Ursula Triebswetter

The Impact of Environmental Regulation on Competitiveness in the German Manufacturing Industry

A Comparison with Other Countries of the European Union

PETER LANG
Europäischer Verlag der Wissenschaften

Bibliographic Information published by Die Deutsche Bibliothek

Die Deutsche Bibliothek lists this publication in the Deutsche Nationalbibliografie; detailed bibliographic data is available in the internet at <http://dnb.ddb.de>.

ISSN 0531-7339
ISBN 3-631-52962-7
US-ISBN 0-8204-7353-7

www.peterlang.de

The Impact of Environmental Regulation on Competitiveness in the German Manufacturing Industry

European University Studies
Europäische Hochschulschriften
Publications Universitaires Européennes

Series V
Economics and Management

Reihe V Série V
Volks- und Betriebswirtschaft
Sciences économiques, gestion d'entreprise

Vol./Bd. 3090

PETER LANG
Frankfurt am Main · Berlin · Bern · Bruxelles · New York · Oxford · Wien

Acknowledgement

This thesis was written at the Queen's University of Belfast, in the School of Economics and Management under the supervision of Prof. David Hitchens. The PhD degree was awarded on 1st of July 2003.

The data used for this thesis were collected in the framework of three projects funded by the European Union where I was part of an international research group. As far as parts of the data were incorporated into the thesis, they were examined with a different research perspective. The angle I took was very narrowly focused on the impact of environmental regulation on competitiveness in Germany instead of the European perspective taken in the original research. The collection of German data was largely my own responsibility either as a researcher at Ifo Institute for Economic Research in Munich or during my Marie Curie fellowship, which led me to the Institute of Prospective Technological Studies (IPTS) in Seville, Spain. There I shared the fieldwork for the cement case in Germany, Spain and the UK with another researcher. All Irish interviews were carried out by researchers from the Queen's University Belfast.

This thesis would not have been possible without the help of numerous persons both in my academic and private surrounding. I am particularly grateful to Prof. David Hitchens who not only supervised my thesis at the Queen's University Belfast, but also co-ordinated the original studies through which my thesis has arisen. Also, I would like to thank Prof. Rolf-Ulrich Sprenger from Ifo Institute for Economic Research in Munich who encouraged me to write the thesis and to choose the research area. Special thanks are contributed to Prof. Karin Wagner from the Fachhochschule für Technik und Wirtschaft in Berlin with whom I closely collaborated in the original, wider cement study. I would also like to thank Josefina Lindblom, Frank Farrell and Michalis Vassilopoulos who were my research colleagues at the Institute of Prospective Technological Studies in Seville for the good collaboration and their support for my thesis. Invaluable research assistance was given to me by Roland Kiewitz, Karin Dressel, Udo Wartha and Claudia Wenzl at Ifo Institute for Economic Research in Munich and by Marion Rückebusch at the Institute of Prospective Technological Studies in Seville. Much help and support was received throughout the thesis from Andrea Möller of the German Economic Institute for Tourism in Munich. Helga Fritzenwenger has very expertly produced the final manuscript. Of course, I am also indebted to the European institutions who made this work possible and to all the companies, which co-operated during this study.

Last, but not least, I want to thank my family and in particular my husband Gernot and my son Julius for their great support and patience.

Contents

Tables

Boxes and Figures

0. Summary

This thesis examines in three case studies (food sector, supply chain of the packaging, food and retail industry as well as cement industry) whether German plants suffer from a negative impact on competitiveness caused by stringent environmental legislation. The findings indicate hardly any evidence of detrimental effects on competitiveness.

The main explaining factors for this result varied from case study to case study. Among the more general and essential factors was the frequent use of clean technology solutions which is less costly both on the resource and financial side than the use of end-of pipe technology. A micro level analysis confirmed that environmental initiatives had often been implemented with both environmental and economic benefits. High environmental awareness went hand in hand with a high number of environmental initiatives in two of the three case studies. Furthermore, environmental pressure was ranked as relatively unimportant compared with other competitive pressures. Finally, the low absolute levels of compliance costs measured as a percentage of turnover, at least in the food and packaging case study, explained why environmental regulation cannot have a great influence on competitiveness in these sectors. In the cement case compliance costs were high. These costs, however, were partially compensated for through the use of alternative fuels.

There were also factors which seemed to be less important for the relationship between environmental regulation and competitiveness. High productivity levels were not among the essential factors explaining our findings. It implies for our case studies that also plants with lower productivity can withstand high compliance costs. There was also hardly any evidence found for a relationship between low age of plant and low compliance costs. Neither did high skill and R&D levels generally help to produce a high number of individual or cost saving environmental solutions. Moreover, small plant size was not found to be a disadvantage when complying with environmental regulation.

1. Introduction

1.1 Background and aim of research

Policymakers and also economists argue that the preservation of environmental goods cannot be let to free market forces because efficient private market allocation is not possible due to the inherent properties of environmental goods. Thus, some form of environmental legislation is needed to protect the environment and human beings from harmful effects of pollution.

Germany is one of the countries in the EU where the environmental debate started very early in the 1970ies and a vast body of environmental regulation has been developed. The stringency of German environmental regulation has been evaluated in cross-national comparisons to be above average in comparison to other industrialised countries (see Jänicke/Weidner, 1995). Moreover, if the trends in environmental quality, emissions and the use of technologies to reduce emissions are compared cross-nationally, then Germany ranks among the worlds best. This is also reflected in the environmental expenditure, which reaches about 3 per cent of German GDP (Wackerbauer, 2002). Only countries like Sweden, Austria, the Netherlands and less clearly the US and Japan reach such environmental performance. Therefore the central aim of this research is to determine the micro level effects on competitiveness of the adjustment, which German firms make in response to this relatively stringent environmental legislation.

There is research evidence that environmental regulation triggers considerable social and economic benefits, including not only lowered illness and mortality, reduced damage to vegetation, but also improved aesthetics and also better recreational opportunities (see Geisendorf et al. (1996) and Hampicke (1996) for extensive German studies and e.g. Carlton/Perloff, 1990 for the US). However, these same regulations have also initiated a debate regarding their direct costs for industry and their indirect and often uncertain impacts on economic growth, productivity and international competitiveness (see the US research by Jaffe et al. (1995), Jorgensen and Wilcoxen (1990) and Portney (1981)). Much attention has also been paid to asymmetrical impacts of regulation on various economic actors. Thus, in the conventional approach, it has been suggested that environmental regulations have differential impacts across large and small firms, new and old firms, new entrants and existing firms, on firms in different geographical locations and that environmental regulation also has the potential to influence international trade (Dean/Brown (1994), Levinson (1996), Pashigian (1984), Kalt (1988), Low and Yeats (1992)). In short, the conventional view fears that the private costs initiated through stringent environmental policy impair competitiveness (Palmer, Oates, Portney, 1995). Conversely some commentators have argued that there are "win-win" opportunities available through environmental regulation, where simultaneously pollution is reduced and productivity increased, the so-called "Porter hypothesis" or revisionist view (Porter/van der Linde, 1995).

What impact could have the stringent environmental regulation on German firms? A priori the answer to this question implies positive and negative aspects. In general, on the positive side, German firms are known to have high productivity. There is a well-trained labour force and e.g. in comparison to the UK the capital stock as well as the R&D endowments in Germany are higher (O'Mahony, 1993). This has a positive impact on productivity levels in Germany (Steedman and Wagner, 1987 and 1989) and one could ex ante assume that German industry is therefore in a favourable position to absorb additional costs arising from legislation. Moreover, with regard to the environment Germany is characterised by two distinctive features, which may facilitate efficient adjustment to environmental legislation. First of all, in Germany there is a well-developed supply of environmental technologies as a consequence of the country's history of strict regulation. Secondly, German consumers are known to be fairly environmentally conscious (see Kuckartz, 2000). Green consumer and also customer pressure within the supply chain of certain production sectors in Germany can be assumed to be comparatively strong and firms will possibly have an early-mover advantage with regard to cost efficient business strategies in this area.

However, there are also factors which make Germany potentially vulnerable to the impacts of environmental regulation with small size being the most obvious one: German industry is characterised by a high proportion of small and medium enterprises as well as family owned companies which might be less able to absorb additional costs of compliance due to their small size and lack of capital (see for instance Bradford, 2000). Small and medium-sized firms often also have problems to define their environmental problems properly and do not undertake important R&D activities themselves. As a result the suppliers of environmental technology strongly dominate the interaction and the selected solutions are often characterised as end-of-pipe technology, which increase production costs (Frank and Swarte, 1986). Unsurprisingly, empirical evidence for Germany shows that small companies often have a problem in implementing clean technologies[1]:

---

1 Clean technologies can be described as a manufacturing process "which allows to reduce the quantity of effluent which pollute the environment and to make the most rational use of raw materials and energy, and all this at a reasonable economic cost" (Rouston et al., 1996). Or as Christie et al. (1995) put it, an "integrated approach to pollution at source and

only 40 per cent of companies with up to 49 employees, but ca. 70 per cent of companies with more than 1,000 employees have installed clean technology measures (Adler, 1997). Since clean technology usually triggers a double dividend, i.e. an increase in efficiency and an improved environmental situation (Kosz, 1997), small firms using end-of-pipe-technology could be disadvantaged in Germany.

Essentially, in this study, the potential competitiveness impacts on German firms caused by environmental regulation are detected by contrasting data collected during in-depth interviews in German firms with the findings in similar (matched) firms in Ireland, the UK and Spain where environmental regulation tends to be less stringent.

1.2 Case studies: Countries, industries and environmental regulation under study

Choice of countries

The competitiveness effects of regulation are represented by sampling firms in Germany where environmental practise can on average be assumed to be rather strict within the EU. In many cases Germany performs well above requirements of EU Directives. Thus the core question is through which environmental measures German firms have adjusted to this regulation, at what costs and how competitiveness was affected. To answer this question the competitiveness effects of environmental initiatives in Germany are analysed by sampling firms in various industrial sectors and environmental problem areas. The results for Germany are contrasted with the findings in similar (matched) firms in competitor countries where environmental regulation tends to be less stringent. In our case Ireland,

---

for prevention of pollution in the first place". Main features which can be used for a specific characterisation of the differences between traditional technologies or products are the following: More economical use of resources and energy, primary recycling, i.e. recycling within the production process, reduction of unavoidable final waste, substitution away from environmentally unfriendly substances, total substitution of products and production technologies by less environmentally dangerous ones, far reaching reduction of the use of end-of-pipe technologies, lifecycle analysis of the production process and the product, recyclability or environmentally more friendly disposal of unavoidable waste and products which are not usable anymore. In principle, one should not think about end-of-pipe and clean technologies as always being obviously distinct. End-of-pipe technologies are not always second-best options. In contrast, clean technology and end-of-pipe technologies should be looked at as cornerstones of a technological continuum. In this technological sense clean technology stands for the ecologically most far reaching technology which is available for technical problem solving, but can also include end-of-pipe technologies.

the UK and Spain have been chosen. At the time of the study Spain and Ireland as cohesion countries were documented to have a softer environmental regime than Germany (Sprenger et al., 1996). Also the UK approach to regulation shows a tendency to be less strict, although the case-by-case evaluation can also embody stricter standards than in Germany. Moreover, firms in the selected sample countries represent contrasts in productivity, skills and R&D input which are likely to be helpful factors in the efficient adjustment to regulation. The study is based on a total of ca. 160 face-to-face interviews undertaken in three industry case studies. The effects of environmental regulation will be analysed from a microeconomic perspective: On the one hand, the consequences of the use of less resources through an increase of capital (more efficient technology) will be measured; on the other hand, also the impact of better organisation and higher awareness (e.g. environmental management systems) will be taken into account.

Choice of industries/choice of regulation

Sample industries and related fields of regulation are chosen in a way to cover a variety of environmental problems, different levels of compliance costs and certain sector specifics which may affect adjustment to environmental legislation in a particular way.

The chosen industries are the food (dairy and meat) sector in Germany, the Republic of Ireland and Northern Ireland; the packaging industry in Germany, the Republic of Ireland and the UK together with the respective supply chain links in the food and retail sector; and the cement industry in Germany, the UK and Spain. Whereas the focus in the dairy and meat industry is on the treatment of waste water and the corresponding environmental initiatives and costs, the issue under concern in the packaging industry is waste minimisation. In the cement industry adjustment to clean air standards and potential consequences on competitiveness are examined. All three areas are regulated to varying degrees in the above named EU countries on a national level. In addition European Directives like the Urban Waste Water Directive, the Packaging Directive and the Directive on Integrated Prevention and Pollution Control (IPPC) have to be taken into consideration.

Another aspect for choosing the industries was the absolute level of compliance costs which influences the competitiveness of the individual sector. In comparison to the average of the total manufacturing sector in Germany, the level of compliance costs[2] is known to be lower than average in several branches of the

[2] measured as environmental investment as percentage of total investment

packaging industry, average in the food industry and above average in the cement industry (Statistisches Bundesamt, 2002; see Table 1.2-1). Moreover, each sector is characterised by different market structures. Whereas the food industry consists mainly of a fairly competitive market with SMEs, the packaging study was chosen as a supply chain example in order to highlight backward and forward linkages arising from environmental pressure, with special attention paid to the monopsony power of the retail sector. The cement industry presents an oligopoly (see Table 1.2-1). These structural specifics can be expected to have different implications for the adjustment to regulation.

In the following the case study contents are briefly presented.

Table 1.2-1 Overview of industries and countries

Sector	Env. focus	Level of compliance costs in %*	Specifics of the sector	Countries
Meat and dairy	Effluent	About average: 3	Consists of SMEs	Germany, Republic of Ireland and Northern Ireland
Packaging (food and retail)	Solid waste	Lower than average: e.g. 1.6 in metal e.g. 1.5 in plastics Higher than average: e.g. 4.3 in paper	Supply chain analysis, Strong demand position of retailers	Germany, Republic of Ireland and the UK
Cement	Air pollution	Higher than average: 3.3	Oligopoly	Germany, Spain, UK
*Figures for Germany in 2000; environmental investment as percentage of total investment.				

Meat and dairy industry

Meat and dairying activities are principally undertaken in small and medium sized enterprises. The aim of the research is to measure the competitiveness impacts of environmental compliance per unit of output and the international difference between such impacts in order to find out whether German firms suffer from a cost disadvantage because of more stringent environmental regulation than in the Irish industries. In order to measure the costs of environmental compliance, the individual abatement measures and their cost and productivity implications were analysed.

The main environmental problem area studied in meat production and dairying is waste water (referred to as effluent) following national legislation and EU Directives. At the time of the study the national German waste water legislation already met the limit values of the European Urban Waste Water Directive; this was neither the case in the Republic of Ireland nor in Northern Ireland.

Hygiene standards and production methods in the industrial process trigger a heavy demand for process and cleaning water which becomes effluent in the end. The use of clean technology can reduce the water demand. In dairying the basic product is milk which is composed of fat, protein, lactose (sugar) and minerals and is used to make a variety of milk products. Milk has an outstandingly high biological oxygen demand (BOD) which is 250 times greater than that of sewage. Milk waste streams also contain valuable materials and only recently their exploitation was started through various clean technology measures. In the meat industry slaughtering of animals and processing of meat releases considerable amounts of organic waste; blood in particular has a very high BOD.

The competitiveness impacts of currently available effluent cleaning measures are measured at plant level. Both the effects of end-of-pipe technologies like effluent plants and clean technology efforts for reuse, recycling and recovering of effluent are measured and internationally compared.

Packaging industry in a supply chain

The focus of the case study in the packaging industry is on how waste reduction and minimisation strategies are undertaken in a supply chain including forward and backward pressures from the fillers (food industry again in the example chosen) and then retailers. The aim is to measure the relative importance of adjustment undertaken for packaging regulation and for market reason.

In Germany the Packaging Ordinance which was enacted in 1991 has meant that the country's waste practise is well above that required by EU standards. Packaging initiatives and other waste minimisation activities and their costs are measured in Germany and then contrasted with similar measures in Ireland and the UK where at the time of the study waste legislation was less stringent than in Germany.

Packaging suppliers, food processors and the distributive sector are considered, given their importance in terms of the implications of packaging regulations and as industries making up one production chain with forward and backward linkages. Packaging itself has an important environmental implication since packaging material accounts for one-third of total household waste in the EU. Paper/board and plastics each represent about 30 per cent of all EU packaging, metals packs represent a further 16 per cent and glass 7 per cent. These materials are all used in the dairy and drinks industries chosen as an example of fillers. The quality of the packaging products delivered by the packaging suppliers can facilitate or hinder the possibilities for waste reduction of the fillers as they attempt to meet environmental and market standards often set by retailers. It is hypothesised that a very limited number of European retailers exert backward pressure on the food and packaging manufacturers. This is not withstanding the market pressure which is sometimes exerted by producers e.g. through intense advertising campaigns.

The cost of individual efforts to reuse, recycle and recover packaging and their impacts on competitiveness are measured at plant level, classified into a common analytical system and internationally compared.

Cement industry

This industry study addresses the cost of clean air standards in Germany relative to Spain and the UK. It examines whether the implementation of techniques for the reduction of air pollution has an impact on competitiveness of cement plants in Germany, Spain and the UK and tries to explain the underlying reasons for this (or lack of) impact on competitiveness.

As in the other two case studies the German cement industry is in comparison to the Spanish and the UK sector subject to the most stringent national clean air legislation. However, contrary to the other two industry studies in the food and packaging area, the German cement industry as a whole does not have a productivity advantage (see chapter 4.3). Simultaneously, Germany is the country in the EU where most small family owned cement plants are found and which do not benefit from the same economies of scale as their larger counterparts abroad. Thus a priori one can argue that German cement factories will suffer from negative competitiveness effects of environmental regulation. But the industry is characterised as an oligopoly with supra-normal profits and one can also argue that in such a business atmosphere environmental investment will be affordable without any damage to competitiveness. Almost all of the European cement plants are in the hand of a few companies and also German companies are internationally active. It is hypothesised that international companies will export their home environmental standard. Branch plants sampled in countries with lower environmental standards are hypothesised to have higher environmental standards than indigenously owned plants.

As a background for the interviews a recently published list of best available technologies (BAT) is used. This list was developed as part of the background reference document for the European cement industry within the framework of the integrated prevention and pollution control Directive (IPPC-Directive, 96/61/EC).

The empirical data encompasses the economic and environmental effects of clean technology measures and end-of-pipe technologies for abatement of NO_x,

SO_x and dust emissions. Capital and running costs of all abatement techniques are measured at plant level and it is examined whether there are factors which can offset part or all of the additional costs of environmental control.

1.3 Research method

The influence of environmental regulation on company competitiveness is a complex issue. As mentioned above there are two major streams of thought in the existing literature. In the "conventional" view it is argued that there is a conflict between competitiveness of firms and their environmental performance (e.g. Whalley & Whitehead, 1994). This view is held because in the past firms have concentrated on end-of-pipe technologies as the major approach towards pollution control, thus generating extra costs for environmental performance improvements. More recently, in the "revisionist" view the notion emerged that improved environmental performance can trigger a competitive advantage by triggering the use of more efficient processes, improvements in productivity, lower costs of compliance and new market opportunities ("first mover advantages") (e.g. Porter & Esty, 1998). Proposals made by the European Commission in 1992 were also based on the assumption that improving the environment and enhancing industrial competitiveness were not incompatible (Commission of the European Communities, 1992), although the use of cost-benefit analysis in evaluating environmental measures was advocated.

These opposite views of the problem show the need for the production of detailed empirical data in order to examine the relationship between economic and environmental performance of companies in depth. Therefore a case study approach generating detailed micro-level data is used. This research method presents an interview-based sample survey technique which allows for a systematic comparison of supply-side features of the firm after controlling for size, ownership and product type. The selection of a sample of plants is undertaken along a spectrum from low to best practise firms defined both by productivity and environmental performance in each industry and country under study.

1.4 Stylised model

Following Wagner/Ngyen/Azomahou (2001) it is assumed as a basic working hypothesis that the interaction between environmental performance (as a result of environmental regulation) and competitiveness is most likely to be indirect and that a priori it is hard to say whether to support the traditionalist or the revisionist view. There are factors influencing either environmental or economic performance, or both "... and (the interaction between environmental and economic performance) should thus be perceived as the outcome of a complex process of interaction and influence" (Wagner/Ngyen/Azomahou, 2001, p.3). Before detailed hypotheses with regard to individual factors of influence are generated, a more general model linking potential factors of influence for environmental performance and economic performance needs to be developed. The stylised model shown in Figure 1.4-1 summarises the factors considered to be most important in the relationship between environmental and economic performance. At the top level the relationship between environmental and economic performance is sketched. This level is understood, later on in the case studies, as the data collected on environmental parameters plotted against economic indicators. The bottom level of the model shows the explanatory factors to be considered as most important in causing a certain level of environmental and economic performance. These factors were found in an initial scanning of the literature from a microeconomic perspective. They include the degree of environmental pressure - which can be regulation and also market pressure - the type of response given by firms, the competitive position of firms and factors internal to firms. Each of these explanatory factors stands as a proxy for a number of influences. As the work progressed, a detailed set of hypotheses was developed. Special attention was paid to the level of influence in the interaction of environmental regulation and competitiveness (see chapter two for a theoretically grounded framework of analysis).

Fig. 1.4-1: Initial stylised model for the relationship between environmental and economic performance

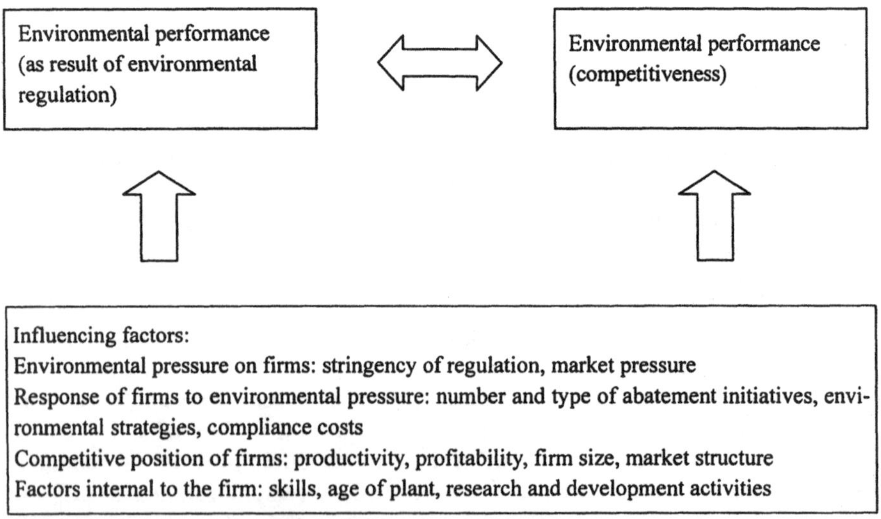

Source: Wagner/Ngyen/Azomahou (2001), own modifications.

1.5 Thesis structure

The thesis is structured as follows:

In chapter two an overview of the literature concerning environmental regulation and competitiveness is given. The possible ways in which regulations could have either positive or negative ways on competitiveness are examined. The empirical evidence on such effects is then shown. The vast majority of the productivity studies shows that the impact on competitiveness varies widely by industry. Chapter two ends with a theoretical framework for analysis and the justification of a set of detailed hypotheses.

In chapter three a brief history of German environmental regulation is presented and the individual regulation for the problem areas at hand in this study, i.e. waste water, packaging waste and air pollution, are described. German legislation is put in an international context, primarily comparisons with other EU countries are drawn. It is shown that German legislation is fairly stringent also on an international level.

In chapter four the structural background of the three case study sectors is given. Emphasis is put on labour productivity and on the competitive strengths and weaknesses of the sample industries/countries. Special attention is paid to the possible influence of environmental requirements on competitiveness in the industries under concern. In most cases the rank order for labour productivity is Germany first, followed by the UK and the Republic of Ireland. In cement, however, Spain shows the highest labour productivity. Concerning the possible impact of stringent legislation the most severe influence can be expected for the German cement industry where costly abatement measures must be undertaken to meet the stringent legislation.

The background to the research method and the measurement of key variables according to the hypotheses are presented in chapter five.

Chapter six contains the full sample description with an emphasis on both economic and environmental aspects and including a review of its representativeness. It is shown that German sample plants show in all cases except cement a strong economic performance and that German plants also record in all cases the highest environmental performance.

The chapters seven, eight and nine deal with the empirical results concerning the impact of stringent environmental regulation in the three selected sectors of the German manufacturing industry. Hardly any negative impact on competitiveness caused by environmental compliance costs was discovered. Overall, in the sample industries the main explanatory factors for this result were found (to a various extent in each case study) in the frequent use of clean technology solutions; high environmental awareness paired with a high number of environmental initiatives; the implementation of environmental initiatives with both environmental and economic benefits on a micro level; the low absolute levels of compliance costs measured as a percentage of turnover (only in the food and packaging case); and the relative unimportance of environmental pressure in comparison with other pressures on competitiveness. Factors which had in fewer cases explanatory power were high productivity levels which were assumed to help to absorb compliance costs; low age of plant which usually also embodies modern and cost saving environmental technology; high skill and R&D levels of which it was expected to produce a high number of individual and cost saving environmental solutions.

The detail to the empirical chapters is as follows:

In chapter seven it is illustrated that German waste water legislation implies relatively higher compliance costs for German dairy and meat plants than in the other sample countries, but hardly any impact on their competitiveness. Overall, there was support for a positive association between achieving average or above average use of waste water initiatives and above average competitiveness both in the dairy and meat industry and both in Germany and the Republic of Ireland/Northern Ireland. Above average economic performance is also associated with an above average score of environmental awareness. For the German dairy and meat sector a significant relationship between younger plant age and a higher number of waste water initiatives was found. A similar positive pattern emerges for skills and R&D and number of waste water initiatives. This was not the case for the Irish economies. More German plants are slightly more efficient users of water, especially in dairying where fewer sample plants were drawing water from (low or no cost) wells. Overall, both fresh water prices and waste water charges were higher in Germany than in Ireland, thus, providing a higher incentive to save water and to reduce compliance costs. More German firms were able to record positive productivity effects triggered by waste water activities, but these impacts were only marginal. In relation to other competitive advantages or disadvantages environmental aspects were only of minor importance both in Germany and the Republic of Ireland/Northern Ireland.

In chapter eight results for the packaging case study within the supply chain of dairy and drinks as well as retailing are given. Waste costs were found to be higher in Germany than in the other sample countries, but they did not cause any damage to competitiveness. The average number of solid waste initiatives was highest in Germany, reflecting the fact that German plants had undertaken a lot of efforts to minimise waste and also to reduce waste costs. However, no consistent relationship between number of environmental initiatives and compliance costs was found. Neither could the hypothesis be confirmed that high productivity plants always incur low waste costs. Only the top performers in the German dairy/drinks and retail industry were able to implement a higher number of solid waste initiatives than the overall German sample. But this was not statistically significant. It could be shown for Germany that increased environmental awareness was associated with increased numbers of solid initiatives undertaken in all sectors of the German sample. There was a trend that younger plant age was linked to lower waste costs in Germany. A positive correlation between skills and R&D and the number of solid waste initiatives could only be found in German packaging plants. The hypothesis that small plant size is linked with higher compliance costs could not be confirmed, partly because small plants undertake significantly less waste initiatives. The most important driver for solid waste initiatives in each of the countries was cost, followed by legislation and customer pressure. In Germany legislation was responsible for significantly more initiatives than in either of the other countries. Initiatives undertaken by German firms were in total more likely to require additional resources (because more initiatives were driven by legislation which frequently caused recycling costs), were less likely to bring about unit cost savings, more likely to trigger general financial savings and more likely to give rise to positive competitiveness effects compared with those undertaken in the other sample countries. The majority of solid waste initiatives, however, was cost driven in every country and these initiatives triggered the largest number of general savings effects for the plants under consideration. Most expenditures were cost neutral and few required capital expenditure. In Germany several companies stated that their environmental initiatives conferred important competitive advantages. No Irish or UK firm quoted environment as a competitive advantage. Environmental factors were mentioned on very few occasions as a competitive disadvantage in Germany. Overall, there was little evidence of packaging regulation as leading to a competitive advantage or disadvantage. Probably the most important reason is that the impact of regulation and the cost of complying with regulation is relatively insignificant in relation to other factors which influence the competitive performance of firms.

In chapter nine empirical results on the impact of stringent environmental regulation for the cement case are presented. In this case Germany again bears the highest compliance costs (much higher than in the other two cases measured as per cent of turnover), but can offset the higher capital costs in several ways, the most important ones being the use of alternative fuels and general reductions in energy consumption. Although German plants were not entirely high productivity plants, they have implemented the highest number of costly abatement measures in the sample. German firms have done this earlier than their counterparts in Spain and the UK. In several cases the German plants have achieved economically and environmentally more beneficial effects than elsewhere. Concerning the purely legislation driven measures which are undertaken only in the German sample and nowhere else, no plant reported a directly negative impact on competitiveness. With respect to competitive advantages German plants most frequently referred to aspects connected to favourable environmental performance like ability to achieve low energy costs, modernity of plant and up to date equipment which enabled them to fulfil strict environmental standards. Concerning competitive disadvantages German plants stressed that environmental requirements were said to be high. But the top environmental performers did not complain about environmental costs.

In chapter ten a summary of the main findings on the impact of environmental regulation on the competitiveness in the German manufacturing industry is given and the general conclusions are presented. The chapter ends with policy implications.

2. Analytical framework: The possible relationships between competitiveness and environmental regulation

This chapter outlines the analytical framework of the study. Theoretically, the research is based on the literature addressed to the question how in principle environmental regulation might impact on competitiveness (section 2.2). It is recognised that regulation may have positive as well as negative effects on competitiveness. Special attention is paid to the role of innovation and diffusion activities caused by environmental regulation (section 2.3). Also, a review of the relevant literature which provides an empirical consideration of the effects of environmental regulation is presented (section 2.4). This background serves for the development of the research hypotheses in section 2.5. In summary, the hypotheses propose that plants are exposed to certain types of environmental pressures, and their ability to respond efficiently to these pressures depends on their competitive position as well as factors internal to the firm. The hypotheses deliberately stress the possibilities and potential consequences of using clean technologies, a focus which so far has been rarely examined in the literature. Hereby, the factors which define competitiveness are also known to influence the adoption of clean technologies. The chapter starts of with a discussion of the meaning and measurement of the concept of competitiveness.

2.1 Meaning and measurement of competitiveness

A study about the competitiveness implications of environmental regulation necessarily needs to begin with a definition and an appropriate indicator of competitiveness.

Definition of competitiveness

In theoretical terms, competitiveness could be defined on the level of industry structure and conditions of market entry. In a market structure of perfect competition any firm which is in operation would be by definition competitive because it can produce at a price level given by the market. Average costs of the firm are not higher than the price level. In a monopoly situation the firm can sell at a price level higher than its average costs. Super-normal profits can be achieved in this situation unless the profits are so high that other firms are encouraged to develop a new product or process with which the market power of the monopoly can be undermined. Similarly, a competitive firm within a contestable market must be able to operate with profits that are just below the level that would trigger the entry of new firms. In fact, most firms are operating under conditions of an oligopoly. They decide strategically and are competitive when they can hold or increase market share in the long run.

However, firms are not only influenced by industry structure and actual competition, but also by potential competition (theory of contestability (Baumol et al., 1982)). Since firms want to prevent potential competitors from market entry, they tend to take lower prices: A competitive firm within a contestable market must be able to operate with profits that are just below the level that would trigger the entry of new firms. A market is the more contestable, the lower any sunk costs are, the easier access to technologies can be gained and the longer existing firms need to react to new competitors. Theoretically, potential competitors could use any deviations from the conditions for a perfectly contestable market for new market entry. If they enter suddenly, attract the entire demand and leave the market without costs, then a so-called "hit-and-run" competition is achieved. Sunk costs can be interpreted as the costs for a failed market entry. Therefore the probability for a market exit is the smaller, the higher sunk costs have been. Famous examples for this would be the gas market, telephone wires or nuclear power; but also market specific knowledge or skills as well as research and development efforts apply.

Also, most economists would stress the need to evaluate the competitiveness impact of environmental regulation by measuring the effect on productivity. For instance, the OECD (1992) defined a nation's competitiveness as "the degree to which it can, under free and fair market conditions, produce goods and services which meet the test of the international markets, while simultaneously maintaining and expanding the incomes of its people over the longer term". Another very broad definition implicitly also refers to productivity and explains competitiveness as follows: Competitiveness "refers to the capacity of an economic actor to produce and sell, and to develop through time this production and selling capacity by putting to good effect acquired skills and assets, know-how, and networks of alliances" (Statistics Netherlands et al., 1996, p. 46). Of particular importance for maintaining competitiveness in the context of this study is the ability to implement technological innovation and the capacity of economic actors to influence and to adjust to market changes consequent on regulatory frameworks and their enforcement. Therefore competitiveness ... "is determined by the ability of a business to respond to changing circumstances in an efficient, innovative and adaptable manner" (OECD, 1998, p. 5).

Indicators of competitiveness

From an empirical standpoint the question arises what indicators are available for the measurement of competitiveness. There are factors related to the input and output side. Concerning the input side (the likely explanations of competitiveness) these are physical and human capital, R&D spending, etc.; on the output side (illustrating the consequences of the relative competitiveness of a firm) the essential factors are profitability, market share, productivity and so on (Jacobson and Andréosso-O'Callaghan, 1996). According to the European Commission competitiveness encompasses three elements - productivity, efficiency and profitability. At the level of an economy, economic indicators of competitiveness include growth, productivity and employment. At the level of a sector of the economy such as manufacturing, competitiveness also includes market share and profitability (Commission of the European Communities, 1996).

Labour productivity has always gained special attention as an indicator of competitiveness. The debate on competitiveness, in particular in the USA, has shown that in the long run increases in living standards measured as GDP per capita, can almost always only be achieved by an increase in labour productivity (see Baumol, Blackman and Wolff, 1989; Dertouzous, Lester and Solow, 1989; Krugman, 1991). Whilst a country can increase its cost competitiveness by decreasing relative wages or the real exchange rate, the only way to increase cost competitiveness whilst holding or improving standards of living is to increase labour productivity (Hitchens et al., 1998). That means that rising living standards can be compatible with constant relative unit labour costs which are defined for any two countries as the ratio of per head total labour costs, defined in common currency terms according to the average market exchange rate, divided by their ratio of comparative productivity levels in real terms (van Ark, 1996).

Given this background, also in this study, labour productivity has been chosen next to other variables on the input and output side as the main indicator of competitiveness and firm performance (see also empirical chapters).

2.2 Possible impacts of regulation on competitiveness from a theoretical standpoint

In this section the different views on the effects of environmental regulation on company competitiveness are presented. Whereas the conventional wisdom is that environmental regulation slows down productivity growth and thereby impairs competitiveness, a more revisionist view supports the positive impacts on competitiveness: New processes and technologies can raise productivity by making more efficient use of resources through innovations (Jaffe, Portney, Peterson, 1995). Recently, the conventional view is being questioned through some general mechanisms that relax the originally expected negative trade-off between environment and competitiveness considerably, but which do not assume the optimistic results of the revisionist side.

2.2.1 The conventional view

The conventional view exclusively focuses on the cost side and hence the negative impact of environmental measures on competitiveness. According to Neumann and Pastowski (1994) this can be explained by three reasons. First of all, the costs of environmental protection are usually much better measurable than the benefits. Still many end-of-pipe technologies are used of which the costs can be easily separated from other production costs. As long as there is no marketable output from emission reduction, the costs of the use of end-of-pipe-technologies can be fully accounted as unproductive, environmental costs. Second, the perception of costs related to environmental protection efforts is stronger than that of environmental damage costs. Environmental damage is most often not perceived in economic categories because e.g. standardised valuation techniques are frequently not existing and firms are not familiar with them. Therefore most of the environmental damage costs are not borne by the firm. And third, the theory of environmental economics seems to regard internalisation strategies of environmental damage costs among the polluters as not promoting Pareto-efficiency when the transaction costs are higher than the damage costs (see also Weimann, 1991). This means that external effects which are spread dispersely in time and space are not taken into account on efficiency grounds. Minsch (1991) even argues that among other things high transaction costs and the asymmetric interest of participants in decreasing transaction costs favour dispersely produced environmental damages and aggravate perception, attribution and internalisation of environmental problems. A further interpretation would be that only so-called "producer-to-producer externalities" where the number of damaged persons is relatively small and the information asymmetry can still be controlled, are considered for internalisation (Fisher and Peterson, 1976). In other words, in the conventional theory, the consideration of environmental protection efforts is almost certain to affect competitiveness negatively when transaction costs are high.

Neo-classical theory suggests that cost incidence of regulation can be measured based on relative slopes of the demand and supply curves in the market subject to higher regulation. For example, in the highly competitive markets of the food industry and packaging industry, an increase in the stringency of regulation through tighter waste water or packaging waste standards can be illustrated as an upward shift of the supply curve. This automatically implies that the regulation causes additional marginal cost of production of a good. Then output prices are raised and sales decline. Ceteris paribus (in particular no changes in demand or technological progress) the prediction is that output levels decline in the sample industries. Also, in industries with some degree of market power (e.g. production of metal cans, see packaging case study or also otherwise oligopolistic markets where they become competitive because of the fear of new entry), an increase of marginal costs prompted by regulation would squeeze super normal profits and would probably reduce output. As regards contestability analysis for our case studies of the food and packaging industry, sunk costs in the form of R&D activities could play a role as a market barrier (the production technologies are standardised so that they can easily be redeployed in another firm of the sector and do not present a type of sunk costs).

However, the consequences for the impact of regulation on the sectors under concern are not straightforward. On the one hand, it could be possible that new environmental regulation might induce further, very special R&D efforts, thus even increasing barriers for market entry. On the other hand, it is also imaginable that regulation triggers activities which are widely copied without a lot of R&D efforts. Then sunk costs are less important, markets become more contestable and output would be likely to be reduced. Of course, independent of market structure and contestability analysis, it can be the case that innovation with the aim to reduce pollution also lowers production costs and increases a firm's cost competitiveness. A rational firm, however, should discover this option without the obligation of regulation. Therefore the standard counterargument against the revisionist view is that in rational economic modelling it cannot be explained why firms do not discover the allegedly efficiency enhancing opportunities of environmental regulation on their own, implying that the revisionist view cannot be of general validity. Moreover, there are two other aspects which are linked to costs and which support the conventional view: Regulation diverts capital resources away from other potentially productivity enhancing projects and will therefore tend to reduce productivity growth. Walley and Whitehead (1996) also stress that regulation takes up a significant proportion of management's time, reducing its availability for other tasks.

At the international level, the cost based approach implies that producers in countries with more stringent environmental controls will be less competitive than those in countries with softer legislation. This concept reflects both the fear of industrial flight where firms relocate production to less regulated markets and that developing countries will become pollution havens which attract polluting industries (see e.g. Jenkins, 1999).

However, within a cost-based framework there are also authors who have argued that environmental costs take up only a small percentage of total costs (Dean, 1992; Jaffe et al., 1995) and that they are only one of many factors affecting production decisions. For instance, studies of investment decisions indicate that key factors for location decisions are political stability, market size and other costs, with environmental regulation rarely if ever mentioned (Sprenger, 1996a).

Taking these different views within the conventional view into account, the debate over the impact of environmental regulation on competitiveness turns into an empirical question. In general terms, a negative impact on the output and employment of firms will be the larger the rise in costs following compliance, the greater the differential cost penalty relative to domestic and foreign competitors, the more significant the compliance costs are in total costs and the greater the degree of price competition between firms and the greater the sensitivity of demand to price increases (OECD, 1993).

2.2.2 The revisionist view

The revisionist view implies that firms are not aware of certain opportunities to which environmental regulation might open their eyes. The savings related to these opportunities could outweigh the costs of compliance. Central to this win-win view where both the environment and the firm benefit is the belief that environmental regulation spurs innovation in a number of ways, not only in the environment industry, but also in the user sectors of environmental technology and even in their network of further suppliers and customers.

In the maybe most well-known variant of the revisionist view, firms can create a type of first mover advantage by the development of environmental technology from which they can benefit in later times when other countries also have to adopt stricter environmental regulations (Sorsa, 1994; OECD, 1996). In the European Union Germany has been particularly successful in exporting environmental technology. This has been attributed to a large extent to national environmental policy which created an early domestic demand (Sprenger, 1996). It is also argued that a lead in the use of environmentally friendly technologies could produce a first mover advantage vis-à-vis those firms which continue to use traditional, more polluting production methods. However, Porter and van der Linde (1995) have also stressed that innovation caused by regulation can directly benefit the user sector of environmental technology and related branches.

First of all environmental regulation might inspire firms to seek ways of increasing resource productivity through changes in production which reduce costs. Regulation may also lead the firm to convert waste into saleable products which create additional revenues (Porter and van der Linde, 1996). The attempt to create a closed loophole economy through waste regulation is one example where legislation triggers a chain of economic and environmental effects along entire production and consumption chains. At this point it shall be stressed that the focus of this thesis is almost exclusively on the measurement of the immediate impact on the user sector of environmental technologies. Only occasionally the effects of regulation outside the chosen industry template are analysed, e.g. the effect of regulation in the tyre sector is picked up in the cement study.

Competitiveness can also be increased by the production of "greener" products. Increased environmental awareness among consumers may allow the firm to charge premium prices for those products in comparison with less environmentally products (Jenkins, 1999).

A further line of thought is related to the reduction of X-inefficiency: Environmental regulation may reduce intra-firm inefficiency and organisational failure (Gabel/Sinclair-Desgagné, 1997) and may shift the firm closer towards its production possibility frontier.

Finally, the favourable outcome of the revisionist view stresses the need for efficient environmental regulation based on market incentives instead of pure command-and-control measures constraining the choice for environmental technologies. E.g. Alpay (2001) could show in a two-country model with tradable emissions permits as the environmental regulatory regime that regulated firms can offset the higher compliance costs through innovation and/or permit revenues by abating more and selling extra permits. According to Wallace (1995) it is critical that the firm has technological freedom to choose the adequate compliance technique, although, of course, consultation between the regulatory body and the firm is useful.

While the supporters of the revisionist view stress the general validity of their arguments, their critics such as Palmer et al. (1995) question the pervasiveness of the innovation offsets. In addition, there is no uniform opinion concerning the time horizon of the innovation offsets. While the supporters of the win-win perspective tend to have the view of very short paybacks of environmental investments, others doubt the prompt return on investment (see also Jenkins, 1999). Again, as in the case of the cost based approach, the differences are essentially empirical.

2.2.3 Position in between the conventional and revisionist view

Lately, economic theory also draws attention to some general mechanisms that relax the negative trade-off between environment and competitiveness considerably, but without expecting a "win-win" situation a priori. According to Xepapadeas and de Zeeuw (1999) positive externalities of the additional R&D efforts due to environmental regulation and the reduction of uncertainty to the firms about policy trends are valid theoretical arguments supporting the Porter hypothesis. The same authors show theoretically that downsizing and modernisation of firms as a consequence of environmental regulation (in the form of an environmental tax) increases average productivity and has positive effects on the marginal decrease of profits and environmental damage.

2.2.4 Conclusions on the possible impacts of environmental regulation on competitiveness

Theory suggests both negative and positive impacts of environmental regulation on competitiveness. Whereas the conventional view stresses the purely cost increasing impact of regulation, the revisionist position argues with innovation offsets caused by environmental regulation. The more moderate view in between the two extreme lines of thinking shows only small positive effects of environmental regulation on competitiveness. Whether one or the other position holds true, is essentially an empirical question. In section 2.5 a set of hypotheses based on the existing theoretical literature will be developed in order to test the impact of stringent environmental regulation on the competitiveness of German manufacturing industry. Appropriate arguments of all three views will have an influence on the formulation of these hypotheses.

2.3 Innovation and diffusion activities due to environmental regulation

All three views on the relationship between environmental regulation and competitiveness - the revisionist, the traditionalist and also the more reserved perspective in between the two extremes - point to the role of innovation and diffusion of environmental technologies either as the response to environmental regulation or as the firm's own decision e.g. following non-regulatory pressure. Thus, in order to understand the potential impact of environmental regulation on competitiveness in greater depth, the general mechanism of innovation and diffusion has to be introduced as a linkage element between regulation and competitiveness. In the following sections, first a general theory of innovation and then the mechanisms of innovations for the environment[3] will be presented. Special emphasis is put on the literature on environmental regulation and innovation. Second, the diffusion of environmental technologies will be treated. A distinction is made between end-of-pipe and clean technologies which will be shown to have differential impacts on competitiveness.

---

[3] Innovations for the environment comprehend "... *all* innovations that have an environmental benefit, compared to their counterparts or in an absolute sense. Conventional innovation is an important source of environmental benefit. Environmental innovations do not constitute a clear recognisable class of innovations. They tend to focus on environmental solutions, and are either end-of-pipe technologies that often are technologically and politically problematic, causing a transfer of problems (environmental damage in another environmental medium or high economic costs) – or pollution reduction at the source through process or product changes. One should judge innovations on their environmental outcomes instead of the purpose for which they are built and used" (Kemp, 2002, p. 1).

2.3.1 Neo-classical and evolutionary innovation theories

Until very recently the areas of environmental and innovation economics have largely developed separately (Hemmelskamp, 1999). Innovation economics itself is in short characterised by neo-classical (Solow, 1957; Lucas, 1988 and Romer, 1990) and evolutionary approaches (see for basic work Nelson/Winter, 1982 and Dosi et al. 1988). In the efforts to integrate innovation aspects in environmental economics the evolutionary branch has been used.

Neo-classical research on innovation examines general equilibrium situations in the economy and the impact of exogenous impulses on these equilibria. It is assumed that the goal of enterprises is profit maximisation and decisions on innovation are the result of consideration of cost/benefits based on subjective probabilities. In the neo-classical models it is shown that economic growth is only partly due to an increased use of production factors, the unknown rest is technical progress which falls like "manna from heaven" (Solow, 1957). Until the 1980ies the models were restricted to the scarcity of production factors and the relative change of factor prices and their impact on innovations (Hicks, 1932 quoted in Erdmann, 1993). However, according to Ruttan (1997, p. 1521) "The major limitation of the microeconomic version is that its internal mechanism - the learning, search and formal R&D processes - remains inside a black box. The model is driven by exogenous changes in the economic environment in which the firm (or public research agency) finds itself". The newer models treat technical progress as endogenous. The new growth theory (Lucas, 1988 and Romer, 1990) regards human capital and learning as the essential sources of productivity growth and assumes the research sector to fulfil this function.

Joseph Schumpeter (1987) can be recognised as the father of the newer innovation theory embodied in the evolutionary approach. He criticises the fact that the (original) neo-classical models consider technical progress as an exogenous variable; he regards it as endogenous variable and also takes the reciprocities between economic and technological development into account. Entrepreneurs are the main motors of innovation and lead to the fact that enterprises do not only compete on price, but also on products and processes. In his view mainly the big companies are the innovators.[4] The strength of the evolutionary model is

---

[4] According to Schumpeter technical change consists of three phases, namely invention, innovation and diffusion. First it starts with technical and scientific research. This may result in an invention. The second phase is the development phase, in which "upscaling" of research takes place, from pilot configuration to pilot plant. This is relatively expensive compared to the first phase. Feasibility studies precede the third phase which is market introduction. When the invention is available for the market, we speak of innovation. The

that it tries to describe what is inside the black box. A behavioural theory of the firm and of its interactions is undertaken in the so called routines. In the early models fixed patterns of production activities, personnel action and R&D activities are used. In later models of the search and selection process some elements of rational choice are included.

2.3.2 Innovation and the environment

Kemp (1993) was the first who tried to integrate innovation theories in environmental economics. He starts its theory of clean or environment-saving technological change building on the Neo-Schumpeterian view, following the work of Nelson and Winter (1977 and 1982). Within this Neo-Schumpeterian, or evolutionary, perspective "... technological advance is seen as a complex, typically evolutionary process, endogenous to economic incentives, the pool of knowledge, and institutional structure. Furthermore, technological change is seen as a result of interactive heuristic search activities; it involves uncertainty and learning in a fundamental way and is characterised by certain paths or trajectories" (Kemp, 1993, p. 79).

Environment-saving technological change should be perceived as regular technological change because clean technology issues cannot be regarded independent from general innovation motives (Kemp, 1993). Recent studies have identified several key factors that are decisive for the general process of technological change and innovation which can also be applied to environmental innovation (see Strasser, 1997 for a review). Among these key factors are

— technological opportunities,
— appropriability conditions,
— firm size and market structure,
— market demand.

The term *technological opportunities* has originally been expressed by Rosenberg (1974). Technological opportunities to develop a particular technology are based on the pool of internal and external knowledge acquired in the company the past. The know-how consists of the basic scientific knowledge and the indi-

---

final phase is diffusion, the process of adoption of the innovation over time. The discussion here concentrates on market introduction, or innovation, and diffusion. Innovations can be product or process innovations and they can be either radical (new techniques or products or important new applications) or incremental (minor improvements to existing techniques and products). Schumpeter himself concentrated mainly on the innovation phase and in particular on the aspect of basic innovations.

vidual skills of the employees, the products and the knowledge embedded in certain organisational forms. Internal knowledge relies on the results of past research activities and the specific knowledge of the employees and also on the capability to draw on external knowledge (Cohen/Levinthal, 1989). Thus the capabilities to develop an innovation can vary among firms. They relate to the size and nature of the knowledge base which is created by the firm's past activities. Simultaneously the accumulated knowledge also determines whether and how a company is prepared to introduce technological innovations for its products and processes (Erdmann, 1993). The external knowledge essentially consists of basic research in universities and institutes. The use of that knowledge can generate a variety of innovations in diverse sectors (see Dosi, 1988 and also Orsenigo, 1991). Concerning the development of clean technologies a growing affinity for science oriented approaches is to be expected because the solution of environmental problems requires collaboration of many disciplines. However, the internal and external knowledge is not used to the same extent in all companies. Subsequently, process innovations are usually developed by special suppliers. Thus e.g. special suppliers of the environment industry mainly develop end-of-pipe technology. Only the knowledge intensive sectors such as chemical and advanced electrotechnical industries have the capabilities to develop environmentally friendly process innovations independently or together with special suppliers. Environmental product innovations are usually developed by the manufacturers of the product (Pavitt, 1984). R&D co-operations, e.g. between universities and companies, are of great importance for innovations.

If a firm engages in the R&D process, it must be able to appropriate the benefits from the innovation. Levin (1986) distinguishes the following *appropriability conditions*: patents, secrecy, technical lead on competitors, learning curve effects and the extent to which a strong market position can be built up. Dosi (1988) complements this list by scale economies. The appropriability conditions vary according to the type of innovation. A process innovation is more driven by learning curve effects, secrecy and technical lead, whereas a product innovation can be appropriated more easily by technical lead and learning curve effects along with additional marketing efforts (Levin, 1986). Patents are additional appropriability conditions in most cases. They are important for some industries, such as the chemical and mechanical equipment industries, as well as for smaller companies. In the area of environmental technology patents have increased world wide since the 1980ies (Adler, 1997). However, this figure relates rather to end-of-pipe technologies than to clean technologies because the environmental relevance of a patent is not a classification criterion in the international patent statistics. Lanjouw and Mody (1993) find that increases in environmental compliance costs as a result of regulation lead to increases in the patenting of new environmental technologies with a one-to-two year lag.

Another important aspect is *firm size and market structure*. Innovation demands a continuous commitment for structured R&D work which might pose a problem for small firms. Concerning market structure and innovation an important question is whether entry to the market is open for new entrants with better technology.

The external or *market demand* for technological change will also be of relevance for a firm's decision to innovate. Early studies on innovation make a distinction between market-pull and technology-push innovations. In the context of environmental technological change market-pull innovations are developed according to customer wishes or as a consequence of environmental regulation.

In the literature on environmental regulation and innovation it is argued that regulation raises the cost of production and thereby strengthens the incentive to engage in cost-reducing R&D (Oates et al., 1993). Thus the cost increase through regulation can be offset or even lead to a reduction of total costs, especially when there are spillovers between innovating firms through the diffusion of knowledge (Kennedy, 1994). Schmalensee (1994), however, notes that the regulation induced R&D is likely to come at the expense of other research efforts which could have been more profitable. Therefore it is not astonishing that cost considerations are often claimed to be tighter for environmental projects than for other innovation projects. Moreover, research and development for clean technologies might be burdened by more insecurities than end-of-pipe innovation processes. The "innovation window" for clean technologies is smaller than for end-of-pipe technologies since in the first case innovation must be integrated from the earliest moments of thinking on research and development activities onwards (Adler, 1997). End-of-pipe technologies are usually geared to the reduction of specific pollutants or a group of pollutants and can therefore be transferred from one industry to another. Clean technologies are much more specific for certain processes in certain industries; thus the market potential is largely restricted. It can be expected that the market for clean technologies will focus on completely new installations or replacement of old plants (see Table 2.3-1 for an overall comparison of end-of-pipe and clean technologies).

These types of investment are connected to the general structural change in the economy. For instance in Germany those industries with a high percentage of environmental investment are characterised by a low degree of capital replacement (Hartje, 1990). Another influence on the demand for environmental technologies can be found in the elasticity of demand. Usually the suppliers of process technology can expect a price inelastic demand for those investments which have to be undertaken because of new regulation. For green consumer goods, however, a price elastic demand reaction can be expected. Only if the green product shows an immediate benefit, e.g. health benefits, a higher willingness to pay can be assumed.

Table 2.3-1 Comparison of End-of-Pipe and Clean Technologies

Economic Parameter	End-of-Pipe Technology	Clean Technology
Total Productivity	Decrease of Productivity	Potential for Productivity Increase
Production Costs	Increasing	Potential for Cost Reduction
Investment Need	Lower	Higher
Sunk Costs	Generally: No	Potential
Information and Access Costs (For Example Licence Fees)	Lower	Higher
Adaptation Costs	Lower	Higher
Compatibility with Current Production Methods	Higher	Lower
Economic Risk (For Example Downtime Due to a Lack of Experience)	Lower	Higher
International Competitiveness of the Economy	Tendency: Negative Influence	Potential for Future Competitive Advantages

Source: Coenen et al., 1995, p. 47, modified.

2.3.3 Diffusion of environmental technologies

General theoretical models of technological diffusion without a connotation to the environment relate to intra- and inter-firm diffusion. With respect to intra-firm diffusion the literature provides two explanations for the use of a new technology which can be a new input, a new form of work organisation or a new management style. The first theory argues that information is transmitted epidemically and firms then decide whether they use the technology. The argument in the second theoretical strand is that there is learning by experience. Most empirical investigations of technology diffusion have found that relatively profitable, small-scale and simple innovations are adopted fastest (Vassilopoulos, 1999).

Concerning the question why new technology does not spread to all users immediately (inter-firm diffusion), different adoption dates can be explained again using the epidemic model. Also the David/Davies approach where only firms for whom it is profitable use the new technology and the number of users changes over time, is frequently quoted for the explanation of inter-firm diffusion. Two other explanations can be found: Firstly, the Bayesian learning approach (number of firms who perceive the use of new technology as profitable), and secondly, a game theoretic approach (Stoneman, 1983). More recent research addresses the theory and empirics of the diffusion of substitute and complementary technologies (Stoneman and Kwon, 1994).

Diverse empirical studies on the use of environmental technology have found out that environmental regulation is the decisive factor for their implementation (e.g. Adler, 1997; Halstrick-Schwenk et al., 1994; Bigleiser and Horrowitz, 1995; Newell et al., 1996). Jaffe and Stavins (1995) found that incentives for the invention and adoption of cleaner technologies are greater under emissions taxes and tradable emissions permits than under command and control regulations. For the analysis of diffusion and acceptability of new environmental technologies and products in the user sector various factors have an influence:

— transfer of information,
— costs,
— risk and uncertainty.

Knowledge and information on available techniques and newer developments is a key prerequisite, if a firm wants to implement environmental technology. Information costs for available and proven techniques are reckoned to be relatively low because the techniques have been used in practise and information is available through conventional information transfer such as fairs, journals, consulting firms or through direct contacts with suppliers. It is much more difficult to get reliable knowledge on newer technologies. In particular small firms often do not know which clean technologies are available (Kemp, 1993). Moreover, the application of technology is almost always specific to a certain location. Tailor-made solutions are therefore required and face-to-face relations between the special suppliers of the clean technology and the potential users must take place (Dielemann, de Hoo, 1993). The quality of these relations is also decisive since the interaction ought to take place between equal partners (Frank and Swarte, 1986). Another knowledge-related factor is the availability of cleaner production expertise in the environmental technology sector. Many suppliers have specialised on putting forward concrete solutions for handling existing waste streams and emissions and are less specialised in identifying areas which require special attention for prevention of waste and emissions (Cramer and Schot, 1989). However, this existing knowledge, if used creatively, can produce a process where alternative technologies can as well be developed.

The *costs* and benefits of a certain innovation for the environment are determined by its price and quality. Costs can involve purchasing costs, implementation costs, financing costs, operating costs; benefits can be found in better environmental performance, cost savings, increased demand and improved public image of the firm. The cost consequences of adopting clean technologies can differ enormously for firms. They do not only depend on the costs of purchasing, but also on the age and the nature of the production process. Hereby sunk costs play an important role. There is the tendency that existing installations invest in end-of-pipe technologies because of sunk costs (the capital costs of an existing installations which are fixed in the short and medium run). In an investment decision only the variable costs and the initial investment cost of an end-of-pipe technology are compared with the total costs of clean technology. Therefore the use of end-of-pipe technology may be a relatively more advantageous solution in terms of net present profits (Hartje, 1990). But the significance of the cost differences between end-of-pipe and clean technologies also depends on the R&D efforts of suppliers and the productivity increases of the users of environmental technologies. Clean technology might be cheaper, if suppliers can decrease costs through innovations and when users achieve cost reductions through e.g. energy savings during the use of clean technologies. The influence of sunk costs on existing installations also depends on both the regulator's and the companies' strategies for compliance with upcoming legislation. If companies can plan investment on a long term basis, the use of clean technology in replacement investment becomes more likely (Faber et al., 1994). Sunk costs, however, are not important in the case of a new installation or a replacement investment so that total costs of both end-of-pipe and clean technologies are compared. Clean technology is chosen when its costs are lower than the costs of a standard technology extended by end-of-pipe measures.

Another important cost factor is represented by the implementation costs related to the use of clean technology and there are also learning and scale effects which need to be taken into account. The implementation costs are the higher, the more process steps are involved in the innovation, the more basic the innovation is and the less experience is available (Maas, 1990). Thus, clean technologies may require a change in production routines (new jobs, rewards) and may meet both managerial and worker resistance (Kemp, 1993). Training is required and maybe even the hiring of new employees with special expertise. Adjustment costs of end-of-pipe technologies are expected to be much lower because only little amendments in the existing technology have to be made. However, learning effects in the use of clean technologies through improved routines over time can lead to a decrease of unit costs. These cost reductions can be the sooner realised the faster production grows due to increased demand. Thus, in the long term, cost structures of firms may change and the higher costs initially associated with clean technology may be overcome (Hemmelskamp, 1999).

Finally, because of the difficulties of obtaining an overview on all available environmental technology and due to the lack of complete information on their performance, users are faced with *technical and economic risks* (Kemp/Soete, 1990). Process-integrated clean technologies that lead to radical innovations, show a high risk (Hartje and Lurie, 1984). The more complex a technology is, the higher is the risk for frequent down time breaks. Also, if a technology is very new, no draw back on existing experience is possible (see Rogers, 1995). Personnel must gain knowledge of and experience with new machines. In particular smaller firms lack expertise in relation to clean technology, e.g. membrane technology (Kemp, 1993). The technical risks associated with clean technology are immediately related to economic risks. The higher the down-time, the greater the costs of non-production. If there is a problem in the use of an end-of-pipe technology, the standard technology can still be used to a certain extent (depending on the enforcement of environmental regulation). Since most companies would show a largely risk averse behaviour, end-of-pipe technologies seem to be advantaged over clean technologies.

2.3.4 Conclusion on innovation and diffusion activities due to environmental regulations

Even when some authors express straightforward views on what the relationship between environmental regulation and competitiveness could look like, these results might alter when the mechanisms of technological change through innovation are introduced in detailed form. Both innovation and diffusion of environmental technology which are spurred by environmental regulation are influenced by a variety of factors which will eventually also influence competitiveness in one way or the other. It became clear that in particular the type of environmental technology (clean or end-of-pipe technology) used in response to regulation will have an impact on the final effect on competitiveness.

2.4 Review of empirical studies on the impact of environmental regulation on competitiveness

Independent of the conventional or the revisionist view, there are not many empirical studies available which have looked at the detailed relationship between environmental regulation and its effects on competitiveness. This might be due to the fact that "... the link between environmental regulatory policy and the allocation of environmental resources is complex, multistep and imperfect" (Gabel/Sinclair-Desgagne, 1997).

A useful typology of available studies is undertaken by Stewart (1993). He separates the studies in research on productivity, location and trade. The micro perspective adopted in this piece of research is closely linked to the productivity studies, but also implements some elements of the other categories of studies.

2.4.1 Productivity studies

Results of productivity studies show that „the effect on measured output, and hence competitiveness, varied widely by industry“ (Stewart, 1993). There are scarce signs that a Porter-type first mover advantage can be spurred by stringent environmental legislation and can lead to a dominant position. In the following examples of systematic studies found in the literature will shed light on how disperse results of productivity studies can be depending on their design and the industries under analysis.

Gollop and Roberts (1983) estimated that the American Clean Air Act reduced productivity growth by 0.59 per cent per year during the 1973-79 period in the fossil fuelled electric utilities industry. Gray (1987) attributes about 30 per cent of the decline in productivity growth in the U.S. manufacturing sector during the 1970's to both workplace and environmental regulation. Another Canadian study by Smith and Sims (1985) did not find a significant effect of pollution charges on the productivity of four brewing firms. In a study of the manufacturing sectors of the U.S., Germany and Canada virtually no impact of pollution investment on productivity growth in Canada was found during the 1967-80 period (Conrad and Morrison, 1989). Also for the manufacturing industry in Canada, in the province of Quebec, Dufour et al. (1998) estimated in a least-squares procedure a reduction in productivity growth caused both by health and safety and environmental regulation. The magnitude of the negative effect of environmental regulation was smaller than in American studies, but comparable to the results of other Canadian studies. A decrease in productivity growth was found, although the study had theoretically also allowed for positive impacts through research and development activities.

A very detailed study on the effect of environmental regulation on productivity was undertaken by Barbera and McConnell (1990) in five U.S. industries which are heavily affected by environmental regulation. It was attempted to measure some of the positive impacts of regulation. The authors analysed the impact on total factor productivity by splitting up this effect into the direct effect concerning the direct costs for abatement capital and into the indirect effect of a shift in the production function. The latter measures the change in the way conventional inputs are used to produce the manufactured good (change in the amount/combination of inputs). Whereas, the direct effect on productivity is negative because direct pollution abatement costs always add to total input costs, the indirect effect can be positive, negative or neutral. A negative impact on total factor productivity can be expected when in conjunction with abatement capital more energy and labour is needed, thus decreasing the productivity of the conventional output. However, abatement requirements could make conventional equipment obsolete more quickly and then lead to new investment and increased productivity. Barbera and McConnell build up a translog cost function and show that the indirect effect varies substantially with industry and that it is in general smaller than the direct effect. Some industries are able to compensate for parts of the productivity loss by changes in other inputs (mainly housekeeping measures). All industries use capital and energy for abatement, but the amount of employment is variable. On average, in the period from 1961 to 1980 the average annual loss in productivity varied from 0.24 per cent in Stone, Clay and Glass to 0.08 per cent in Non-ferrous Metals. In Stone, Clay and Glass the productivity loss is about a 55 per cent reduction in comparison what productivity growth would have been as a whole in this period in the absence of regulation. Abatement requirements partly lead to a decrease in relative employment (e.g. in Chemicals) and partly to an increase. In Chemicals costs go down in the early periods, the firms may have shifted to low-cost production methods and processes as a result of environmental regulation. However, the elasticity of cost with respect to abatement becomes larger and positive in the late 70ies when regulatory pressure also increased in response to the fast growth in agricultural and petro-chemicals. The total abatement effect is negative in all five industries, but it appears that the environmental regulation may have had a smaller effect on total factor productivity than others have predicted for the aggregate level, beginning to decrease in the early 1970ies.

Another study by Gray/Shadbegian (1995) found a clearly negative relationship between a plant's pollution abatement costs and productivity. They examined three US industries (pulp and paper, oil refineries, steel mills) and came to the result that the heavier the pollution and regulation was, the lower the productivity was. Raising environmental performance did not trigger productivity benefits large enough to outweigh the measured compliance costs.

Repetto (1995) tested the hypothesis that superior environmental performance leads to lower profitability than inferior environmental performance within the same industry. The findings for 50 finely defined US industries show that there was no tendency for good environmental performers to be less profitable. It was concluded that there is no evidence that superior environmental performance puts firms at a market advantage or that it adversely effects market performance.

Boyd and McClelland (1999) have undertaken one of the most recent and most profound studies on the impact of environmental constraints on productivity. They have tested the Porter hypothesis of a possible win-win potential to reduce input use and pollution output, while maintaining given productive output levels and have also allowed for the measurement of losses of production from environmental constraints. Within a framework of overall production efficiency measurement they evaluated the effect of pollution levels and related resources/production changes and the specific capital requirement for a sample of

US integrated paper mills using a distance function approach. The plant as economic unit is looked at as making decisions on their desired outputs as well as on the pollution outputs. Thus the extent to which the decisions regarding the pollution output constrain the plant's efficiency can be measured. The most obvious constraint is that a plant may use scarce investment for pollution abatement activities. Different investment types are taken into account in order to measure the potential constraints on productivity. Results show that the input use by the paper industry could be reduced anywhere between 2 per cent and 8 per cent without adverse effects with energy and fuel use showing the greatest potential for reductions. Simultaneously constraints from the legal obligation for environmental investment reduced productivity by about 1 per cent to 4 per cent. From the total level of production inefficiency of 17.1 per cent in 1990 9.4 per cent are due to environmental constraints and 7.6 per cent of the inefficiencies are due to other differences between plants. The 9.4 per cent of general efficiency loss due to environmental obligations can be split into 2.7 per cent due to the abatement capital constraint and in 6.7 per cent of other indirect effects of environmental constraints.

Aden and Kyu-Hong (1999) examined the intensity of regulatory exposure in Korea. They find a positive association between plant size and productivity and plant-level abatement expenditures in their study of the manufacturing industry (see for similar results also Hettige et al. 1996). In a survey of two high-polluting sectors (textiles and plastics/resins/petrochemicals) they applied a two-stage least squares estimation of a responsive regulation type model. It was also found that domestically owned plants spend more on abatement than firms with some level of foreign ownership. This, however, can well be a specific feature of Korea where the country's large firms do not get a lot of public support and want to protect themselves from public criticism.

The only study which clearly supports the Porter hypothesis is undertaken by Murty/Kumar (2001) and analysed the effects of waste water regulation in India on the productive efficiency of firms. Using panel data of 92 water polluting firms for the period between 1996 and 1999 it was found that the technical efficiency of firms increases with the intensity of environmental regulation which induces water conservation efforts.

To summarise, most of the systematic research evidence comes to at least mixed findings concerning the relationship between environmental regulation and competitiveness on a productivity level. Clear proof of the Porter hypothesis is scarcely found.

2.4.2 Location studies

Studies on locational impact of environmental regulation examine whether a movement from nations with stringent standards to those with lower standards is observable (Stewart, 1993). US and also international studies have found out that the costs of environmental regulation are only of minor importance in the decision making process concerning the siting of new production facilities (see for instance Birdsall and Wheeler, 1992). This result may reflect the small percentage of environmental costs of overall production costs. Although studies on relative toxic intensity of sectoral output showed the greatest growth in dirty production in developing countries starting in the early seventies, the hypothesis of migration of dirty industry to poorer countries with less stringent environmental standards cannot be proven (Lucas, Wheeler, Hettige, 1992). There might be also a compositional effect of industrialisation where in the early stages high-polluting activities represent a relatively high percentage of output. Birdsall and Wheeler (1992) even found that between 1960 and 1990 relatively closed economies in developing countries produced more pollution-intensive output than economies which were more open to trade. However, this study did not control for the adoption of low environmental standards in these rather closed economies. In a study on chemical and mineral-processing firms in the US (Leonard, 1988), the author concludes that there is no general industrial flight recognisable in foreign direct investment. However, some displacement effect, for heavily regulated industries with stagnant markets making environmental investment difficult for them, can be discovered. According to Leonard's study there are no documented examples of healthy American industries which moved abroad by force of environmental regulation. Earlier research of Leonard (1984) indicates that relocating industries have been hit harder by regulation concerning workplace safety than by environmental legislation. An update of Leonard's research by Ferrantino (1995) for the early 1990s again found no shift of U.S. investment in pollution-intensive industries towards developing countries. A more recent study by Eskeland and Harrison (1997) which examined outward investment from the U.S. and the pattern of foreign investment in four developing countries found no evidence for support of the pollution haven hypothesis. A joint ESCAP/UNCTC study from 1990 on the activities of multinational companies found that environmental costs were only 0.54 per cent of output value and concluded that factors other than environmental control costs are much more important in determining locational decisions (cited in World Development Report of UNCTC, 1992).

Overall, location studies do not find a major movement from countries with stringent environmental standards to those with laxer standards.

2.4.3 Trade studies

The locational studies are closely linked to the trade studies. If a country has to face high compliance costs with the production of certain goods, the country would try to export goods with relatively cheaper compliance costs and tend to import the goods with relatively higher abatement costs. Kalt (1988) tested this hypothesis based on 1977 compliance costs for several US industries and found substantial trade effects. Using a Heckscher-Ohlin analysis he discovered that in 1977 US exports were much less pollution-intensive than imports, a result which was not true ten years ago. Kalt (1988) attributes this difference in the composition of exports and imports to a regulatory-based comparative disadvantage. Still, there is little econometric evidence that this had any significant impact on overall US export levels. Looking at the exports of "dirty" industries (those with highest pollution control costs) in several countries, Low and Yeats (1992) find that differences in environmental standards had a significant effect on trade patterns. Developing countries experienced during the time period the study took into account (1965-88), a more rapid growth of exports of dirty goods than exports generally. However, other variables which affect location decisions like wage rates or taxes were not considered. Ratnayake (1996) did not discover a significant impact of environmental regulations for New Zealand's international competitiveness. A study by Albrecht (1998) found an improvement in exports and competitiveness in countries with a proactive strategy on the regulatory settings of the Montreal Protocol which forbid CFCs. This was at the level of final products like refrigerators.

To summarise, the older trade studies are mainly based on data from the 1960's and 70's and show that there are no significant effects for most industries. There are also alternative explanations for the economic performance of developing countries. The newer studies from the 1990ies find even less evidence to suggest that stringent environmental standards lead to a loss of competitiveness.

A conceptual study by Alanen (1996) examines the competitiveness impacts of environmental cost internalisation between different sectors of economic activity using a framework which encompasses both productivity and trade elements. Her view is "... that although the observable effects at an economy-wide level have been small at present levels of environmental protection, competitiveness impacts are likely to vary greatly between different sectors, and that marginal companies may be affected in any sector" (Alanen, 1996, p. 11). It is argued that the ultimate competitiveness effects on a sector depend on several factors related to products, production processes and markets of different sectors. In a conceptual framework called the environment-competitiveness matrix competitiveness impacts of environmental protection efforts are described along two axes: cost C representing price changes, and (non-price) differentiation D, in particular through merchandise (content and image of products), shaping sub-markets in which competition is less intense. Box 2.4-1 presents a summary of some factors impinging on the competitiveness impacts of environmental cost internalisation.

Box 2.4-1 Summary of some factors impinging on the competitiveness impacts of environmental cost internalisation

Dimensions of competition	**Subdimensions**	**Determining factors**
Cost (proxy for price)	Magnitude of environmentally related net cost increase	Type and extent of environmental externality (e.g. pollution intensity)
		Technological aspects (e.g. investment cycle, innovation potential)
	Competitive significance of the cost increase	Ability to absorb costs (e.g. cost structure, profit margins)
		Ability to pass on cost to customers (e.g. price elasticity of demand, prices of competing products)
Differentiation	Feasibility of competition based on differentiation	Nature of product (e.g. degree of homogeneity, price based competition)
	Differentiating importance of environmental characteristics	Customer environmental concern (e.g. destination of exports, sensitivity issue)

Source: Alanen (1996).

The advantage of the presented framework is that the impacts of environmental cost internalisation can be separated from other factors affecting competitiveness, something which has been difficult in previous empirical studies. However, according to Alanen (1996) it is necessary to identify additional influencing factors and to examine their relative explanatory power and the correlation between individual factors. This concept had significant influence on the formulation of hypotheses for this thesis (see section 2.5).

2.4.5 Conclusions on empirical studies on the impact of environmental regulation on competitiveness

The vast majority of the productivity studies quoted above has examined the effects of environmental regulation on overall productivity and has shown that the impact on competitiveness varied widely by industry, being positive, neutral and negative. The different research frameworks, however, make it difficult to compare results properly.

In both location and trade studies no significant relation between environmental regulation and industry migration or trade performance was identified. Even when trade patterns have changed over time, it is not clear which phenomenon of the many possible ones is responsible.

A shortcoming of all studies is that no systematic search for the impact of the type of environmental abatement measure was undertaken. In most cases the impacts of end-of-pipe technologies were measured, but not those of clean technologies. It should also be noted that much of the evidence has been U.S. based, with only little attention paid to the European case.

2.5 Development of hypotheses for the case studies

Based on the mechanisms found in the theoretical literature on environmental regulation and competitiveness as well as on the factors influencing technological change, appropriate hypotheses for the case studies of this thesis were developed. The focus of the case studies was on the effects on productivity and competitiveness of compliance with environmental legislation. Special attention was paid to Germany where regulation has been known to be relatively stringent. Taking the preceding arguments into account, it is not clear whether or not environmental regulation has a negative impact on productivity and competitiveness.

The adopted baseline here is that average productivity levels (as indicator of competitiveness) and the sources responsible for international differences in these productivity levels are hypothesised to be causally associated with the direct and indirect productivity effects of the solutions to the environmental regulation and other environmental pressures. The hypotheses deliberately stress the possibilities and potential consequences of using clean technologies, a focus which so far has been rarely examined in the literature. Hereby, the factors which define competitiveness are also known to influence the adoption of clean technologies. As mentioned earlier the competitiveness factors are related to the input and output side of the firm (Jacobson and Andréosso-O'Callaghan, 1996). Concerning the input side (the likely explanations of competitiveness) the decisive factors are physical and human capital, R&D spending, etc. E.g. Griliches (1986 and 1995) found for the US in the period of the 1970s that firms which spend a high percentage of their R&D on basic research achieve a higher level of productivity and reach a higher level of output relative to their other inputs. The listed input factors also influence the take-up of clean technologies (OECD, 1985a, 1987 and 1995; Vassilopoulos, 1999). On the output side (illustrating the consequences of the relative competitiveness of a firm) the essential factors are profitability, market share, productivity and so on. The latter provide the resources for the take-up of cleaner production methods and in their function as indicators of competitive performance they simultaneously illustrate the consequences of the adoption of clean technologies (OECD, 1987 and 1995).

In summary, the hypotheses express that plants are exposed to certain types of environmental pressures, and their ability to respond efficiently to these pressures depends on their competitive position as well as factors internal to the firm. This framework for research is also depicted in Figure 2.5-1 below. Although the case study sectors under concern have different properties, the sample data can be analysed according to common hypotheses concerning the ability to adjust efficiently to environmental regulation.

Influence of plant competitiveness (productivity) on adjustment to regulation and subsequent compliance costs

The proportional cost of environmental compliance relative to turnover incurred by the firms is likely to be a negative function of the productivity level (Hitchens et al., 1998, 1999, 2000, 2001). This would be because firms with the capabilities to achieve high productivity will also find it easiest to respond to environmental challenges without the penalty of reduced output and employment. It implies for international comparisons that countries/regions with relatively high productivity (levels of competitiveness) will have relatively low compliance costs. Also, within a country/region those firms with above average productivity could be expected to have relatively low compliance costs. According to this hypothesis compliance costs in Germany could a priori be expected to be low because the country is a high productivity economy.

Competitiveness and environmental initiatives including environmental awareness

A favourable environmental performance is hypothesised to be linked to characteristics of high productivity (best practise) plants. Just as these high productivity plants would be more aware of any gains in efficiency than the average, their management and workforce can be expected to undertake a higher number of environmental initiatives because they show a higher environmental awareness. Still, these plants are hypothesised to be able to keep compliance costs low for a number of reasons also related to best practise firms:

— *Type of environmental initiative and compliance costs*: In general, compliance costs could be assumed to increase with the number of environmental initiatives. High productivity plants, however, will not only react more actively to environmental pressure with a higher number of environmental initiatives than other plants with lower productivity, but also prefer clean tech-

The preceding analysis and findings yields the following hypotheses:

Fig. 2.5-1 Framework for research
Stringent environmental regulation in the areas and industries of ...

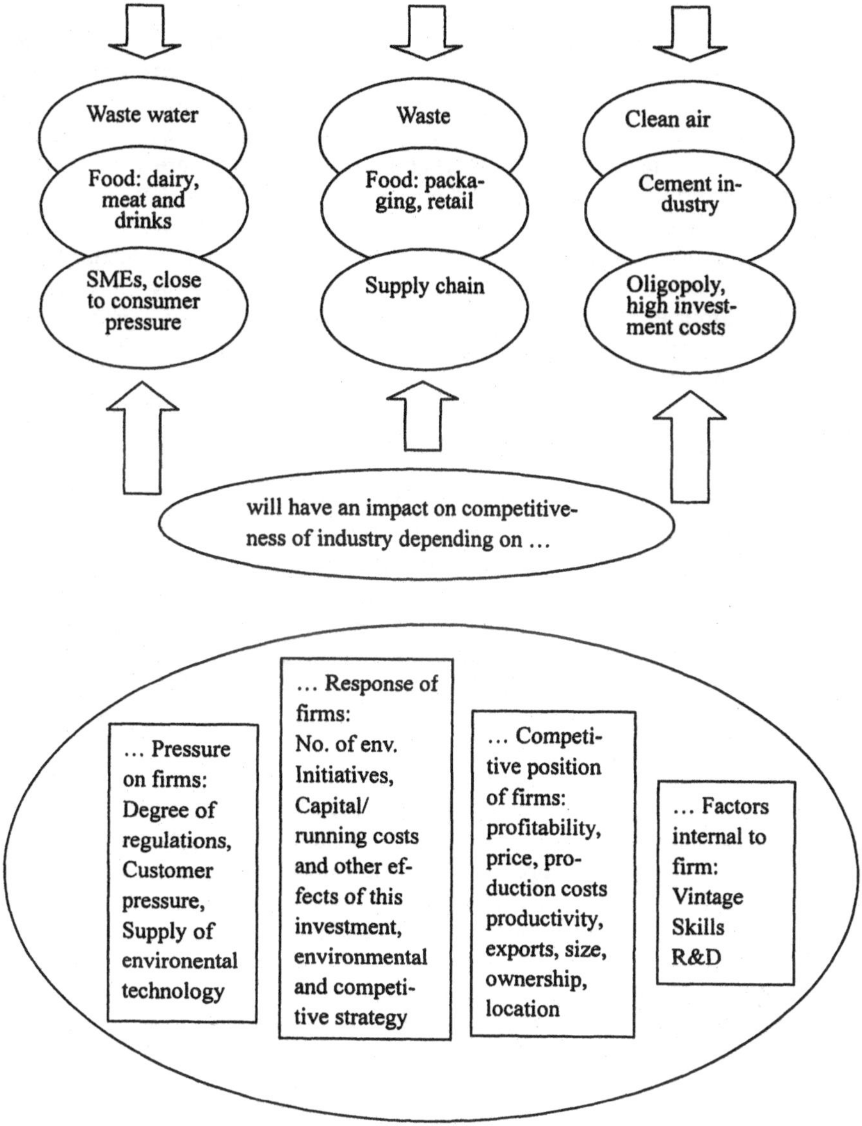

nologies to end-of-pipe technologies. This is because the latter is normally associated with additional costs and losses in productivity, whereas clean technology can improve productivity, reduce compliance costs and simultaneously bring emissions down (Coenen et al., 1995).

— *Skills and R&D*: The type of environmental activity is also anticipated to be connected with the skill and R&D endowment of plants. Plants with a well trained labour force and a strong R&D basis will also integrate environmental activities in their work routines and come up with a higher number of individual, clean technology solutions and are likely to be more successful in reducing environmental costs than their counterparts lacking these skills (interview with Bundesinstitut für berufliche Bildung, 2000; Jenkins, 1999). Regulation could lead firms to increase their R&D efforts, thus leading to a decrease in production costs (Oates et al., 1993) and stimulating productivity growth (Dufour et al., 1998; see Kennedy (1994) for a formal argument). In comparison to the UK Germany is said to have a better skill base and the R&D activity is also higher (Daly et al., 1985; Prais, S.J., 1981; Steedman and Wagner 1987 and 1989; Mason et al., 1994). Therefore Germany might have relatively good preconditions for the use of clean technology.

— *Plant age*: The age of a plant and its machinery is likely to have an impact on its environmental performance and on compliance costs (Alpay, 2001). Modern plants with younger machines are more productive and less polluting than plants using older machines. They can reach better environmental standards because the newest technology also embodies best environmental technology and are allowed to achieve environmental goals at a lower cost than with old equipment which would need expensive retrofitting.

Influence of plant and firm size

Plant size is assumed to have an influence on the adjustment to regulation and related environmental pressure (Henriques and Sadorsky, 1996). Large plants are hypothesised to undertake more environmental initiatives because they often have more financial and human resources for the solution of environmental problems than small and medium sized plants and have also better opportunities to research cost efficient environmental measures. Large multiplant firms could possess certain economies of scope across several markets in complying with a regulation; that is large diversified companies may be able to spread compliance costs across several markets (Arnold, 1995). Thus the proportional cost of compliance relative to turnover is likely to be a negative function of the size of plants/firms.

Influence of history of regulation

Plants in countries with a history of stringent environmental legislation will be in a better position to develop or gain access to environmentally and economically beneficial solutions than their counterparts in other countries with a history of less stringent regulation. This is because their supplier network is more experienced with solutions to environmental problems and can offer a more differentiated supply of environmental goods including clean technology. Germany has a very long history of environmental regulation and a very developed supply of environmental goods of which it could potentially benefit when adjusting to legislation (Wackerbauer, 2002).

Influence of green market pressure in supply chain relations

Not only regulatory pressure forces firms to react, but also environmental pressure coming from the market place (Hitchens et al., 2000; Dobilas and Macpherson, 1997; Henriques and Sadorsky, 1996). Relations within the supply/production chain are likely to be both an influence upon, as well as being influenced by, the level of environmental standards, e.g. a manufacturer may find it easier to increase the environmental standards of its products if it has a reliable and competent base of suppliers to draw on. A manufacturer may be forced to upgrade product and process environmental standards by pressure coming from the customers of plants in the three sectors under study.

Influence of green consumer pressure

Plants which operate in a market where final demand for clean products and processes is large and consumers are prepared to pay a premium, will be more involved in environmental measures than other firms which do not have similar customer groups and are not rewarded for their environmental activities (Sprenger, 1996). Currently green consumer pressure can be expected to be higher for Germany than for Spain or the UK where the general level of environmental awareness among the population is just beginning to grow. The demand for environmentally sound products is expected to grow in the future because environmental awareness of consumers will increase (Steger, 1993; Welford, 1995).

Influence of market structure (packaging and cement study only)

The industries researched in this study (food, packaging and cement) operate in different market structures. This will influence a firm's strategy towards the environment. In a relatively competitive market like the food industry, constant striving for new cost efficient environmental solutions is required because chances to pass on costs to customers are small. In an oligopoly like the cement industry individual incentives for renewal and updating always depend on what other members of the oligopoly will undertake. On the one hand, adjustment to new environmental requirements may therefore be delayed in an oligopolistic market. On the other hand, if an industry achieves super-normal profits, it is to a certain extent protected against additional costs arising from environmental compliance and the pursuit of wider management goals which includes environmental protection is possible (Hettige et al., 1996). Thus, market power will allow to pass on environmental cost increases (Alanen, 1996). Also, where a firm is acting in a protected local market, it can pass on any higher cost arising from environmental regulation than if it was active in a global market with competition from countries with laxer regulation (Levêque, 1993). Monopsony power in the retail sector (packaging supply chain study) could be used to exert environmental pressure and force suppliers to adopt environmental initiatives.

Influence of ownership and location

Ownership: Ownership is likely to be of influence on the environmental response a plant or a firm gives to environmental regulation. The industries under consideration in this study are to varying degrees composed of multinational enterprises. Multinational companies are likely to show better environmental performance than other enterprises because they export both the environmental and the efficiency standard of their home country (at least in their new production facilities, see Epstein and Roy, 1997 and 1998). Therefore, where multinational branch plants are sampled in those parts of the EU with lower environmental standards/enforcement, they are expected to undertake a higher number of environmental initiatives and are in a better position to reduce environmental costs than indigenous companies producing the same products.

Location: Location can affect the cost of compliance. The assumption that environmental regulation has differential impacts between regions and countries would be related to differences in competitiveness between Member States of the EU which are based on the underlying differences in productivity and skills/R&D (Barras et al., 1997). These factors are related to the efficient adjustment to regulation. Therefore plants in peripheral locations are likely to be less advanced on environmental matters, to undertake less environmental initiatives and to be potentially subject to higher environmental costs because of lack of access to suppliers, language barriers, higher transport costs etc. than plants in centrally located regions.

Possible employment impact of environmental regulation

The direction of the employment implication of adjustment to stringent environmental regulation is not clear a priori. The available macro-econometric evidence has shown that the overall employment effects of environmental policies are likely to be small.

Summary: Influence of high standards and strict enforcement

Given the above hypotheses, although high environmental standards and strict enforcement may represent a short term cost and burden to the firm, they could in the medium and longer term push firms on to a higher growth path by forcing them to make product and process changes which yield higher competitiveness. If this happened it would represent part of the so-called "double dividend", i.e. gains in environmental performance would also be accompanied by increased economic performance.

The above listed hypotheses will be empirically tested for the effects of environmental regulation in the food industry, the supply chain of the packaging, food and retail industry, and the cement industry (see chapters seven, eight and nine).

3. German environmental regulation: history and case study regulation in an international context

Germany has a history of about 30 years of environmental policy. The stringency of the regulation behind this policy has been evaluated in cross-national comparisons to be above average in comparison to other industrialised countries. Concerning emission levels and the use of environmental technologies Germany ranks among the world's best. In the European Union only Sweden, Austria and the Netherlands reach such environmental performance (see Jänicke&Weidner, 1995).

This chapter sets the scene for the empirical analysis given in chapters seven, eight and nine. The purpose is to present a brief history of German environmental policy indicating that German legislation has indeed been more stringent over a long time relative to other countries, in particular in the EU (section 3.1). Sections 3.2 to 3.4 discuss the individual pieces of environmental regulation as they apply to the three case study industries chosen for this thesis (food sector, supply chain of the packaging industry, retail and food sector, cement industry). The view of a more stringent regulatory approach in Germany than in the other sample countries Ireland, Spain and the UK is confirmed. As far as possible, comparisons are also drawn relative to the remaining EU member states in order to gain a clear picture about the overall stringency of the selected fields of German environmental policy within the EU. Section 3.5 provides a summary and the conclusions to the chapter.

3.1 Brief historical overview of German environmental policy

This section presents a short historical overview of German environmental policy from the 19th century until today and discusses its stringency relative to other countries.

3.1.1 Developments from 19th century until end of World War II

Environmental regulations have a long tradition in Germany. The first systematic piece of legislation was the Prussian Industrial Statute of 1845 which emphasised protection of health and property in the vicinity of emitting premises. Increasing industrialisation gave rise to further legal statutes, but there was no central agency responsible for management, implementation, control and sanctions concerning environmental matters. Thus, the environmental situation was bad and worsened as industrialisation progressed in the 20th century. In particular, shortly before and during the Nazi period the defence industry contributed heavily to pollution (Brüggemeier and Rommelspacher, 1992). Although there was already in 1936 a law on air hygiene, the environmental situation did not improve. This was the case because industry was largely protected against this law.

3.1.2 Post World War II period

In the years following the Second World War social and economic issues were in the forefront of political discussions and the environment continued to be neglected. The years of the "economic miracle" caused massive damage to water and air (Wey, 1982). Then, in the early 1960ies public discussion about pollution began in the highly industrialised and polluted state of North Rhine-Westphalia and led to legislative and organisational measures, in particular in the field of air pollution control.

However, it was mainly stimuli from abroad, primarily from the US and also from Japan which led to a comprehensive discussion of environmental problems in Germany in the early 1970ies (see Weidner, 1995). This was in response to the publications in the US by Rachel Carson (1962), Paul and Anne Ehrlich (1970) and the Club of Rome (especially Meadows, "The limits to growth", 1972). Also the foundation of the US Environmental Protection Agency, the Council on Environmental Quality and the enactment of the National Environment Protection Act had an impact on the origins of similar institutions in Germany.

Japan was influential for Germany because of pollution-related diseases.[5] This was not widespread in Germany; still the degree of awareness was high enough that the Social Democrat-Liberal Coalition Government initiated in 1969 the origins of environmental policy in Germany. One of the first measures taken in-

---

[5] Japan uses some environmental policy instruments which are without precedence worldwide or at least rare; among the most important ones are compensation payments for health related damages, industry participation in the costs of public environmental measures and total emissions control for the reduction of air and water emissions in selected polluted areas (Weidner, 1992). Japan is one of the few countries where almost all areas of environmental protection are regulated not only through specific laws and standards, but – unlike Germany – also by environmental goals (OECD, 1985 and 1994). Japanese standards are quite strict (in many cases stricter than in Germany), e.g. BOD and COD daily averages of 120 mg/l, dust between 0.05 and 0.07g/Nm^3 according to type and size of facility, NO_x between 60-150 ppm at a gas boiler or 450-550 ppm at a solid firing boiler (Barret and Therivel, 1991). Simultaneously there is also a strong institutional base which is responsible for enforcement of environmental regulation. This is internationally an exception. According to the OECD (1994, p. 46) "Japan's air pollution monitoring network is probably the most sophisticated to be found in OECD countries".

volved the transfer of responsibilities relating to pollution control from the Ministry of Health to the Ministry of the Interior. The reason for that can be found in a recommendation of a project group on governmental and administrative reform (PRVR, Projektgruppe Regierungs- und Verwaltungsreform beim Bundesminister des Innern, 1969). This project group thought that the Ministry of the Interior was especially capable to handle environmental tasks because it would not have particular interests in this area and would be able to produce a trade-off between health, economic and other interests regulation (Müller, 1986).

3.1.3 Principles of modern environmental policy

The period between 1969 and 1974 is described by Müller (1989, p. 23) as a phase of "very active policy design" by the administration, and compared internationally, this environmental policy was "progressive and demanding for that time" (Weidner, 1995, p. 6). During this period several pieces of clean air legislation were passed quickly, among those the Federal Air Quality Protection Act of 1974.

In 1971 the Government published an Environment Programme for the coming years and anchored the three basic principles of modern environmental policy: the principle of precaution, the principle of co-operation and the polluter pays principle. Relative to other European countries Germany was with this approach quite advanced (see Box 3.1-1).

Box 3.1-1 Principles of modern environmental policy in Germany relative to other countries

The *precautionary principle* implies that already the assumption of danger from emissions leads to a policy reaction. It also implies that economic aspects must not play a central role in the design of environmental policy. The design of emission limit values particularly reflects the implementation of the precautionary principle. Especially, the existence of national emission limit values shows that a country aspires far-reaching precaution in its environmental policy. In Germany the precautionary principle gained an important position in 1976, when the Environment Programme was updated. With respect to the precautionary principle Germany has served as a model at EU level and also internationally. In the EU in the 1980ies only Austria, Germany and Sweden applied the precautionary principle in the regulation of SO_2. During this time also Japan, the US and Switzerland developed an environmental policy oriented at the precautionary principle (Hohmeyer et al., 1988).

Co-operation principle

The co-operation principle aims at involving all stakeholders in the design of environmental policy. It reflects the decision structures of a country as well as the intensity of public discussion about environmental problems and their solutions. In Europe Austria, Germany and Switzerland are the forerunners.

Polluter-pays-principle
The polluter-pays-principle is decisive for the success of environmental policy. It means that the polluter has to pay for measures to reduce pollution (no subsidies). The polluter is also liable for damages. Canada, Sweden and Japan (compensation payments for health-related damages) were the forerunners in this area with extensive liability programmes for environmental damages. The German charging policy in the area of waste water has been an early way to realise the polluter-pays-principle in the 1970ies. But unlike in Canada and Sweden liability questions are less stringently regulated in Germany. According to the OECD (2001) the polluter-pays-principle is well established and implemented in Germany, particularly for traditional air pollutants. The full cost of environmental services in the area of waste and water management are increasingly covered.

3.1.4 Grassroots for the environment movement

Between 1974 and 1978 there was a period of environmental stagnation due to the economic recession related to the oil crisis. During that time, however, public environmental pressure grew enormously. The nuclear energy question led to a large anti-nuclear movement and provided the grassroots for the environment movement and the Green Party later on (Poguntke, 1992). Already in 1978 the Green Party became a member of the Parliament of the City of Bremen. In 1983 the Green Party was elected to join the Bundestag. The green pressure led to far reaching activities, in particular concerning vehicle regulation and large combustion plants and the development of an ecological action programme.

3.1.5 First Minister of the Environment

At the end of the Social Democrat-Liberal Period broad parts of the society supported an even more stringent legislation than that proposed by the government. The Conservative-Liberal Government with its new Minister of the Interior, Friedrich Zimmermann, pushed through several strict pieces of environmental legislation. There was strong pressure for clean air policy because of the "Waldsterben" and the Ordinance on Large Combustion Installations was quickly passed. In 1986, after the nuclear catastrophe in Chernobyl, the Ministry of the Interior got sharply criticised for its poor information politics concerning radiation levels. During this crisis the New Ministry of the Environment, Nature Conservation and Nuclear Safety was founded and the first full-time Minister of the Environment was appointed.

Under the first Minister of the Environment Klaus Töpfer important goals such as the development of a concept for protecting the North Sea and the Baltic were met, and also laws on environmental liability were extended. However, there was no progress on many other issues, e.g. eco-taxes and environmental charges, vested interests were hard to overcome in terms of energy policy and there was sharp criticism of the planned construction of hazardous waste incinerators.

3.1.6 German reunification

On the first of July 1990 the environmental union between the Federal Republic of Germany and The German Democratic Republic was confirmed by a jointly designed environmental framework law ("Umweltrahmengesetz") where the East German standards were equalised to the West German ones. The general ordinances of the Federal Ambient Pollution Prevention Act (Bundes-Immissionsschutzgesetz) (e.g. regarding noise, air pollutants etc.) were completely transmitted to the new Länder directly after the reunification on October 3rd 1990 and are now fully implemented (Sprenger et al., 1991).

After the reunification also the European environmental law had to be applied in the new Länder. However, due to the catastrophic state of the environment, the economic and social situation as well as due to implementation problems, the EU law provided transitional arrangements which were in principle not longer valid than December 31, 1992. In the critical areas - water, air and waste - the transition period lasted until the end of 1995 because short term success in these areas could not be expected. E.g. if the limits for drinking water quality had to be met, there would not have been any drinking water available. Thus EU directives 75/449 (surface waters), 80/68 (ground water) and 80/778 (drinking water) had to be fulfilled only until December 31, 1995.

The effectiveness of environmental policy in East Germany has progressed much more slowly than expected and the planned end of the clean-up by the year 2000 could not be realised. In part this is due to the desolate environmental situation which needs more resources, but also due to the general poor situation of the economy and the chronically high unemployment.

3.1.7 Socialdemocratic-Green Coalition since 1998

The latest milestone in the history of German environmental legislation was set, when in 1998 the Socialdemocratic-Green Coalition formed the new government of Germany. The new government has a stringent environmental programme on its schedule. One of the most prestigious projects is an ecological reform of the tax and subsidy system. So far an energy tax has been introduced and a stepwise increase of it has already started. This has been the most publicly notable activity until now.

3.1.8 Conclusions and future needs of environmental policy

Within the EU Germany is with Sweden, Austria and the Netherlands one of the most advanced countries in environmental terms. In Germany particular progress has been achieved in clean air policy, construction of sewage plants and ban of leaded gasoline. Despite this progress the Council of Environmental Experts (SRU, 1987 and 1994) has criticised the lack of transparency in standard setting in German environmental policy and the general lack of use of economic instruments instead of regulatory intervention. So far most areas of German environmental policy are still dominated by the traditional approach of standard setting where no obvious economic incentive for pollution reduction is given. Here and in some other areas Germany can still learn from examples and experiments undertaken abroad. Concerning the use of economic instruments in environmental policy the United States have most experience, e.g. bubble concept, emission banking etc.[6]

In Germany there is also scope for further environmental improvements with respect to some challenging national and international problems (e.g. CO_2, transport related pollution). Given the environmental progress already achieved, the marginal abatement costs at point sources are increasing substantially. Further environmental progress will depend on more cost-effective environmental policies as well as improved integration of environmental aspects in sectoral and economic policies (OECD, 2001). Transposition of the EU Directives on Integrated Prevention and Pollution Control (IPPC) and Environmental Impact Assessment should contribute in this regard.

3.2 Waste water regulation for the dairy and meat industry in Germany, the Republic of Ireland and Northern Ireland

This section presents the detailed waste water regulation for the dairy and meat industry in Germany and compares its stringency with that in the Republic of Ireland as well as Northern Ireland. German legislation is not only found to be on average more stringent, but charges related to waste water treatment are also higher in Germany than in the other two sample countries. The section starts with the main environmental problems of the dairy and meat industry. Concerning the individual country legislation, first the regulatory standards are discussed. Then complementary charging systems are presented for each case study country.

---

[6] Still, the United States is said to have a highly regulated system which is even more comprehensive and detailed than in Germany (Rehbinder and Stewart, 1985; Salzwedel, 1982).

3.2.1 Environmental problems in the dairy and meat industry

Environmental problems in the dairy industry

Water is the environmental medium which is most strongly polluted by the dairy industry (Hegenbart, 1992). Not only organic material, but also disinfectants produce a considerable danger for the environment. Whey - as a sort of liquid waste - can be recycled for the larger part. E.g. sweet whey can be processed into milk powder although this requires like any drying process a lot of energy input. However, sour whey presents a problematic substrate which can be used in a microbiological way at the utmost. Since waste water in the dairy industry nearly always is identical with otherwise usable and saleable products, there should be a major incentive for pollution avoidance. Thus, there is a large potential for an as complete as possible use of all materials (see Table 3.2-1).

Table 3.2-1 Effects of sewage technologies in the dairy industry

Application area	Media to be recycled	Technology	Material to be recycled or removed
Dairy Industry	whey, low fat milk	reverse osmosis	water
	whey, low fat milk	ultrafiltration	proteins
	material from ultrafiltration	reverse osmosis	milk sugar

Source: Bauer, J.: Integrierte Umwelttechnik, Abwasser, p. 23, Landsberg/Lech 1992.

A further main problem area in the dairy industry concerns the energy input. The cooling chain from the producer to the consumer guarantees perfect products, but is also a potential to improve the ecological performance of the products.

Environmental problems in the meat industry

As for dairy processing operations the main environmental problems associated with the meat industry are the high consumption of water, the discharge of high-strength effluent and the high energy consumption.

Hygiene requirements necessitate the use of large quantities of fresh water. Wa ter is used for watering and washing livestock, cleaning process equipment and work areas and washing carcasses. Moreover, abattoir effluent contains high amounts of blood, fat, manure and cleaning agents. It is typically characterised by a high BOD_5 and/or COD need, fat, nitrogen, phosphorous and salt (sodium). Concerning energy consumption thermal energy, in the form of steam and hot water, is used for cleaning and sterilising and for rendering. Electricity is used for the operation of machinery and for refrigeration, ventilation, lighting and the production of compressed air. Noise, odour and solid wastes may also be issues in the meat sector. The main environmental issues of the entire meat processing chain are summarised in Table 3.2-2.

Table 3.2-2 Environmental issues at abattoirs

Process	Environmental issues
Reception of livestock Truck washing Cattle washing	Effluent containing manure wastes High water consumption Noise
Stunning and bleeding	Effluent with high organic load, especially if blood is discharged
Hide treatment (pigs)	Energy consumption for hot water used in scalding Generation of putrescible by-products Effluent with a high content of organic matter
Splitting and evisceration	Energy consumption for equipment sterilisation Generation of putrescible by-products Effluent with high organic load
Refrigeration	High energy consumption Fugitive losses of refrigerants, e.g. CFCs or ammonia
Cutting and boning	Electricity consumption Generation of putrescible by-products Energy consumption for equipment sterilisation
Casing and offal processing	Effluent with very high organic load Very high water consumption
Rendering	Effluent with very high organic load Potential for odour generation High energy consumption
Cleaning	High water consumption Consumption of chemicals Large volumes of effluent with high organic load

Source: United Nations Environment Programme (2000).

3.2.2 Waste water regulation for the dairy and meat industry in Germany

In contrast to the case-by-case approach in the Republic of Ireland and Northern Ireland the German water protection regulation has traditionally been a limit value approach. In the mid seventies this system was accomplished by a charg ing system in certain areas depending on the difference between so called direct and indirect dischargers distinguished in the German water management act (see Box 3.2-1).

Germany already applies the standards of the EC Urban Waste Water Directive to food processing. This implies that biological oxygen demand (BOD) and chemical oxygen demand (COD) pollution must not be higher than 25 or 110 mg/l on a national level. In the EU only Germany, Austria, Flanders (Belgium) and France follow an approach of national emission standards for the implementa-

Table 3.2-3 Overview of regulatory approaches for the implementation of the Urban Waste Water Directive

Parameter	Austria	Belgium (Fl.)	Denmark[1]	F. Rep. of Germany	Finland	France	Greece	Italy	Ireland	Lux.	Netherlands[2]	Portugal	Spain	Sweden	United Kingd.
Nature of Standards	National +	sectoral	Individ.	Nat.+ sect	Indiv.	National + sect. (partly)	Regional	National framework	Individ			Nat. framework	National framework	Individ.	
pH [-]	6,5-8,5	6,5-9,0	n.a.	n.a.	Na	5,5-8,5 5,5-9,5	6,0-9,0	5,5-9,5	n.a.	n.a.	6,5-8,5	6-9	5.5 - 9.5	n.a.	n.a.
Temperature [°C]	30	30	n.a.	n.a.	Na	30	35	3 (streams) 30 (lakes) 35 (sea)	n.a.	n.a.	30	incr. ≤ 3	incr. ≤ 3	n.a.	n.a.
BOD_5 [mg/l]	20	15-100	0.2-1812	25	6-50 (BOD_7)	30-100	60	40	n.a.	n.a.	10-60	40	40-300	n.a.	n.a.
COD [mg/l]	75-120	120-600	0.12-1022	110-150	Na	125-300	180	160	n.a.	n.a.	300-600	150	160-500	n.a.	n.a.
Total N [mg/l]	5 (NH_4-N)	60-100 (NKj-N)	1.3-383	25 10 (NH_4-N)	Na	30	15 (NH_4-N) 0,6 (NO_2-N) 20 (NO_3-N)	15 (NH_4-N) 0,6 (NO_2-N) 20 (NO_3-N)	n.a.	n.a.	15-30	15	15-50 (NH_4-N) 10-20 (NO_2-N)	n.a.	n.a.
Total P [mg/l]	1-2	10	0.02-26	2	0,4-1,5	10 (Phosp.)	10 (Phosphorus)	10	n.a.	n.a.	3-100	10	10-20	n.a.	n.a.
TSS [mg/l]	—	60-100 (SS)	n.a.	n.a.	Na	35-100	40	0,5	n.a.	n.a.	n.a.	60	30-300	n.a.	n.a.

Source: Haskoning (1996).

tion of the Urban Waste Water Directive (see also Table 3.2-3 for an overview). The case-by-case approach in the Republic of Ireland and Northern Ireland implies that licences would be currently sometimes more and sometimes less strict than the German values (see also sections 3.2-3 and 3.2-4 below).

The relevant German water protection legislation consists of

— the water management act ("Wasserhaushaltsgesetz") in its first version from 1957 including an amendment valid since 1987 prescribing higher demands on water quality and the wastewater clearing process according to the technical standard for both direct and indirect dischargers,
— the law on waste water charges ("Abwasserabgabengesetz"), (1976, latest amendment in 1998), which is only valid for direct dischargers.

Box 3.2-1 Direct and indirect dischargers

Indirect dischargers are those firms which do not have a permit to discharge into waters, but which discharge their waste water indirectly into waters via municipal sewage plants where the sewage is treated together with waste water of other origins and is finally discharged into waters. For the use of a public sewage plant the firm needs a permit. The basis for this permit is the waste water statute on the municipal level. If the waste water contains dangerous substances, the waste water of indirect dischargers must be pre-treated at the discharger's site and a second permit for discharging into the municipal sewage plant according to the water management act is required.

Direct discharging means that waste water is not discharged into the public sewage system, but directly into waters. Essentially communities, waste water associations or industrial plants with an effluent treatment plant of their own fall in this category. Thus direct discharging does not mean that waste water is discharged into waters without any treatment. Also direct dischargers need a permit.

National emission limit value approach in Germany

Paragraph 7a water management act is of special importance because it is a framework regulation for direct and indirect dischargers, but with various appendices each regulating a different industry branch. The appendices indicate the limit values for direct dischargers and - in the case of indirect dischargers – also for municipal sewage plants. There is an appendix of its own both for the milk and meat processing industry as well as municipal treatment plants. For the direct discharging of sewage into waters, i.e. after the treatment process on the firm level the following requirements are set for the dairy and meat industries (according to the state of technology):

Table 3.2-4 Requirements for the direct discharge of sewage for dairy and meat industry (pollution concentration limits according to § 7a Federal Water Management Act, Various Appendices)[1)]

Appendix of Water Manag. Act	Industry Branch	COD	BOD_5	NH_4-N	Nitrogen	Sulphate (SO_4^{2-})	Fluoride (F^-)	Phosphorous
		in mg/l	in mg/l	in mg/l	total in mg/l	in mg/l	in mg/l	total in mg/l
No. 3	Dairy[2)]	110	25	10	18	–	–	2
No. 10	Meat[2)]	110	25	10	18	–	–	2

[1)] Single qualified spot check or 2 hour mixed check.

[2)] Sewage from small direct dischargers is excluded, i.e. the limit values in appendix 3 are not valid for dairies with an amount of untreated sewage which is less than 3kg BOD_5 per day (for meat processing plants: less than 10 kg BOD_5 per week).

Source: § 7a Wasserhaushaltsgesetz; Appendices 3 and 10, 2000.

The effluent quality before the treatment process is of course much worse.

Table 3.2-5 shows that the limit values for municipalities which gather the sewage of indirect dischargers vary according to the treatment capacity of the public sewage plants.

Table 3.2-5 Requirements for the discharge of sewage from municipalities*

Capacity of Sewage Plant measured in	COD mg/l	BOD_5 mg/l	NH_4-N mg/l	Phosphorus total mg/l
kg of BOD5 per day of untreated sewage				
< 60 kg/d	150	40	–	–
60 until < 300 kg/d	110	25	–	–
300 until 1200 kg/d	90	20	10	–
1200 until 6000 kg/d	90	20	10	2
6000 kg/d	75	15	10	1

*Single qualified spot check or 2 hour mixed check.

Source: Bundesministerium für Umwelt, Naturschutz und Reaktorsicherheit und der Länderarbeitsgemeinschaft Wasser (Eds.): Mindestanforderungen an das Einleiten von Abwasser in Gewässer - § 7a Wasserhaushaltsgesetz; Milchverarbeitung, Hinweise und Erläuterungen zu Anhang 3 der Rahmen-Abwasser VwV, Bonn 1994.

Charging system in Germany

The cost of waste water consist of the waste water charge for direct dischargers, sewage fees for indirect dischargers and of a strong polluter surcharge also for indirect dischargers.

Waste water charge

Only direct dischargers must pay a waste water charge for discharging sewage into waters (including groundwater) according to the law on waste water charges ("Abwasserabgabengesetz"). Communities will pay for indirect dischargers which are connected to the public sewage system.

The total amount of the waste water charge which has to be paid by a certain discharger varies corresponding to the pollution concentration and the total quantity of the pollutant per year (expressed in pollution units; a unit quantity of each specified pollutant is set equivalent to a predetermined number of pollution units). The Federal authorities have set maximum limits for COD, mercury, cadmium, toxicity to fish, organic halogen compounds (AOX) and for additional heavy metals (chromium, nickel, lead and copper) (see Table 3.2-6).

The charge is currently 70 DM per damage unit per year and has developed as follows:

- since 1. January 1981 12 DM,
- since 1. January 1982 18 DM,
- since 1. January 1983 24 DM,
- since 1. January 1984 30 DM,
- since 1. January 1985 36 DM,
- since 1. January 1986 40 DM,
- since 1. January 1991 50 DM,
- since 1. January 1993 60 DM,
- since 1. January 1997 70 DM.

When the pollution reduction achieved by environmental measures go beyond the maximum limits for the above named pollutants by 20 per cent, the investment cost of the firm can be set against the unit charges (according to the degree of the additional environmental measures the waste water charge can be lowered up to 80 per cent).

Thus, companies have a price and profit incentive to invest in equipment to treat effluent in order to avoid the damage unit charges. In particular, in the period when the new law on the waste water charge was notified, i.e. before the law was in effect, and also during the first years when the waste water charge was in effect, it set strong incentives for an accelerated construction of treatment plants in the manufacturing sector (Sprenger et al., 1980). Thus, the technical goal of the waste water charge has been achieved. However, since a clear information system was not provided alongside the introduction of the waste water charge, clear quantitative estimations of the incentive function of the waste water charge can hardly be given (Sprenger et al., 1994). The development in the last years which is characterised by increased possibilities for the firms to set their effluent investment against the waste water charge, restricts the economic incentive function of the waste water charge. More and more, the function of the waste water charge is seen in supporting the enforcement of the water management act instead of being an economic incentive for avoidance of pollution (Ewringmann and Scholl, 1996).

Table 3.2-6 Weighting of pollutants according to the Law on Waste Water Charge

Pollutant	kg of Pollutant (M) Equivalent to a Pollution Unit	Concentration Value	Quantity per Year
COD	50 kg	10 mg/l	250 kg
Phosphorous	3 k	0,1mg/l	15 kg
Nitrogen	25 kg	5 mg/l	125 kg
AOX	2 kg	100 ug/l	10 kg
Mercury	20 g	1 ug/l	100 g
Cadmium	100 g	5 ug/l	500 g
Chromium	500 g	50 ug/l	2,5 kg
Nickel	500 g	50 ug/l	2,5 kg
Lead	500 g	50 ug/l	2,5 kg
Copper	1000g	100 ug/l	5 kg
Fish toxicity	3000 m^3 of effluent divided by dilution factor*	Dilution factor of 2	–
*A dilution factor is chosen for which effluent is not toxic anymore in a fish test.			

Source: § 3 Law on Waste Water Charges, 1976, last amendment in 1998.

In 1990 the revenues from the waste water charge were about DM million 321. This was only 3.3 per cent of the waste water capital expenditures in 1990. The revenues are used by the Länder and are dedicated to measures increasing water quality. According to state law states can pass on the charge to indirect discharg ers. Thus, it depends on the tax base of municipal waste water fees whether indirect dischargers have a similar burden as direct dischargers.b[7] Empirical evidence from a nation-wide survey, however, does not suggest that the waste water

[7] The strategies of communities for calculation of waste water fees vary a lot. The tax base can be only the amount of fresh water used or various methods including pollution criteria. Thus regional differences in the fees can be explained.

charge is passed on to indirect dischargers. Competitive distortions between direct and indirect dischargers (in favour of the latter) are then even increased because identical water intensive production technologies and/or products within the same sector have to bear different cost burdens according to the companies' position as a direct or as an indirect discharger. Partially whole sectors are to a large extent organised as indirect dischargers so that not only intrasectoral, but also intersectoral distortions due to the waste water charge can exist.

Sewage fees

Indirect dischargers only have to pay local sewage fees for the use of public sewage plants, but no waste water charge. Regional differences are widespread. E.g. in the state of Sachsen-Anhalt in 1997 the sewage fees were with 6.36 DM/m^3 almost twice as high as in Baden-Württemberg where the fee was only 3.38 DM. In 1997 the weighted average value for the sewage fee on the federal level was 4.61 DM/m^3 (Bundesverband der deutschen Gas- und Wasserwirtschaft (BGW), 1997). However, in 1999 the upward trend for the waste water fees could be stopped and the average value on a federal level could be reduced to 4.43 DM/m^3.

Strong polluter surcharge

The German system tends to provide an incentive to become an indirect discharger because the waste water charge in the sense of a "non-compliance penalty" can be avoided. The number of 4,000 direct dischargers and 350,000 indirect dischargers in the German manufacturing industry also reflects this tendency (Rügemer, 1995). This tendency is restricted by high sewage fees for indirect dischargers which can vary widely between the states or even within the states and by the obligation to pay a strong

polluter surcharge. Indirect dischargers of which the sewage has an extremely high BOD_5 or COD_5 load have to pay this surcharge. The food industry is hit hardest within the manufacturing industry. In particular, slaughterhouses and breweries have to pay the surcharge. The total sum of sewage fees and strong polluter surcharge can urge an indirect discharger - depending on the amount and toxicity of its waste water - to construct an own sewage plant and become a direct discharger.

3.2.3 Waste water regulation for the dairy and meat industry in the Republic of Ireland

Regulatory approach in the Republic of Ireland

The major part of industrial effluent produced in Ireland consists of readily biodegradable organic matter arising from food processing. Until 1993, licensing of industrial waste water was undertaken by the Local Authorities under the Water Pollution Act of 1977, and subsequent measures in the following decades. Maximum values are specified for pH, BOD, suspended solids, ammonia, phosphates, fats and greases, but a flexible approach regarding water quality and volume flow provides for regional variability. Consequently different standards may be set, although the typical values are about 20 mg/l for BOD and 30 mg/l for suspended solids.

In 1992, the integrated pollution control system (IPC) was introduced via the Environmental Protection Act. Then the licensing function for certain industrial activities became a task of the environmental protection agency, while other activities remained within the remit of the local authorities. IPC relates the maximum emission limit values to the following factors:

— best available technology for environmental protection in a certain activity (BATNEEC, i.e. best available technology not entailing excessive costs),
— the receiving capacity of the waters,
— whether it is a new and greenfield establishment or an existing plant (softer standards tend to be set for older firms which may be unable to survive a sudden increase in costs),
— the inclusion of the maximum emission value within an overall management plan to identify ongoing environmental improvements.

Table 3.2-7 Thresholds for IPC licensing in the dairy and meat sector in the Republic of Ireland

Industrial Category	Threshold for IPC licensing
Manufacture of dairy products	50 million gallons milk per annum
Slaughter of animals 1 sheep = 1 unit 1 pig = 2 units 1 head of cattle = 5 units	1500 units / day

Source: Forbairt (1996).

Of all water licences in force at the end of 1998, about 25 per cent were IPC licences and ca. 75 per cent were still given by local authorities. About a third of each type of licences were for discharges direct to waters and two thirds were with consent to discharge to sewers (Environmental Protection Agency, 2001). An IPC licence is needed for the following dairy and meat plants:

Up to now under the IPC system emission limit values for discharges to water from the slaughter of animals have been established, but not for the manufacture of dairy products (see Table 3.2-8 below). The emission limit values for slaughter of animals seem quite stringent, although for each parameter deviant values may apply in co-ordination with the Licensing Authority.

However, when the interviews for this case study were conducted in 1995, no Irish meat plant was subject to this new IPC regulation. The licences in the Irish sample plants were still granted by the local authorities under the Water Pollution Act.[8] They included specific standards to be met by the discharger, depending on the local water quality (similar to the Northern Irish approach, see below) and usually also included monitoring requirements (Forbairt, 1996).

Concerning the transposition of the EU Directive on Urban Waste Water Treatment into Irish law regulations made in 1994 provide the framework for the construction and upgrading of sewage systems and treatment plants (O'Leary & Carty, 1998). The regulations impose a schedule for the provision of collecting systems and treatment capacity based on the population equivalent of the urban area and the type of water body into which the treated waste water is discharged. Until 2005 about 80 per cent of the national generated urban waste water load shall be treated in secondary treatment plants.

Charging system in the Republic of Ireland

The licensing system is supplemented by some charging approaches which vary widely, but broadly speaking they represent a partial revenue raising system (Hitchens et al, 1998). According to Scott and Lawlor (1994) the full capital costs of treatment are not yet recovered.

Evidence from a survey conducted with 33 local authorities in the mid 1990ies suggests that only in a few cases were capital contributions for the use of public sewage treatment plants paid by industry (Lawlor, 1996 and Table 3.2-9 below). The OECD (2000) also reckons that these capital cost contributions vary widely and globally cover about one-third of total capital costs. No abstraction charge designed to reflect resource scarcity is in place. Self-treatment of effluent was found to be quite common, especially in the food industry. There was no clear

---

8 Only when the effluent volume exceeds $5m^3$/day, a licence from the local authority is needed.

evidence that firms decided to self-treat their effluent because of charges for treatment through public services.

Table 3.2-8 Emission limit values for discharges to water* concerning the slaughter of animals in the Republic of Ireland

Constituent Group or Parameter	Limit Value	Notes
PH	6-9	4
BOD	>90% removal or 40 mg/l	1,4
Toxic units	5	2,4
Total Nitrogen (as N)**	>80% removal o or 15 mg/l	4,5
Total Phosphorus (as P)**	>80% removal o or 2 mg/l	4,5
Total Ammonia (mg/l as N)	10	4
Oils, Fats and Grease (mg/l)	15	4
Fish Tainting	No Tainting	3,4
Mineral Oil (Interceptor) (mg/l)	20	4

* All values refer to daily averages, except where otherwise stated to the contrary, and except for pH which refers to continuous values. Limits apply to effluents prior to dilution by any uncontaminated streams, e.g. cooling water, stormwater etc.

** Only applicable to waters subject to eutrophication. One or both limits apply, depending on the sensitivity of the receiving waters.

Notes:

1. The daily raw waste load for BOD is defined as the average daily mass arising for treatment over any three month period. Calculations of the removal rates for BOD should be based on the differences between the waste loads arising for disposal and those discharged to the receiving waters. The amounts removed by treatment (chemical, physical, biological) may be included in the calculation.
2. The toxicity of the effluent shall be determined on an appropriate aquatic species. The number of Toxicity Units (TU) = 100/96 hr LC50 in percentage vol./vol. So that higher TU values reflect greater levels of toxicity.
3. No substances shall be discharged in a manner which, or at a concentration which, following initial dilution causes tainting of fish or shellfish, interferes with normal patterns of fish migration or which accumulates in sediments or biological tissues to the detriment of fish, wildlife or their predators.
4. Consent conditions for these parameters for discharge to municipal treatment plants can be established with the Licensing Authority, and different values may apply.
5. Reduction in relation to influent load. Total nitrogen means the sum total of Kjeldahl-nitrogen plus nitrate-nitrogen

Source: Environmental Protection Agency (1996), BATNEEC Guidance Note for the Slaughter of Animals.

In the 1990ies there was significant support from the EU, mainly the Cohesion Fund, for investment in sewerage and waste water treatment facilities in order to comply with the Urban Waste Water Directive. A report by the European Commission in 1999 noted that Ireland's progress in the implementation of the Urban Waste Water Directive has been slow because of the long lead time taken to bring projects from the design stage through to completion (Berg, 1999). Still, the EC concluded that a radical change in water resources management is underway, but criticised the absence of direct domestic user chargers for water. A survey in 1996 indicated that only 31 of the 88 local authorities levied charges for domestic waste water services (Barrett et al., 1997).

Table 3.2-9 Republic of Ireland charging for waste water services

Type of charge	No. of respondents to survey	Basis of charge
Capital contributions from industry	12	Half were volume related
Industrial effluent operating charges	16	Volume related
Industrial effluent monitoring charges	28	Mainly related to the frequency of monitoring
Total no. of respondents	33	

Source: Lawlor (1996).

3.2.4 Waste water regulation for the dairy and meat industry in Northern Ireland

In Northern Ireland a mixture of licences and charges is in place. Charges are only paid by indirect dischargers. However, the licenses are used to limit the discharge of those companies directly discharging into rivers and waters (Hitchens et al. 2000).

The consents are granted by the Department of the Environment in Northern Ireland and the agreed limit values in the licence vary according to the existing quality of water and volume flow. As in the Republic of Ireland maximum values are specified for pH, BOD, suspended solids, ammonia, phosphates, fats as well as greases, and the flexible approach regarding water quality and volume flow provides for regional variability.

When effluent is discharged through the public sewage system, a variant of the so-called Mogden formula is used to calculate the charges to the firm.

Charge per $m^3 = R + V + B\ (x/480) + S\ (y/340)$ with

480: regional standard for COD in mg/l,

x: company's COD,

340: regional standard for suspended solids (mg/l),

y: company's suspended solids,

R: cost for receiving and conveyancing (10 pence per m^3),

V: volume charge required for screening and settlement (8 pence per m^3),

B: cost of biological treatment (10 pence per 480 COD mg/l, i.e. the average strength effluent),

S: cost of sludge treatment (8 pence per 340 suspended solids mg/l, i.e. the average strength effluent).

This charging system has only been implemented in the mid 1990ies. The Water Executive of the Department of the Environment is gradually bringing all industrial indirect dischargers within the remit of this system.

The Northern Irish charging system represents a revenue raising procedure, i.e. total capital and running costs are estimated for the whole region and then proportioned amongst firms according to use.

To date the British integrated pollution control (IPC) legislation is also enacted in Northern Ireland. Certain industrial processes must be authorised under the Industrial Pollution Control (NI) Order 1997.[9] The processes with the greatest pollution potential are known as Part A processes for which a single authorisation covers releases to air, water and land. In Northern Ireland there are two other categories of process - Part B and Part C processes. Authorisations for Part B and Part C processes cover releases to air only. The Chief Inspector of the Industrial Pollution and Radiochemical Inspectorate is the enforcing authority for part A and part B processes, while district councils are responsible for regulating part C processes. Authorisation of existing processes under the Order is being carried out under a phased programme, while new prescribed processes need to apply before starting operation (http://www.ehsni.gov.uk/).

3.2.5 Comparison of German, Irish and UK waste water regulation for the dairy and meat industry

Both in Northern Ireland and the Republic of Ireland a mixture of licences and charges is used with direct dischargers not being charged. Licences are used to limit direct discharge. In contrast to both the Irish and Northern Irish effluent

[9] This was after the empirical data for the food case study were collected so there was no direct impact from the new legislation.

control system the German approach uses a national emission limit regardless of variations in local conditions. Germany also operates maximum discharge limits for direct dischargers and is reported to use a generally tougher effluent standard. Both direct and indirect dischargers are subject to a charging system.

Since in the two Irish jurisdications are given on a case-by-case basis, it is hard to compare the average rigour with the German systems. For consents given on a local basis in the Republic of Ireland typical values have been reported to be about 20 mg/l for BOD and 30 mg/l for suspended solids. The Northern Ireland system relies on measurement of COD, not BOD, because COD values can be more readily reproduced, while BOD values would be very sensitive to natural conditions. The parameters of 480 mg/l of COD and 340 mg/l of suspended solids in place at the time of the interviews conducted in Northern Ireland in 1995 and 1996 were said to be reduced over time since the Water Executive had an incentive to raise charges and to earn profits after its privatisation. It is, however, noteworthy that Republic of Ireland firms have commented that the licence systems in the two Irish economies show similar levels of stringency, when they have bought Northern Ireland subsidiaries in the food industry (Hitchens et al., 2000).

In summary, Germany not only operates generally tougher effluent standards, but also charges those companies using the public sewage system more heavily than in the other two sample countries (see results of case study presented below). Northern Ireland lies in between Germany and the Republic of Ireland, where charging according to effluent volume just started.

3.3 Packaging regulation for the packaging, food and retail industry in Germany and the Republic of Ireland/Northern Ireland

This section analyses the packaging regulation in Germany, Northern Ireland and the Republic of Ireland. For the empirical part it is noteworthy that at the time of the interviews only the German DSD system was in place. This renders the German approach stricter than in the Republic of Ireland or the UK where the packaging waste management systems were not yet enacted. Still, their design and their progress until today is important for the comparative approach of this study.

The section starts with the environmental problems related to packaging waste. Then the individual packaging waste schemes for the case studies countries are presented.

3.3.1 Environmental problems related to packaging waste

The main environmental issues during all stages of the lifecycle of packaging materials are energy consumption, resource efficiency as well as air and groundwater pollution. The intensity of the environmental burden depends on the type of material used.

Scarcity of landfills increased as a result of the growing amounts of waste and this factor was among the essential reasons for new waste legislation. Waste avoidance, recycling and environmentally friendly disposal of waste within the production process are now the top elements on the waste agenda in the EU, also concerning packaging waste. In the following sections the packaging waste schemes for the sam,ple countries will be analysed.

3.3.2 Packaging regulation in Germany

The German Packaging Ordinance of 1991 was already well ahead of the EU Directive on Packaging and Packaging Waste (94/62/EC) which requires that member states decide how best to meet minimum targets for recycling and recovery. The Directive states that by 31 December 2001, recovery of a minimum of 50 per cent and a maximum of 65 per cent by weight of all packaging waste has to be achieved. Such recovery includes recycling of 25–45 per cent of pack ackaging waste, and recycling of a minimum of 15 per cent of packaging waste from each individual packaging material.

Germany has - together with Austria, the Netherlands and Sweden – set the most stringent targets for the recycling of packaging waste within the EU (see Table 3.3-1 for the state of the art in the year 2001).

Since the start of the packaging waste scheme in Germany two Ordinances have been enforced (1991 and 1998) and a further draft Ordinance has been already designed in 2001. In the following these legal developments will be analysed.

The Packaging Ordinance of 12.6.1991

In the Packaging Ordinance ("Verpackungsverordnung", 12.6.1991) collection and sorting quotas for all packaging materials are agreed. All of the sorted waste must then be recycled, thus delivering minimum recycling quotas of 64 per cent and 72 per cent depending on the type of packaging (see Table 3.3-2) .

Table 3.3-1 Recycling targets within the EU

	Global targets			Recycling targets for packaging materials						
Country	Recovery	Recycling	Recycling of each material	Glass	Paper/ cardbord	Plastics	Steel	Aluminium	Beverage/ composites	Reuse/prev. target
Austria[1]	50	25	15	93	90	40	95	95	40	
Belgium	80[2]	50	15							x
Denmark[3]				65	55	15				
Finland	61	42	15	48	53	45[4]	25	25		x
France	50-65 75[5]	25-45	15							
Germany	65	45	15	75[6]	70[6]	60[6]	70[6]	60[6]	60[5]	x
Greece										
Ireland	50-65	25-45	15	45	31	10	5	25		
Italy	50-65	25-45	15							
Luxembourg	55	45	15							
Portugal	25[7] 50[8]	25[8]	15[8]							x
Spain	50-65	25-45	15							x
Sweden				70	40/65[9]	30[9]	70	70[9]		
The Nether- lands	65	45[10] 65[11]	15	90[11]	85[11]	35[11]	80[11]			x
UK	58		18							

Note: Targets are to be achieved by 2001 if not stated otherwise.

[1] Austria: Material-specific targets for individual company compliance, expressed as percentages of packaging taken back and own packaging waste arising which must be recycled.

[2] Belgium: Targets have to be achieved by 1999.

[3] Denmark: Targets of waste management plan, other targets are included in voluntary agreements (see country report).

[4] Finland: Target for plastics applies to recovery.

[5] France: Target for household packaging waste to be achieved by the end of 2002.

[6] Germany: Material specific recycling targets apply to sales packaging.

[7] Portugal: Target to be reached by 2002.

[8] Target to be reached by 2006.

[9] Sweden: Recycling target for: corrugated cardboard 65 %; paper/cardboard 40 %; aluminium drinks containers 90 %; PET drink bottles 90 %.

[10] Netherlands: Mandatory target to be achieved in 1998 defined in the Packaging and Packaging waste decree.

[11] Voluntary target defined in the Covenant II to be achieved by 2001.

x Particular reuse and/or prevention targets are described above.

Source: ARGUS (2001).

Table 3.3-2 Recycling requirements according to the German Packaging Ordinance of 1991

Material	Since 1.1.93	Since 1.7.95
Glass	42 %	72 %
Tin plate	26 %	72 %
Aluminium	18 %	72 %
Paper, board	18 %	64 %
Plastics	9 %	64 %
Composites	6 %	64 %

Sources: BGBl. Teil I, Jg. 1991, Nr. 36 vom 20. Juni 1991, Verordnung über die Vermeidung von Verpackungsabfällen; Ifo calculations.

For beverage packaging, detergents and paints the Packaging Ordinance of 1991 foresees a compulsory deposit. However, in the case of beverages, this deposit need not be paid as long as the share of reusable beverage containers does not decrease in the respective distribution area (i.e. state level) relative to 1991 (the year of the first Packaging Ordinance) and simultaneously the target of 72 per cent for reusable beverage containers is not forgone on the federal level (72 per cent was the level when the Packaging Ordinance was agreed in 1991).

The Foundation of the DSD System

The immediate consequence of the Packaging Ordinance of 1991 would have been that each company had to take back its own package material. Due to the practical unfeasibility of such an idea and because of the legal possibilities foreseen in the Packaging Ordinance, the companies involved founded the DSD ("Duales System Deutschland GmbH") on 28.9.1990 - a private disposal company which is responsible for the collection, sorting and reprocessing of used sales packaging materials of its participating companies.

The system is called dual system because it is introduced as a second system of waste management next to the municipal waste disposal systems. The foundation of the DSD was initiated by ca. 400 companies from the retail sector, the manufacturing industry producing consumption goods, the packaging industry and raw material producers. In 2001 about 600 companies are involved. The DSD serves the goals of the Packaging Ordinance (achievement of given recycling quotas) and shall also secure the financing of the disposal. The DSD makes the participants pay a licence fee for every product disposed, the licence symbol is the so-called green dot. Precondition for obtaining the green dot is a recycling guarantee for the respective material given by the packaging industry and/or the disposal industry. Thus the responsibility for the actual recycling does not stand with the fillers.[10] In 2001 the recycling guarantee for glass is given by one company, that for paper by seven companies. Six companies guarantee for the recycling of aluminium, nine companies for tinplate, one for plastics, one for beverage containers and one company for other composites (www.gruener-punkt.de). That implies that fillers simultaneously contract with the DSD GmbH (licence contract) and with the companies giving the recycling guarantee (recycling contract). Thus, the licence fees are based on the contracts negotiated with recycling companies and shall cover costs for the collection, sorting and recycling of materials. In October 1994 DSD fees for aluminium and paper were increased, those for glass and plastics decreased. Next to the material and weight related fee a piece-related fee was introduced in order to cover the relatively high sorting effort necessary for small packaging (see Table 3.3-3 for the current fee structure valid since 1.1.1995). The current fee structure does not reflect all developments of the DSD activities: E.g. the increase of the recycling costs in the mid nineties, the fact that the DSD was close to bankruptcy several times due to various reasons (e.g. free riders, recycling costs were higher than the revenues from fees, high payments to municipal disposal). However, since 1.1.1999 a discount of 9.5 per cent on the total sum of licence fees can be paid because innovation and new technology have led to lower costs. Also, the revised Packaging Ordinance of 1998 has forced more companies to become licence takers and pay the green dot fee[11].

The big retailers used their monopsony purchasing power so that almost all their suppliers licensed their products, i.e. the DSD fees for sales packaging are paid by the suppliers to a large extent (unless there is scope for the suppliers to in-

---

[10] This implies that fillers can obtain the green dot without giving a recycling guarantee themselves and therefore the backward pressure on the packaging industry for innovative recycling techniques may be weakened. There is only competition between the different types of materials, the competition between individual firms for innovative recycling techniques is limited. Companies which develop new techniques can use them only efficiently when they gain the respective orders from the companies giving the recycling guarantee (Bundeskartellamt, 1993). Related to this issue, in April 2001 the European Commission has decided that the license contracts of the DSD system do not conform with EU legislation. The DSD system was accused of making companies pay the green dot fee even when the packaging is not disposed by the DSD system, but by other disposal companies. This would imply that the DSD system uses its position of being nearly a monopoly covering 80 % of the German disposal market. The European Commission requires the DSD system to change its license contracts accordingly (Süddeutsche Zeitung, 21./22.04.2001). Recently, also the Federal Cartel Authority has supported the position of the European Commission.

[11] In the past the free riders like butchershops, bakeries, etc. have caused annual losses for the DSD of about 800 mill. DM (Süddeutsche Zeitung, 23. September 1997). This amount is fairly high given the estimate of total compliance cost to German industry of DM 4 bill.

Table 3.3-3 DSD licence fees in Germany since 1.1.1995

First part: Fees according to weight	
Packaging Material	Fees in DM/kg
Glass	0.15
Paper	0.40
Tin Plate	0.56
Aluminium	1.50
Plastics	2.95
Cartons	1.69
Mixed materials	2.10
Natural materials	0.20
Second part: Fees per piece	
1. Packaging with Volume	
Volume	Fees in Pf/piece
< 50 ml and < 2 g additively	0.10
< 50 ml, portioned packaging per 15 pieces	0.20
< 50 ml and > 2 g	0.20
50 – 200 ml and < 3 g additively	0.30
50 – 200 ml > 3 g	0.60
> 200 – 400 ml	0.70
> 400 ml – 3 l	0.90
> 3 l	1.20
2. Packaging without Volume	
Surface	Fees in Pf/piece
< 150 cm^2 and < 2 g additively	0.10
< 150 cm^2, portioned packaging per 15 p.	0.20
< 150 cm^2 and > 2 g	0.20
< 150 cm^2 – 300 cm^2 and < 3g, add.	0.30
150 cm^2 – 300 cm^2 and > 3 g	0.40
> 300 cm^2 – 1,600 cm^2	0.60
> 1,600 cm^2	0.90

Sources: DSD (1995a and www.gruener-punkt.de).

crease prices; this is currently a rare opportunity for food producers, see empirical part of this study). Especially food retailers reacted to the 100 per cent obligation to take back or reuse packaging material with the installation of reusable containers and stacks. Moreover, many retailers developed catalogues listing the characteristics of the materials they favoured. Through the demand power of the retailers transport packaging has been diminished to a great extent. Also wrappings need to be taken back to 100 per cent by the retailers. Many producers reduced the wrappings, the rest can be given back by consumers directly in the shops, but usually this is done when the whole waste of the product is disposed either in the DSD containers or in the households' garbage.

With respect to the ecological effects of the DSD fees, there is an incentive to lightweight packaging (as far it is still possible) and also a tendency to substitute e.g. paper for plastics due to the differentiated fee structure. In general the fees do not reflect the varying degrees of recyclability of the materials because they are not sufficiently subdivided (e.g. no difference is made between the various types of plastic) (Öko-Institut, 1995).

Due to the DSD system collection of waste was very successful. Recycling quotas for all materials were in 1999 even higher than required by the Packaging Ordinance of 1998 (see Table 3.3-4). In 2001 there are more than 18,000 licensed companies.

Table 3.3-4 Collection, sorting and recycling in Germany in 1993, recycling quotas for 1999 in brackets

Material	Collection quotas	Sorting quotas	Recycling quotas
Glass	63%	97%	62% (88%)
Tin plate	36%	98%	35% (105%)
Aluminium	13%	54%	7% (87%)
Paper, board	66%	83%	55% (70%)
Plastics	39%	74%	29% (108%)
Composites	31%	84%	26% (66%)

Sources: DSD, 1993 and 2001 (www.gruener-punkt.de).

On average 49 DM of licence fees are paid in Germany per inhabitant for all sales packaging (Süddeutsche Zeitung, May 21, 1997).[12] That means that total running costs can be estimated to be about 3.92 bn. DM (see also DSD annual report for 1999). These fees are used to subsidise the recycling companies. There is no official figure for the initial set-up costs available.

[12] For plastic packaging this amounts even up to 75 DM per inhabitant (Süddeutsche Zeitung, October 7, 1997).

The Packaging Ordinance of 21.08.1998

The Packaging Ordinance of 1998 amended several aspects of the 1991 version. These amendments will be presented in the following:

Collection and sorting targets were abolished for ease of administration. Instead minimum targets for recycling concerning the total amount of packaging were introduced (see Table 3.3-5). Recycling efforts going beyond the minimum targets are planned to be either in the form of energy recovery or recovery of resources. Energy recovery was not allowed in the previous version of the Packaging Ordinance.

Furthermore, since 1.1.1999 at least 60 per cent of plastics must be recycled in the sense of recovery of resources. 60 per cent of this recycling quota for plastics must be secured through processes which replace new material of the same kind or where the plastics is still available for similar use. Resource recovery currently can only easily be undertaken with monomaterials. In 2000 about 590,000 tonnes of plastics were recycled and a saving of primary energy of 34.4 MJ/kg plastics was achieved. This is close to the goal of the DSD system of achieving up to 40 MJ/kg plastics (Süddeutsche Zeitung, 03.05.2000).

Table 3.3-5 Recycling requirements according to the German Packaging Ordinance (21.08.1998)

Material	Since 1.1.96	Since 1.1.99
Glass	70 % (previously 72 %)	75 % (previously 72 %)
Tin plate	70 % (previously 72 %)	70 % (previously 72 %)
Aluminium	50 % (previously 72 %)	60 % (previously 72 %)
Paper, board	60 % (previously 64 %)	70 % (previously 64 %)
Composites	50 % (previously 64 %)	60 % (previously 64 %)
Plastics*	50 %	60 %

*60% of the recycling quotas for plastics must be secured through processes which replace new material of the same kind or where the plastics is still available for similar use. This rule will be examined again until 01.01.2000.

Source: Verordnung über die Vermeidung und Verwertung von Verpackungsabfällen, 21.08.1998.

As a further aspect of change, manufacturers and retailers who did not already participate in an organised system for take back and recycling of sales packaging (i.e. mostly small shops which produce and sell simultaneously like bakeries) must give proof of their disposal. If in the future the quotas are not fulfilled, obligatory connection to dual systems will be the consequence.

As foreseen in the Packaging Ordinance of 1991 manufacturers and retailers are only freed from participating in dual systems with respect to beverage packaging, when not less than 72 per cent of all beverages (except milk) are packed in returnable containers on a state or federal level. If the quota falls below this margin, a compulsory deposit for non-reusable beverage containers will be introduced. However, the threat of a deposit refund will be restricted: If some types of beverages have not fallen under the 72 per cent margin for reusable packaging, there will be no deposit on them. Due to this change in the legislation the deposit would only have to be paid for those beverages with an especially high increase in throwaway packaging, e.g. beer packed in cans.

The remaining waste (i.e. waste which cannot be recycled) is disposed by the municipal authorities. However, there is a new provision that also the private disposal industry can intervene, but only upon agreement with public disposal companies and only if the private operators possess a label/certificate showing the specific expertise of the company. So far, in practise, there is not much competition because municipalities want to use their capacity.

Finally, with respect to milk, new life cycle analysis has found out that a pouch made of polyethylene is as environmentally friendly as a reusable glass bottle. The Packaging Ordinance of 1998 took this into account and raised the quota for reusable milk packaging from 17 per cent to 25 per cent.

Second Draft Ordinance for Changing the Packaging Ordinance, 07.02.2001

The major planned change in the second draft change of the Packaging Ordinance of February 2001[13] is a general separation of ecologically beneficial and ecologically detrimental beverage packaging. Moreover, it aims at simplifying the existing Ordinance through the introduction of a deposit for all environmentally unfriendly beverage packaging irrespective of the former quota of 72 per cent for reusable packaging. After 1995 the quota of 72 per cent reusable packaging for all beverages has been endangered several times, in particular because the share of non-reusable beer cans has expanded by 20 per cent (Breitenacher et al., 1997).

According to § 3 of the draft, ecologically beneficial beverage packaging consists of reusable packaging, beverage cartons and beverage pouches made of polyethylene (the latter is used for packaging of milk) (BMU, 2001). This approach acknowledges the results of a study carried out by the Federal Environ-

[13] This draft was accepted by the German Bundestag on 02.05.2001.

mental Agency in August 2000 in which no essential ecological advantage or disadvantage was found between the existing reusable glass bottle systems and beverage cartons for mineral water, beverages without CO_2 and wine. The study also showed that throw-away glass packaging as well as cans made of tinplate and aluminium have significant ecological disadvantages in the area of beverages with CO_2 content. The total environmental burden stemming from environmentally unfriendly beverage packaging accounts for about 0.1 per cent of the total pollution in Germany or corresponds to a pollution level caused by some 10,000 to some 100,000 citizens (Umweltbundesamt, 2000).

§ 8 of the draft states the introduction of a compulsory deposit refund for beverages in environmentally unfriendly packaging. The deposit is at least 0.25 Euro incl. VAT or at least 0.50 Euros incl. VAT for containers with a volume larger than 1.5 litre. Currently the perspective is to enact this new regulation from 1st of January 2003 (www.bmu.de/fset800.php). It would imply that the differentiation between single beverage market segments will not apply anymore. According to the opinion of the ministry of the environment consumers would understand this differentiation only with difficulties and wonder why e.g. soft drinks would be exempt from a deposit. If the new draft id adopted, the costly surveys concerning the extent of the quota for reusable packaging could be dropped by the government.

It is assumed that the deposit for environmentally unfriendly beverage packaging will set an incentive for fillers, retailers and consumers to offer/demand a larger amount of beverages in environmentally friendly packaging. The deposit system would enable a sorted take back of used packaging and facilitate high quality recycling afterwards.

In terms of costs it is reckoned that net costs of about 265 million DM would arise for the installation and operation of a deposit refund system (joint calculation of the ministry of the environment and the ministry for economic affairs, BMU and BMWi, 2001; Table 3.3-6 below).

These costs are related to the need for 83,000 automatic take back machines of which each unit is estimated to cost about 25,000 DM. About half of these machines would be installed in discounters and hypermarkets with more than 800 m^2 sales area where more than 80 per cent of the throw-away beverage volume is sold in Germany.

If put in relation to the current packaging amount of about 14.4 billion units, an average additional burden of 1.84 Pfennig per packaging results. In relation to Germany's total number of inhabitants a per capita burden of 3.31 DM per year follows.

Table 3.3-6 Overview of costs for the installation and operation of a deposit refund system

Total investment need		**2.08 billion DM**
Annual costs*:	Depreciation	297.1 million DM
	Interest payments	166.4 million DM
	Operating costs	166.4 million DM
	Logistics	275.0 million DM
	Gross costs	904.9 million DM
Minus	Revenues for secondary resources	90.0 million DM
	Net costs	814.9 million DM
Minus	Savings on licence fees for DSD system	550.0 million DM
Additional costs incurred through deposit refund system		264.9 million DM
*These costs do not include the costs for manual take back of used beverage packaging because they are very hard to estimate.		

Source: BMU and BMWi (2001).

The introduction of a deposit would make the DSD lose about 550 million DM of revenues through forgone licence fees. On the one hand, this reduces the cost of establishing a deposit system; on the other hand, one cannot assume an equally high cost reduction in the DSD due to reduced activities. On the contrary, the specific costs for collection and sorting of the remaining packaging licensed by DSD will increase. However, the ministries do not fear a major increase in the licence fees of the DSD system because there is still a potential for cost reduction (see e.g. already existing discount on licence fees). Only for glass, where the deposit on throw-away glass is assumed to lead to a decrease by one third in the collection of glass, some structural consequences for DSD can be envisaged. Possibly the frequency of collection and maybe the number of containers will be adapted to the new situation.

3.3.3 Packaging regulation in the Republic of Ireland

In Ireland the Waste Management Act of 1996 is geared to facilitate the achievement of the packaging waste recovery and recycling targets laid down in the European Parliament.

All companies, i.e. raw material manufacturers, converters, packer/fillers, distributors, wholesalers, retailers and also importers are affected, if turnover net of exports exceeds IR £1 million and packaging supplied to the domestic market is greater than 25 tonnes (this is more stringent than the initial threshold values for Great Britain and Northern Ireland). Each company is obliged to take back packaging of a similar type to that which it produces. Packers and fillers must ensure recovery rates of at least 40 per cent. Firms elsewhere in the supply chain are required either to recover material themselves or to send back collected material to a supplier. However, if firms have not found after their best efforts a source of take back for a certain material, they are then permitted to resort to landfill.

The EU Packaging Directive allows some delays for Ireland:

— by 30 June 2001, recovery of a minimum of 25 per cent by weight of all packaging waste instead of 50 per cent; [14]
— by 31 December 2005 (instead of 2001), recovery of a minimum of 50 per cent and a maximum of 65 per cent by weight of all packaging waste: such recovery must include recycling of 25 per cent of packaging waste and recycling of a minimum of 15 per cent of packaging waste from each individual packaging material.

The Foundation of Repak

Repak is the Republic of Ireland's counterpart to the German DSD system and other consortia elsewhere. Repak ensures that members' legal requirements with respect to registration and licences are met. Companies which are participating satisfactorily in the Repak waste recovery scheme are exempted from obligations under the Irish Waste Management (Packaging) Regulations of 1997.

Repak Ltd is at present the sole approved body under the Waste Management (packaging) Regulations, 1997. Repak Ltd is a non-profit company established by industry to co-ordinate and finance the recycling of packaging waste. Repak operates a recovery scheme in relation to packaging waste. Member companies are required to:

— provide information to Repak in relation to packaging output and waste recovery activities
— prepare a written plan for the recovery of packaging waste arising on their own premises
— direct packaging waste arising on their own premises to recycling
— make a financial contribution to Repak to enable Repak to secure the collection/recycling of packaging waste from households.

---

[14] Actually the Irish Government has set a target of 27 per cent of all packaging to be achieved by 30 June 2001 (Commission of the European Communities, 1999).

Repak's first report in May 1998 of its activities since it became operational in July 1997 indicated that:

— membership of Repak now stands at some 500 companies

— contributions from member companies exceeded £1.2 million by end 1997

— Repak has paid some £800,000 to Kerbside Dublin and to Gandon Enterprises (i.e. Rehab Recycling) in 1997 to support the waste collection/activities of those bodies

— member companies collected, separated and recycled 30,000 tonnes of packaging waste arising on their own premises

— Repak have financed directly the collection/recycling of 21,000 tonnes of waste to date by its contributions to Kerbside Dublin and Rehab.

The current Repak fee structure was introduced in January 2000, it is fixed for a three year period and costs are reviewed annually. It is based on the principle of "Shared Responsibility", which ensures that the cost of recovery and recycling of packaging waste is spread across the supply chain.

Fees are generated twice yearly via statistical submissions made by each Member. Five parties are charged on the quantity and type of packaging handled and the sum of these figures equate to 100 per cent of the cost of recycling 1 tonne of the material in question.

Fees for Year 2000 & 2001 membership, (based on previous year's statistics) are assessed as Participation Fees, at a flat fee per tonne, irrespective of the material type, and are charged depending on the activity, or activities, of the Member. Only the brandholder/importer has to pay a material specific fee (see Tables 3.3-7 and 3.3-8 below).

Table 3.3-7 Repak fee structure according to activity in the supply chain

Activity in Supply Chain	Cost/Tonne		Fee Type
	IEP	EURO	
Manufacturer	1	1.27	Participation Fee
Converter	1	1.27	Participation Fee
Brandholder/Importer	See Table below		Material Specific Fee
Distributor/Wholesaler	1	1.27	Participation Fee
Retailer	2	2.54	Participation Fee

Source: www.repak.ie

Table 3.3-8 Repak material specific fee structure for brandholders and importers

Material	Cost 2000		Cost 2001	
	IEP/tonne	Euro/tonne	IEP/tonne	Euro/tonne
Aluminium	61.56	78.17	45.00	57.14
Composite A (Paper/Plastic)	36.31	46.10	36.31	46.10
Composite B (Steel/Aluminium)	96.04	121.95	60.00	76.18
Glass	4.40	5.59	4.40	5.59
Paper	10.89	13.83	10.89	13.83
Plastic	36.31	46.10	46.00	58.41
Steel	96.04	121.95	50.00	63.49
Wood	5.17	6.57	5.17	6.57

Source: www.repak.ie

Material specific fees depend on the material supplied and are charged at the brandholder/ importer stage in the supply chain. These will vary from year to year. Between 2000 and 2001 fees for packaging made of aluminium, for composites made of aluminium and steel as well as for steel packaging have been reduced substantially. Only for plastic packaging they have been increased. Importers are responsible for activities preceding their activity (i.e. packaging material manufacturer). Some companies' operations span across all five activities (e.g. where a retailer imports packaging or packaged products).

Credits for minimisation, reuse, reduction and recycling of packaging will result in lower fees for the Member.

3.3.4 Packaging regulation in Northern Ireland

In order to describe the Northern Irish situation the British regulation has to be examined because it provides the standard which Northern Ireland is most immediately trying to attain.

British legislation

The UK has adopted a unique approach to fulfil the EU requirements of packaging waste recovery and recycling targets. The "shared producer responsibility" for packaging waste is defined very narrowly and refers only to the industries which produce or use packaging. It does not include a partnership with the consumers or local authorities as in other countries. Responsibility for recovery and recycling of packaging waste is divided between the commercial enterprises belonging to the packaging chain which are raw material producers, packaging manufacturers, packer/fillers and sellers (ARGUS, 2001).

The Producer Responsibility Obligations (Packaging Waste) Regulations 1997 came into force on 6 March 1997 and transposed elements of the EC Directive on Packaging and Packaging Waste 94/62/EC, including targets for recovery and recycling of packaging waste to be met by 2001.

The Regulations place on certain businesses three main obligations:

— to register with the Environment Agency and provide data on the packaging handled by the business;

— to carry out specified tonnages of recovery and recycling of packaging waste, calculated on the basis of the tonnage handled by the business, the national recovery and recycling targets for the year and the percentage obligation for the activity that the business carries out on packaging;

— to certify that the necessary recovery and recycling has been carried out. In addition, consumer information obligations are placed on businesses that sell packaging to the final user or consumer.

Any business handling more than 50 tonnes of packaging and with financial turnover of more than £2 million will be obligated if it performs one or more of the following activities within the packaging chain: manufacturing raw materials for packaging; converting materials into packaging; filling packaging; selling packaging to the final user; importing packaging or packaging materials into the UK. Certain packaging is exempt: packaging which is exported; any packaging that is reusable; data on wood and other packaging must be included up until 2000, even though it is not included in the recovery obligations.

The activity obligations associated with each of these activities are:

— raw material manufacturing 6 per cent;

— converting 9 per cent;

— packing/filling 37 per cent;

— selling 48 per cent.

In 1997 the Department of the Environment planned the following recycling and recovery targets for the period 1998 to 2001:

Table 3.3-9 British recycling and recovery targets, 1998-2001

	1998/99	2000	2001 planned	2001 revised
Minimum recycling target for each material	7	11	16	18
Recovery target for obligated businesses[1]	38	43	52[2]	56
Overall target recovery rate of UK packaging waste	32	40	50	51

Notes:

[1] Since these rates relate only to those firms which fall within the terms of the legislation on material recovery, this explains why their rates exceed the required overall UK target rate. In 1998/99 the targets applied to businesses with turnovers of more than 5 million Pound Sterling or a weight of output of paper, metal and glass greater than 50 tonnes. By 2000 the legislation is to apply to businesses with greater than 1 million Pound Sterling of turnover, and by 2001 to any firms with wood output of greater than 50 tonnes are also included.

[2] of which a minimum of 26 per cent must be recycled.

Source: Department of the Environment (1997), Department of the Environment, Transport and the Regions (2000).

According to the Environment Act (1995) two courses of action are allowed. If obligated to adhere to the legislation a company will have a choice to either carry out recovery and recycling obligations itself – individual compliance – or to join a registered compliance scheme. Membership of such a scheme will exempt the company from its obligations entirely. Such schemes will assume responsibility for meeting recovery and recycling obligations on behalf of their members, although they will not take on responsibility for actually removing waste from the members' sites. Members of schemes will need to supply output data to the scheme.

Individual compliance required

- Registration with the appropriate agency by 31 August 1997;
- Payment of the registration fee of £750;
- Provision of data (metric tonnes) on packaging handled/created;
- Provision of this fee and the data every year by 1 April in order to remain registered;
- The meeting of the legal recovery and recycling obligations;
- By 31 January each year the company must show certificates of compliance to the Agency.

Participation in a compliance scheme required

— The individual compliance schemes such as Valpak (first collective scheme set up in the UK) to register by 31 August 1997, or by 1 April in subsequent years;
— Demonstration of the ability both financially and technically to deliver the obligations of its members;
— Payment of a fee and provision of single page data form for each of its members by 31 August 1997;
— Once a scheme is registered, its members are exempt from their legal obligations.

Unlike in Germany, where only one scheme is operated, in the UK exist next to the large-scale cross-sectoral national compliance schemes, such as Biffpack, Cleanapack, Recycle UK, Valpak, and Wastepac a number of smaller compliance schemes. E.g. the scheme called Impact is specifically aimed at smaller companies which will become obligated for the first time when the producer responsibility threshold under the Regulations is reduced from 2000 onwards.

To exemplify the fee structures of the compliance schemes Valpak will be presented in more detail. The fee consists of four elements:

i) There is a once-only payment which depends on the joining company's UK turnover, excluding exports. It is distinct from the total turnover threshold in the regulations, which refers to the audited accounts of a business (see Table 3.3-10).

Table 3.3-10 One time joining fee in the Valpak scheme

UK turnover excluding exports	Joining fee (special rates – expire April 2001)
£2m to £5m (GB)	£650 (plus VAT)
£2m to £5m (Northern Ireland)	£1,000 (plus VAT)
£5m to £50m (GB & NI)	£1,150 (plus VAT)
£50-£100m (GB&NI)	£3,500 (plus VAT)
Over £100 m (GB & NI)	£5,500 (plus VAT)

Source: Valpak (2001).

ii) By law, all companies affected by the packaging waste regulations have to pay an annual registration fee. Companies based in Great Britain have to register with the Environment Agency with an annual fee of £460. Companies based in Northern Ireland have to register with the Environment and Heritage Service with an annual fee of £760. The relevant registration fee is included in our joining fee in the first year.

iii) Companies also have to pay a material levy. This represents the cost incurred by Valpak in obtaining evidence of compliance on behalf of the companies. The levy is calculated according to the tonnage and materials which make up the participating company's net recovery and recycling obligations. Every member pays the same cost per tonne for each obligated material.

As of April 1998, the Valpak levies were expected to be in the following ranges:

Table 3.3-11 Material levies per tonne according to Valpak, April 1998

Material	Fee
Aluminium	£ 6-20
Steel	£25-90
Glass	£2-90
Paper	£2-30
Plastics	£6-160

iv) Valpak passes on its operating costs through a flat annual administration cost of £250, plus a graduated charge per tonne of net obligation. This decreases as the tonnage increases.

Currently the Government is proposing to change the mechanism according to which the registration fee paid each year by members of compliance schemes is determined. The previous sliding-scale system (as shown above for Valpak) shall be replaced with a flat fee of £460 per scheme member per year, regardless of the size of the compliance scheme. According to the UK government's view the flat fee more accurately reflects the enforcing authorities' monitoring costs and takes account of concerns about a system that was perceived by most compliance schemes to be unfair and anti-competitive (Department of the Environment, Transport and the Regions, 2000). So far competition between schemes has hardly taken place on the grounds of their performance and operational efficiency, but on the grounds of a regulatory fee differential.

In order to ensure fair competition the Office of Fair Trading examines the effects whenever new compliance schemes are created.

The UK landfill tax

In addition to the Packaging Waste Scheme the UK has introduced a landfill tax in order to give further incentives for waste minimisation. The landfill tax applies to waste disposed of by way of landfill, at a licensed landfill site unless the waste is specifically exempt. The landfill tax had to be paid from 1st of October 1996 onwards.

A company is liable to pay landfill tax if it is the licence holder for a landfill site. If a company operates a landfill site, any waste deposited in it, including waste which it has produced, will be liable to tax unless exempt from the tax. This includes in-house sites where producers dispose of their own waste. The tax is chargeable by weight and there are two rate measures: A lower rate of £2 per tonne for inactive (or inert) wastes and a standard rate of £7 per tonne applies to all other taxable wastes.

Northern Ireland Legislation

Government policy on environmental regulation in Northern Ireland is that environmental standards should be the same as those in Great Britain. However, at the time of the set-up of the Producer Responsibility Obligations the Northern Ireland recycling rate for households lagged 10 per cent behind the 20-30 per cent in the rest of the UK. Also, Northern Ireland then produced about one-and-a-half times as much household waste per head.

The present regime in Northern Ireland is set out in The Pollution Control and Local Government (Northern Ireland) Order 1978. In 1999, however, the Northern Ireland Waste and Contaminated Land Order (November 1997) was implemented. This Order embodies most aspects of the Environmental Protection Act that apply to Great Britain and has led to a registration of obligated firms with the Environment and Heritage Service. For the year 2000 Northern Ireland was expected to meet the recycling and recovery targets set for the UK as a whole.

As in the rest of the UK, the landfill tax is regarded as additional economic incentive to move to alternative means of waste management. In urban areas of Northern Ireland suitable spaces for landfill sites are especially scarce. That is why in the Belfast area an incinerator is planned.

Expected Difficulties in Applying the Recycling Regulation to Northern Ireland

Northern Ireland has a much smaller number of firms in absolute terms lying above the threshold values for turnover and weight of output. This implies that the burden for each obligated firm in Northern Ireland may be relatively more onerous than in Great Britain.

Originally the first organised compliance scheme, Valpak, had no plans to establish itself in Northern Ireland. By now, a variety of compliance schemes are UK-wide in operation. Also, Northern Irish plants use certified recycling plants in the Republic of Ireland.

Some of the start-up problems of the recycling industry in Northern Ireland were cushioned in an interim period. Until 1999, companies which had plants both in Northern Ireland and Great Britain had to apply the recycling targets only on one side of the Irish Sea, not on the other.

3.3.5 The comparative rigour of the packaging waste management systems in the sample countries

In the following the comparative rigour of the packaging waste management systems will be analysed by comparing three indicators of stringency which are

— recycling levels (as a result of recycling targets),

— he annual costs of compliance and

— the material specific fee related to the packaging waste management systems.

Comparison of recycling levels

German practice in 1995 through the existing DSD, 'green dot', system was with 80 per cent rates of material collection and 64–72 per cent (depending on material) rates for recycling already well above the EU standards required by the end of 2001. Also, the recycling figures shown for the mid-1990s in Table 3.3-12 place Germany well ahead of the Republic of Ireland, the UK and Northern Ireland. While the legislation in both Irish economies has been designed to achieve the EU targets with respect to recovery and recycling rates, the Republic of Ireland does not have to realise those standards until 2005. Thus their immediate objective for 2001 is considerably lower than for Northern Ireland or Germany.

Table 3.3-12 Comparative rates of recycling in the mid-1990s (as % of national consumption levels of materials)

Country	Glass, 1996	Glass, 1997	Paper, 1994
Germany	79	76	60
Republic of Ireland	46	37	12
United Kingdom	22	24	34
Northern Ireland	20	–	–

Source: Eurostat Yearbook 2000 and Statistical Office of the European Communities 2000.

Comparison of annual costs of compliance schemes

Another indicator of the strictness of the various regulatory frameworks is obtained by comparing their annual actual or projected costs, i.e. administrative plus the running costs for the firms (see Table 3.3-13).

Table 3.3-13 Actual/projected total annual cost* of packaging recovery and recycling systems

	£ mn, using market exchange rates	£ per employee in the industrial sector (in parenthesis: per person in total employment)
Germany	1 700 (1996)	132 (48)
United Kingdom	600 (2001)	85 (23)
Republic of Ireland	4–5 (2001)	14 (4)
Note: *Initial set-up cost, recurrent administrative cost and annual compliance cost to the firms (e.g. fees paid).		

Source: See text and also Office for National Statistics (1997).

There is the problem that such estimates usually do not allow for the expenditure made (e.g. on clean technologies) in order to avoid regulation changes in the future. The estimated annual compliance cost for German industry is £1.7 billion (i.e. DM 4 billion) compared to a projected cost of only £600 million for the UK system which is to be in place by 2001. These costs consist of administrative expenses, the costs of membership of collective responsibility schemes or of the disposal of material at the level of the individual firm. This may be an underestimate of the total compliance cost imposed on German industry, given that in 1996 the licence fees paid within the DSD were themselves almost DM 4 billion (and to this would have to be added the start-up and recurrent administrative costs of the system).

IBEC (1997a) estimated that by 2000 the total cost to the Republic of Ireland industry of the REPAK scheme would be £4–5 million and they judged that 'the costs per company would be lower than in any other EU country'. Certainly, on a pro rata basis, estimated cost appears to be no more than one-third that projected for the UK.

However, this comparison is misleading. Informed sources in the Republic of Ireland have accepted that the IBEC figure is likely to be an underestimate. Moreover, like is not being compared to like. REPAK member firms still bear their own responsibility for material recovery and hence will have to pay for this over and above their fees to REPAK (such payments were not considered in the IBEC estimate).

In the context of this study it is worth stressing that one consequence of the traditionally more stringent regulation of packaging in Germany has been that the infrastructure for the handling and processing of material for recycling and recovery is much more strongly developed than in the Republic of Ireland or the UK.

Comparison of material specific fees

In contrast to the Irish economies the German DSD system gave with its material-specific fees from the beginning onwards price-related incentives to substitute between material types (e.g. from plastic to paper) albeit this seems to be a fairly blunt instrument in practise. In the case of the Irish economies (i.e. considering the likely impact of the UK system on Northern Ireland) the incentive effects have in the beginning appeared to be more generalised – towards light weighting (which, indirectly, may then favour certain material types). However, after the first year, also in the UK material specific fees were introduced.

Table 3.3-14 shows that fees in Germany are many times higher than in the Republic of Ireland and the UK. Economic incentives for the reduction of packaging waste are therefore higher in Germany.

In summary, Germany operates a more stringent packaging waste scheme system in terms of recycling targest and licence fees than Northern Ireland and the Republic of Ireland.

Table 3.3-14 Comparison of material type fee structure in Germany, Republic of Ireland and UK

Material	DSD cost 2001, D		Repak Cost 2001, IR		Valpak Cost 2001, UK	
	DM/tonne	EURO/tonne	IEP/tonne	EURO/tonne	£/tonne	EURO/tonne
Aluminium	1 500	750	45.00	57.14	6-20	9-30
Composite A (Paper/Plastic)	2 100	1 050	36.31	46.10	n.a.	n.a.
Composite B (Steel/Aluminium)	2 100	1 050	60.00	76.18	n.a.	n.a.
Glass	150	75	4.40	5.59	2-90	3-120
Paper	400	200	10.89	13.83	2-30	3-40
Plastic	2 950	1 475	46.00	58.41	6-160	9-213
Steel	560	280	50.00	63.49	25-90	33-120
Wood	200	100	5.17	6.57	n.a.	n.a.

Source: own compilation from compliance schemes data.

3.4 Clean air regulation for the cement industry in Germany, Spain and the UK

This section discusses the main differences between the legislative approaches for the cement industry in the three sample countries Germany, Spain and the UK. It shows that on average Germany operates the strictest national emission limit values. After a brief overview of the main environmental issues of the cement industry, the individual country regulations are presented.

3.4.1 Overview of environmental issues in the cement industry

The cement industry is an energy intensive industry. Various fuels can be used to provide the heat required for the process with coal being the traditionally used primary fuel. In the EU in 1995 the most commonly used fuels were petcoke (39 per cent) and coal (36 per cent) followed by different types of waste (10 per cent), fuel oil (7 per cent), lignite (6 per cent) and gas (2 per cent) (see CEMBUREAU, 1997). Waste has been used as a fuel for more than 10 years and to varying degrees in the Member States.

The clinker burning is the most important part of the cement manufacturing process in terms of the key environmental issues which are energy use and emissions to air. The main emissions are nitrogen oxides (NO_x), sulphur dioxide (SO_2) and dust. Other emissions to be considered are carbon oxides (CO, CO_2), volatile organic compounds (VOCs), polychlorinated dibenzodioxins (PCDDs) and dibenzofurans (PCDFs), metals, and noise (EIPPC Cement BREF, 2000).

3.4.2 Clean air regulation for the cement industry in Germany

In Germany the precautionary principle has led to the concept of "Stand der Technik" which is also feasible on economic grounds. Emission limits for clean air are laid down according to the general regulations of the Technical Instruction Air (TA Luft) of 1986. German cement plants are legally subject to the specific ordinances of the Federal Ambient Pollution Prevention Act (Bundes-Immissionsschutzgesetz). If a cement plant uses for its clinker burning process in addition to normal fuel also waste fuel, then the regulations of the 17th Ordinance on the Federal Ambient Pollution Prevention Act (1990, last updated in 1999) concerning incineration installations for waste and similar burnable materials also apply (Schneider et al., 2000). Air emission limits are valid on a uniform federal level.

As can be seen in Table 3.4-1 German dust emission limits are among the most stringent emission limit values in the EU, and the relative stringency is particu-

Table 3.4-1 National dust emission limits for the production of cement within the European Union

Dust mg/Nm³	Data bas.on	New/modified or existing plant	kiln stack	clinker cooling	cement grinding	other point sources
Austria	Na (a)	new/modified	50	50	50	50
		existing	50	50	50	50
Belgium	P	new/modified	50	50	50	
		existing	50-150	50-400	50-150	50-300
Denmark	P	Existing	50 (b)	50 (b)	50 (b)	50 (b)
Finland	P	new/modified	50	50	30-50	30-50
		existing	50 (c)	50	30-50	30-50
France	Na	new/modified	50	100	50	30
		existing	50 (d)	100 (d)	50 (e)	30
Germany	Na	new/modified	50	50	50	50
		existing	50	50	50	50
Greece	Na/R	new/modified	100	100	100	
		existing	150	150	150	
Ireland	Na	new/modified	50	100	75	50
		existing	50	100	75	50
Italy	Na/P	Existing	50	50	50	50
Luxembourg	P	Existing	30 (f)			
Netherlands	P	Existing	15 (f)	10 (f)	10 (f)	10 (f)
Portugal	Na	new/modified	50	100	75	50
		existing	100	100	75	50
Spain	Na	new/modified		170/100 (g) 100 (h)	300/250 (g) 75 (h)	300/250 (g) 50 (h)
		existing	400/250 (g) 100 (h)	170/100 (g) 100 (h)	300/250 (g) 75 (h)	300/250 (g) 50 (h)
Sweden	P	Existing	50 (i)	50	50	20
United Kingdom	Na (j)	new/modified	40 (k)	50 (k)	40 (k)	50 (k)
		existing	(l)	(l)	(l)	(l)

Na=National law; R=Regional law; P=Typical permit

a) Daily averages and reference condition of 273 K, 101.3 kPa, dry gas and 10% O_2.
b) Limits under discussion. Reference condition of 273 K, 101.3 kPa, dry gas and 10% O_2.
c) Existing plant must meet 50 mg/Nm³ by January 1, 2001. Monthly averages and reference condition of
d) 10% O_2 and dry gas.
e) Existing plant with emission <150 mg/Nm³ must meet limit for new plant by 2001.
f) Existing plant must meet limit for new plant by 2001.
g) Daily average values.
h) Current limits. i) Limits under discussion. j) Daily average value. A limit value of 90 mg/Nm³, including start/stop and CO-trips, applies for monthly averages.
k) IPC Guidance Note S2 3.01. l) Benchmark releases m) Benchmark releases are, in particular, not applicable to existing plant but are a factor in considering appropriate limits.

Source: EIPPC Cement BREF, 2000 [Based on [Cembureau report, 1997] and information provided by experts of the Technical Working Group set up in order to support the production of the BREF]

larly conspicuous in comparison to Spain where dust legislation is in many provinces still quite soft. In Germany both new and existing plants are not allowed to emit more than 50 mg of dust per Nm^3.

NO_x limits were set with a provision for updating and after conclusion of a large demonstration project on NO_x abatement in the German cement industry, the updated limits are now 500 mg/Nm^3 for new installations and 800 mg/Nm^3 for existing installations. Within the EU these are now the second most stringent limit values, only in Denmark and Sweden regulation is tighter (see Table 3.4-2).

SO_2 regulation in Germany is also tighter than the average in the European Union. However, SO_2 is on average not as much a widespread problem in Germany as e.g. in the UK where different raw materials (chalk) produce much higher SO_2 levels.

Concerning emission limit values for metals and other emissions Germany does not seem to hold an above average stringent position (see Table 3.4-3). However, according to TA Luft the limits required for total organic compounds (TOC) are divided into three classes of danger. This is of particular importance when alternative fuels are burnt which is quite common in Germany (see also Appendix A.3 on environmental regulation of alternative fuels).

In connection to the EU Directive on Integrated Prevention and Pollution Control (IPPC) the background reference document (EIPPC Cement BREF, 2000)for the cement sector also mentions emission levels associated with the use of best available techniques (BAT). These associated levels are in the range of 200 - 500 mg/m^3 for NO_x, 200 - 400 mg/m^3 for SO_2 and between 20 - 30 mg/m^3 for dust.

However, these suggested values for the future are not the main focus of this research; the focus of this work is to evaluate the impact of differences in existing national environmental legislation. Although the current regulation in Germany is already quite strict, also German cement plants would have to invest in further pollution abatement equipment in case the BAT associated emission levels will be implemented in the near future in the member states.

Table 3.4-2 National SO_2 and NO_x emission limits for the production of cement within the European Union

SO_2, NO_x mg/Nm³ PCDD/Fs	Data based on	new/modified or existing plant	SO_2 normal situation	SO_2 S-rich raw materials	NO_x	PCDD/Fs ng TEQ/Nm³
Austria	Na(a)	new/mod. existing	200 200	400 400	500 1000	
Belgium	P	new/mod. existing	1000 1000		1800 1800	
Denmark	P(a)	existing	5/250/450(b)	no limit	1200/2500/850(c)	no limit
Finland	P(d)	existing	150-400		1200-1800	
France	Na	new/mod. existing	500 500(e)	1200/1800(f) 1200/1800(e,f)	1200/1500/1800(g) 1200/1500/1800(g)	
Germany	Na	new/mod. existing	400 400	400 400	500 800	
Greece						
Ireland	Na	new/mod. existing	400 400	700 700	1300 1300	n.a. n.a.
Italy	Na/P	new/mod. existing	 600		 1800	10000(h) 10000(h)
Lux.	P	existing	100(i)		800(j)	0.1(k)
Netherl.	P	existing	(l)		1300(j)	0.1
Portugal	Na	new/mod. existing	 400		 1300	0.1 0.1
Spain	Na	new/mod. existing	2400/6000 (m) 600(n) 2400/6000(m) 600(n)	2400/6000(m) 1800(n) 2400/6000(m) 1800(n)	 1300-1800(n)	
Sweden	P	existing	—	<200	<200	0.1
UK	Na(o)	new/mod. existing	200(p) (q)	 600-2500(r)	900(p) 500-1200(q)(s)	

Na=National law; R=Regional law; P=Typical permit

a) Daily averages and reference condition of 273 K, 101.3 kPa, dry gas and 10% O_2.
b) 5 for semi-dry process, 250 for wet process and 450 for wet process with wet scrubber and heat recovery. Limits under discussion.
c) 1200 for semi-dry process, 2500 for wet process and 850 for wet process with wet scrubber and heat recovery. Limits under discussion.
d) Monthly averages, reference condition of 10% O_2 and dry gas
e) Existing plant must meet limit for new plant by 2001.
f) 1200 mg/Nm³ if ≥ 200 kg/h; 1800 mg/Nm³ if < 200 kg/h.
g) 1200 mg/Nm³ for dry process with heat recuperation, 1500 mg/Nm³ for semi dry and semi wet processes, and 1800 mg/Nm³ for wet and dry processes without heat recuperation.
h) General rule for any kind of industrial emission.
i) Half hour average. j) Daily average value.
k) 6 hour average.
l) 90 kg/h as daily average, maximum 375 tonne/year.
m) Current limits. n) Limits under discussion.
o) IPC Guidance Note S2 3.01.
p) 'Benchmark releases'
q) Benchmark releases are, in particular, not applicable to existing plant but are a factor in considering appropriate limits.
r) Limit values reflect the actual levels of releases. Daily averages and reference condition of dry gas and actual O_2 content.
s) Actual releases, daily averages, not all plants currently have limits.

Source: EIPPC Cement BREF 2000 [Based on [Cembureau report, 1997] and information provided by experts of the Technical Working Group set up in order to support the production of the BREF]

Table 3.4-3 National metals and other emission limits for the production of cement within the European Union

Metals, others mg/Nm^3	Measuring interval	Σ (Cd, Tl, Hg)	Σ (Se, Te, As, Co, Ni)	Σ (Sb, Pb, Cr, Cu, Mn, V, Sn)	TOC	HCL	HF	NH_3	CO
Austria	3 years(a)	0.1/0.2(b)	1.0(c)						
Belgium	6 months	0.2	1(d)	5	75(e)	30	5		
Denmark									
Finland									
France	1 year	0.2	1.0	5.0(f)					
Germany	3 year	0.2	1.0	5.0(g)	(h)	30 (i)	5(i)		
Greece									
Ireland									
Italy	varies	0.2	1.0(j) 1.0(k)	5.0(l) 5.0(m)	(n)	30	5	250	
Luxemburg(o)	1 year	0.2	1	5(p)	30	30	5		
Netherlands	1 year	0.15	1.0(q)	1.0(q)	40	10	1		
Portugal	6 months	0.2	1.0	5.0	50	250	50		1000
Spain	varies	0.2(r)	1.0(r)	5.0(r)					
Sweden	1 year	(s)	(s)	(s)					
UK									

a) Half hourly averages and reference condition of 273 K, 101.3 kPa, dry gas and 10% O_2.
b) Group comprises Cd, Tl, Be. The limit 0.1 applies individually, 0.2 applies for the sum
c) As, Co, Ni, Pb.
d) As, Co, Ni.
e) CH_4 excepted.
f) Zn also included.
g) Cyanides (soluble) given as CN, fluorides (soluble) given as F, Pt, Pd and Rh are also included.
h) Limits are required for organic compounds divided into three classes of danger: (from TA-Luft) CLASS I: 20, CLASS II: 100, CLASS III: 150.
i) Emission limit applies for sum of this and other compounds.
j) Se, Te only.
k) As, Cr(VI), Co, Ni.
l) Sb, Cr(III), Mn, Pd, Pb, Pt, Cu, Rh, Sn, V.
m) Σ(Cd, Tl, Hg, Se, Te, Sb, Cr(III), Mn, Pd, Pb, Pt, Cu, Rh, Sn, V)
n) Limits are required for specific organic compounds (ca. 200) divided into five classes of danger: CLASS I: 5, CLASS II: 20, CLASS III: 150, CLASS IV: 300, CLASS V: 600.
o) Half hour average.
p) Σ(Cd, Hg, Tl, As, Co, Ni, Se, Te, Sb, Cu, Pb, Cr, V, Fluorures)
q) Σ(Sb, As, Pb, Cr, Co, Cu, Mn, Ni, V, Sn, Se, Te)
r) Limits under discussion.
No limit specified in the permit. Actual levels are reported initially and are taken into account in the issuing of the permit. Selected elements are reported annually.

Source: EIPPC Cement BREF 2000 [Based on [Cembureau report, 1997] and information provided by experts of the Technical Working Group set up in order to support the production of the BREF]

3.4.3 Clean air regulation for the cement industry in Spain

Spanish cement legislation is based on national framework regulation, but each autonomous province has the right to set regional limits. According to the Spanish cement federation Oficemen[15] the national cement regulation has not been updated for many years and thus has virtually stayed unchanged since 1975 (Ferraro Garcia, 1982). Already in 1975 differences between Spain and West Germany, the United States and France have been marked (see Table 3.4-4 below). These differences have widened as many other countries have moved on to a more stringent environmental policy.

Table 3.4-4 Emission limit values for dust in the cement industry in 1975 in mg/Nm³

Country	Kiln	Clinker cooling	Other points
France	150	150	150
Japan			
General limits:			
< 40,000 m³/h	400	–	–
> 40,000 m³/h	200	–	–
Limits for polluted areas:			
< 40,000 m³/h	200	–	–
> 40,000 m³/h	100	–	–
Spain:			
Existing installations	400	200	300
New installations	200	120	170
Sweden	250 (1)	–	250
Switzerland	155 (2)	155	155
United Kingdom			
< 1,500 t/day clinker	460	230	230
1,500-3,000	230-460	230	230
> 3,000	230	230	230
United States,			
In g/t of raw material	150	50	–
West Germany	120	75	75

Source: Ferraro Garcia (1982).

In 1998 Spanish NO_x emission limit values were still fixed between 2400 and 6000 mg/Nm³ (see Table 3.4-2 above). Current dust emission limits still range

[15] Personal communication with Oficemen in Madrid during April 1999.

between 100 and 400 mg, with new plants being treated in a stricter way than existing plants. Also SO_2 limits are currently still relatively soft.

However, there is tremendous regional variation in Spain with a tendency of the North being more progressive than the South[16]. E.g. the government of Catalunia is already advanced in transposing the new IPPC Directive which proposes relative stringent emission values associated with the use of special best available techniques in the cement industry (see also chapter 9 on the empirical results). Moreover, there are examples of voluntary agreements between individual cement plants and the environment ministry of Catalunia where favourable environmental performance is documented.

The use of alternative fuels in the Spanish cement industry is starting to grow because waste markets are just established or starting (see also Appendix A.3 on the issue of alternative fuels). Measurement of metals and other emissions related to the burning of waste is under discussion.

3.4.4 Clean air regulation for the cement industry in the United Kingdom

In the UK permits are given on a plant by plant basis and plant size has always been a criterion in fixing limit values with larger plants being subject to stricter limits (see e.g. Table 3.4-4 for British limits in the past).[17]

The emission limit values of today quoted in the Tables 3.4-1 and 3.4-2 are understood as benchmark values, i.e. these limits are among the strictest in the industry, but are not applicable for existing plants, even when they are a factor in considering appropriate limits. According to the Environment Agency there are about six plants in the UK which are of a very clean standard.[18] The use of alternative fuels in the UK is currently not as widespread as in Germany.

3.4.5 Comparison of German, Spanish and UK legislation for the cement industry

The three main pollutants of the cement sector are dust, NO_x and SO_x emissions. Especially the German dust and NO_x emission limits both for new and existing plants are among the most stringent emission limit values in the EU. Also, Ger-

---

[16] Personal communication with the ministry of the environment in Catalunia during February 2000.

[17] One could argue that large plants can reach uniform limit values more easily than small plants because emissions proportionately grow with production capacity and gas volumes in the cement industry (Ferraro Garcia, 1982).

[18] Personal communication with the Environment Agency in Reading in May 1999.

many follows a national emission limit value approach. In comparison, Spanish legislation is in many provinces still quite soft. In the UK permits are given on a plant by plant basis. Any emission limit values are understood as benchmark values, i.e. they are among the strictest in the industry, but are not applicable for existing plants. The burning of alternative fuels is more widespread in Germany than in the UK and Spain. It leads in all countries to additional maximum emission limits for heavy metals.

3.5 Conclusions on the stringency of environmental regulation in Germany

This chapter has set the scene for the empirical chapters of this study. In summary, Germany has a long history of stringent environmental policy and it could be shown that in all three case study industries environmental regulation is indeed more stringent than in the other sample countries. In the food case, Germany not only operates generally tougher effluent standards, but also charges those companies using the public sewage system more heavily than in Northern Ireland and the Republic of Ireland. In packaging, recycling levels in Germany are in the mid nineties well ahead of the Republic of Ireland, the UK and Northern Ireland. Material specific fees for the individual packaging materials are much higher in Germany; economic incentives for the reduction of packaging waste are therefore higher in Germany than in the other sample countries. Also in the cement industry, Germany has on average the most stringent emission levels in place. Spanish and UK plants may sometimes be regulated as strict as German plants. But in Spain there is wide regional variation and in the UK benchmark levels are not used for existing plants. Given this background the question is whether Germany is disadvantaged by this stringent regulation. This is debated in the next chapter on a sectoral level. The empirical analysis in the chapters seven, eight and nine then focuses on the impact of stringent environmental regulation on competitiveness in representative samples of the individual case study countries under concern.

4. Structural background of case study industries

This chapter provides the main structural aspects of the case study industries in the sample countries. The purpose is to give an overview on the state of competitiveness (productivity) of the sample sectors and to analyse the possible influence of environmental requirements on competitiveness, thus serving as background for the empirical evidence presented in chapters seven, eight and nine of this study. For each case study the likely impacts of environmental regulation on industry competitiveness are first analysed from a general perspective and then in detail for Germany. It is found for Germany that the expected impact of environmental regulation on competitiveness is likely to be small in the food and packaging case, but larger in the cement industry. The chapter starts with the competitive situation in the dairy and meat industry (section 4.1), then presents the packaging industry in a supply chain (section 4.2) and proceeds to the cement industry (section 4.3). Conclusions are given in section 4.4.

4.1 Structural background of the dairy and meat industry in Germany, the Republic of Ireland and the UK

This section presents in detail the key structural features of the dairy and meat sectors in Germany, Ireland and the UK. The purpose is to give an overview over the essential points relating to competitiveness, important trends and the possible influence of environmental regulations on competitiveness. It is found that the impact of waste water regulation on competitiveness in the German food industry is likely to be small. Interviews for the empirical part of this case study were undertaken in 1994 and 1995. Therefore the data on the structural issues mainly concentrate on the 1990ies. Newer trends are sketched as far as EU data were already available.

4.1.1 Structural background of the dairy industry

This sector consists of primary milk products, i.e. pasteurised, powdered and condensed milk, as well as the processing of milk into by-products, both intermediary (such as casein) and final consumer products (such as cream, butter, yoghurt, cheese and icecream).

Production value, company size and leading European companies

The dairy industry within the EU is marked by a high absolute amount of value added, only slightly lower than that of the meat sector, with France, Germany, Italy and the UK exhibiting the highest levels of value added. In 1998 total employment was about 300.000 and the production value amounted to Euro 85 billion. In general, the market is now saturated and new markets can only be opened through product differentiation. Consumer preferences are being influenced by factors such as calorie content and perceived influence on fitness and health.

In terms of processing the biggest producers are Germany, France and the UK. The first two show high production levels for all the main types in the subsector, while the UK specialises in the production of fresh milk based products and drinking milk. Highest export quotas are achieved in Denmark, the Republic of Ireland and the Netherlands.

The dairy sector is distinguished by the presence of companies of different sizes, from small cottage industries that produce only cheese, to large multinationals with high production outputs and fully automated production cycles. Consequently production levels vary depending on the type of company. Intensive capital production techniques have been adopted in every subsector in order to cut costs and provide higher standards of hygiene with the exception of cheese production, which is carried out mostly in small to medium sized enterprises.

EU milk production has been affected by a considerable decrease in livestock numbers (by more than 20 per cent for the EU as a whole). This is due to the reform of the Common Agricultural Policy (CAP) which is only partially compensated for by productivity growth and the effect of increases in the Italian, Spanish and Greek quotas (Statistical Office n of the European Communities, 1997). The milk quota system has also brought about strong competition between milk processing companies in the raw material supply stage, and this has led to mergers and take-overs which have reduced the number of enterprises in this subsector (especially smaller firms).

From 1985 to 1991 the number of dairy enterprises dropped by about one-quarter, although there is a significant difference in the level of fragmentation of the different EU countries, from Italy with 2 400 enterprises and a turnover of Euro 12 billion in 1992, to Germany with 315 enterprises and an Euro 15.6 billion turnover. The UK had around 670 enterprises in that year, but experienced an almost 50 per cent drop until 1994. In 1998 the number of German dairy enterprises had declined to almost half of the number which had existed in 1994 (see Table 4.1-1).

The leading European companies were for yoghurt: Danone, Yoplait, Müller, Zott, Milram, Northern Food, Unigate, MD Food; for cheese: Campina Melkunie, Draft, Coberco, Friesland Frico Domo, Entremont, Besnier, MD Food; for milk: Campina Melkunie, Dairy Crest, Unigate, Waterford, Parmelat, MD Food.

Of the survey countries, the Republic of Ireland had the largest dairy industry in relative terms (that is, relative to the size of total manufacturing output) (Statistical Office of the European Communities, 1997).

Table 4.1-1 Number of dairy firms in the sample countries, 1985–1998

Country	1985	1988	1991	1994	1998
EU	5 472[a]	6 114[a]	3 911[a]	4 362[a]	–
Germany	489	408	315[b]	354[b] (1992)	165
Republic of Ireland	90	84	31	–	–
United Kingdom	643	653	670	339[c] (1992) 323[c] (1994)	–
				323[c] (1994) 323[c] (1994)	

Notes:
[a] EU15 including East Germany from 1991 onwards.
[b] Including East Germany from 1991 onwards.
[c] UK figures include only producer dairies with a milk intake of over 180 tonnes p.a. in 1992. The 1994 data exclude producer processors.

Source: ASSILEC, quoted in the Statistical Office of the European Communities (1995, 1997, 2000).

Comparative productivity in the dairy industry

The productivity results (gross value added (GVA) compared by using purchasing power parities (PPPs)) shown in Table 4.1-2 imply a fair degree of variability in performance. In 1990 there was not much difference between West Germany and the UK, with the Republic of Ireland lagging further behind. By 1991, however, the UK had fallen behind whilst Germany was still relatively closely positioned. In 1998 the UK has caught up and even more so the Republic of Ireland relative to the 1991 level. The relative labour cost data shown in Table 4.1-3 are shown to be highest in West Germany whilst they are considerably lower in the UK and Republic of Ireland.

Factors which influence competitiveness in the German dairy industry

Regarding international competitiveness the Germany dairy industry suffers despite the high productivity levels and quality standards from some undeniable comparative disadvantages. Partly, this is due to the predominant organisational form of German dairies as co-operatives. The co-operative dairies process 60 per cent of all milk delivered to German dairies. However, the average production of co-operatives is about 20 per cent lower than in large private enterprises.

Table 4.1-2 Gross value added in dairying at market prices per employee, 1990, 1991 and 1998 (compared using PPPs, West Germany = 100)

Year/industry	Exchange rate	PPP
1990		
West Germany	100	100
United Kingdom	121	105
Republic of Ireland	111	83
1991		
West Germany	100	100
United Kingdom	107	88
Republic of Ireland	–	–
1998		
Germany	100	100
United Kingdom	102	91
Republic of Ireland	95	91
Note: All sectors defined according to the NACE. A single PPP, judged to be 'representative', was applied to each sector: 'Milk etc.'.		

Source: Statistical Office of the European Communities (1994, 1995a, 2000).

Table 4.1-3 Total labour costs per employee in the dairy sector, West Germany, Republic of Ireland and United Kingdom, 1975–1991 and 1998 (Euro at market exchange rate; West Germany = 100)

Year	West Germany	Republic of Ireland	United Kingdom
1975	100	56	57
1979	100	58	54
1987	100	81	63
1989	100	72	69
1990	100	69	87
1991	100	74	–
1998	100 (Germany)	73	60

Source: Statistical Office of the European Communities (1981, 1984, 1993, 1995a, 2000).

For example, other European companies in France, Denmark or the Netherlands have a substantially higher milk intake which can be up to five times the amount of the average volume processed in German firms and thus take advantage of economies of scale. Concerning international marketing strategies these large French, Danish and Dutch dairy corporations systematically "europeanise" their operations and products. In contrast to these efforts German firms are less marketing-oriented and mainly concentrate on the home market (Maurer/Drescher, 1995). Currently foreign competitors are entering the German market and building up high-capacity plants. At the moment the most modern plants are situated in East Germany. These firms are expected to be among the most competitive dairies within Europe. All the East German dairies which were visited for this study had experienced substantial growth.

Prospects of the dairy industry at the EU Level

The trends up to 2000 and close beyond are likely to include the following:

- increasing extent of extra-EU trade competition (more imports allowed because of the World Trade Organisation e.g. from Australia, New Zealand, Argentina);
- import growth exceeding export growth;
- continued concentration and rationalisation;
- declining employment;
- adjustment to the anticipated impact of forthcoming EU enlargement to include large dairy producers in Eastern Europe (higher competition);
- a moderate acceleration in EU consumption and output growth rates (Statistical Office of the European Communities, 1997 and 2000).

Important regulations for the dairy industry

In addition to the effluent regulation described above in detail, milk processors have been mainly affected by both hygiene and production quality regulations (e.g. EC 856/84; 1447/84 or 1527/88) as well as packaging regulation (see also case study on the EU Packaging Directive in this book).

4.1.2 Structural background of the meat industry

The meat sector consists of slaughtering and meat processing. Final products comprise cut and processed meat, frozen meat and meat cans.

Production, employment and leading European companies

Total meat production in the EU was 38.9 mio. tonnes in 1999. The biggest meat producers in the EU were France and Germany (see Table 4.1-4).

The meat sector is characterised by the presence of a number of large companies as well as small and medium sized enterprises. Whereas the industry is still fragmented, the Common Agricultural Policy as well as central marketing strategies lead to growing concentration.

Details for the sample countries show the following: Total employment has slightly grown at falling turnover in the German slaughtering industry between 1997 and 1998, but in meat processing the number of employees was reduced at constant turnover hinting at a growing labour productivity in this industry branch. In the Irish meat processing industry (incl. slaughtering) both employment and turnover were stable in the same period. Concerning the United Kingdom turnover has strongly declined due to the Creutzfeld Jacob disease (see Ta bles 4.1-5 to 4.1-7). The export share in the Republic of Ireland was with 52 per cent remarkably high in comparison to Germany.

The largest companies on a European level belong to multinationals like Unilever (UK/NL) and Nestlé (CH). Further large companies with meat as a core business are Südfleisch (D), Nordfleisch (D), Socopa (F), Arcadie (F), Vital Sogesviande (F), Moksel (D) and Cremonini (I) (Statistical Office of the Commission of the European Communities, 2000).

Table 4.1-4 Total meat production in selected EU countries in 1,000 tonnes

Country	1994	1995	1996	1997	1998	1999
Germany	6 095	6 089	6 162	6 236	6 464	6 704
France	6 864	7 077	7 335	7 470	7 493	7 450
Ireland	1 070	1 106	1 132	1 148	1 190	1 290
Spain	4 258	4 428	4 558	4 703	5 120	5 450
Unit. Kingdom	3 921	4 118	3 830	3 925	4 059	3 781
EU 15	36 042	36 598	37 101	37 177	38 571	38 945

Source: Zentrale Markt- und Preisberichtsstelle (2000).

Table 4.1-5 Selected output and employment data for slaughtering and meat processing firms in Germany

Indicators	Slaughtering (without poultry)			Meat processing		
	1997	1998	1999	1997	1998	1999
No. of plants	200	219	228	816	1 024	1 062
No. of employees	16 668	16 459	17 430	86 344	84 106	85 586
Turnover, mio. Euro	7 061	6 266	6 350	11 250	11 665	11 440
Export share in %	8.8	10.6	11.0	2.6	2.9	3.0

Source: Zentrale Markt- und Preisberichtsstelle (2000).

Table 4.1-6 Selected output and employment data for slaughtering and meat processing firms in the Republic of Ireland

Indicators	Republic of Ireland		
	1997	1998	1999
No. of employees	12 300	12 800	12 900
Turnover, Mio. Euro	2 061	2 123	2 057
Gross value added in Mio. Euro	148	147	–
Export share in %	–	–	52.9

Source: Central Statistics Office, Census of Industrial Production, 1998, Early Estimate of Turnover for 1999.

Table 4.1-7 Selected output and employment data for slaughtering and meat processing firms in the UK

Indicators	United Kingdom		
	1997	1998	1999
No. of employees	115 000	114 000	113 000
Turnover, Mio. Euro	17 908	13 379	–
Gross value added in Mio. Euro	–	3 375	–
Export share in %	–	10.3	–

Source: UK Meat and Livestock Commission; Statistical Office of the European Communities 2000.

Comparative productivity in the meat industry

Table 4.1-8 considers the relative labour productivity performance of the sectors in the sample countries. The chosen indicator is the level of gross value added measured in factor cost per employee (although some of the Republic of Ireland data was only available measured in market prices). These comparisons should be regarded as approximations, especially where average market exchange rates have been used. For most years Germany lies at the top with the UK and the Republic of Ireland being close. The relative labour cost data shown in Table 4.1-9 confirm West Germany to be at the top whilst the UK and Republic of Ireland lie further behind.

Concerning slaughtering in Germany the predominant organisational form used to be municipal slaughterhouses. Many of the small municipal slaughterhouses have closed down because they are not competitive anymore and a lot of investment would be required in order to meet European hygiene standards. Private companies benefiting from economies of scale now dominate the market.

Table 4.1-8 Gross value added (GVA) factor cost per employee in the meat industry 1989, 1991 and 1998 (West Germany = 100)

Year/industry	GVA at factor costs	GVA at market prices
1989		
West Germany	100	100
United Kingdom	80	77
Republic of Ireland	108	51
1991		
West Germany	100	100
United Kingdom	—	91
Republic of Ireland	—	34
1998		
Germany	100	100
United Kingdom	—	101
Republic of Ireland	—	97

Note: Data for the Republic of Ireland and the United Kingdom required some estimation to adjust from factor cost to market prices. For 1991 purchasing power parities were used (PPPs), otherwise converted into Euro using market exchange rates. The comparisons using PPPs attempt to make allowance for differences in product price levels in the sample countries when comparing real levels of output.

Source: Statistical Office of the European Communities (1981, 1984, 1993, 2000). Production Census Data in the various countries for 1991 and 1998.

Table 4.1-9 Total labour costs per employee in the meat sector, West Germany, Republic of Ireland and United Kingdom, 1975–1989 and 1998 (Euro at market exchange rate; West Germany = 100)

Year	West Germany	Republic of Ireland	United Kingdom
1975	100	63	51
1979	100	63	52
1987	100	63	66
1989	100	63	66
1998	100 (Germany)	69	67

Source: Statistical Office of the European Communities (1981, 1984, 1993, 2000).

Factors which influence competitiveness in the German meat industry

International competitiveness on a pure cost basis is not an easy subject for the German meat industry due to three factors. Firstly, other countries like the Netherlands can raise the animals for a cheaper price (mass production). Secondly, fees for veterinarians vary not only European-wide, but also on a national level due to locally different ways of calculating the fees. Thirdly, often the disposal of unusable animal waste is in the hands of local firm having a monopoly. Thus the disposal costs per animal can vary between about 5 per cent and about 40 per cent of the total cost of slaughtering the animal.

East German food processing faces particular problems created by the sudden demise of its formerly substantial farming sector. Thus most of the meat has to be imported which raises prices due to transport costs.

Market volume has fallen steadily since 1992. The German meat processing industry could adjust only by reducing prices. Simultaneously, however, the production costs have risen. As a consequence a strategy of rationalisation is pursued by buying ready components (standard meat, cut meat chunks), less variety in the product assortment as well as closing down of business parts. Often, several shifts are worked in order to improve the cost situation. The temporarily lower input prices were compensated by a cost increase in the disposal of packaging, distribution and labour costs.

Prospects of the meat industry at the EU Level

A major shock for the meat industry was the BSE crisis. Following the announcement made by the British Secretary of State for Health on 20 March 1996 that there might be a link between BSE and variant Creutzfeldt Jakob Disease (vCJD), there was a significant loss of confidence in beef and in all beef in general, when in various countries of the EU cases of BSE were discovered. Total UK domestic usage of beef fell by 19 per cent in 1996 but had risen to 2 per cent above its 1995 level in 1999 with other types of meat gaining considerably. E.g. in the UK, lamb and poultry consumption were each about 10 per cent above its 1995 level (Department for Environment, Food and Rural Affairs, 2001). In Germany first cases of BSE were discovered in fall 2000. After an immediate sharp drop of beef consumption, in summer 2001 beef consumption was back to its original level before the crisis in Germany.

Future trends for the meat industry as a whole are likely to include the following:

- — changing consumer preferences: increasing demand for poultry since it is relatively cheap, easy to prepare and contains less fat than red meat;
- — increasing demand for certified meat products (e.g. according to type of breeding, methods of slaughtering, type of fodder etc.);
- — growing extent of extra-EU trade competition (more imports allowed because of the World Trade Organisation e.g. from Australia, New Zealand, Argentina,);

— continued concentration and rationalisation;

— declining employment;

— adjustment to the anticipated impact of forthcoming EU enlargement to include large meat producers in Eastern Europe (higher competition);

— a moderate acceleration in EU consumption and output growth rates (Statistical Office of the European Communities, 2000).

Important regulations for the meat industry

Just as the dairy industry also the meat sector has been subject to hygiene, production quality and packaging regulation. In addition, already in 1991 scare as to the safety of meat led to the EC Directive 92/102 on the identification and registration of stock farms and their livestock. The emergence of the mad cow disease in subsequent years led to export bans and changes in production methods.

4.1.3 The likely impact of effluent regulation on competitiveness in the dairy and meat sector

Meat and dairy processing typically consume large quantities of water and energy and discharge significant quantities of effluent. In order to keep the likely impact of effluent regulation on competitiveness small, factories will therefore try to reduce both water and energy consumption and will also try to reduce the pollutant load in effluents. Classical end-of pipe measures in the food industry are effluent treatment plants which typically increase production costs and can thereby impair competitiveness. However, there is also a large the potential for the cost reducing use of clean technologies which will be shown in the following.

Water consumption

Since water is used so extensively in meat and dairy processing, water saving measures are common cleaner production opportunities for these industries (UNEP, 2000). Efforts to reduce water consumption also result in reduced effluent volumes. Concerning water consumption there are a number of clean technology solutions available which decrease costs. The first step is to gain knowledge about water use patterns by installing water meters and regularly recording water consumption. The next step is to identify all wasteful practises, e.g. hoses left running when not in use. Automatic control of water use is preferable to relying on operators manually controlling water use. In a third step, essential process functions can be optimised. Here the bottleneck for water reduction will be hygiene requirements. Careful investigations are therefore necessary.

Water reuse is a popular strategy for water reduction. Waste waters that are only slightly contaminated could be used in other areas. However, this option requires screening, filtering or in-line bacterial control. Not all authorities will allow this strategy. Certainly the reuse of water should never impair product quality and hygiene.

Full recycling of water is more possible, the cleaner the treated effluent is. As quality standards become tighter and approach drinking water quality, there is a strong incentive to take advantage of the investment that goes into effluent treatment. Treatment techniques must then contain techniques such as membrane filtration in order to remove dissolved solids.

Effluent

Opportunities for reducing the pollutant load of effluent in the food industry focus on avoiding the discharge of polluting substances, such as blood or scraps of meat in abattoirs or fat in dairies, to the effluent stream. This means capturing materials before they enter drains and utilising dry cleaning methods wherever possible. E.g. blood recovery can decrease organic loads by approximately 40 per cent in abattoirs (Jones, 1974). According to UNEP (2000), most gains in reducing the pollutant load of effluent in the food industry can be made by improving cleaning practises. Box 4.1-1 gives an overview of possible methods for reducing the pollutant load of effluent which trigger a cost saving potential. Naturally, however, the final cost burden of an individual plant depends on its efforts and possibilities of reducing both the volume and the pollution load of effluent as well as on its cost scheme for fresh and waste water.

Box 4.1-1 Checklist of ideas for reducing effluent loads in meat processing

Maximise the segregation of blood by designing suitable blood collection facilities and allowing sufficient time for bleeding, typically seven minutes.
Sweep up solid materials for use as by-products, instead of washing them down the drain.
Fit drains with screens and/or traps to prevent solid materials from entering the effluent system.
Use offal transport systems that avoid or minimise the use of water.
Use water sprays with a pressure of less than 10 bar for carcass washing to avoid removing fat from the surface.
Use dry cleaning techniques to pre-clean process areas and floors before washing with water.
Segregate high-strength effluent streams, such as rendering effluent and waste waters from casings and paunch washing and treat them separately.

Source: UNEP (1999).

Energy

Simple plant optimisations and housekeeping measures can provide substantial energy savings almost immediately without capital investment. Additional savings can be made through the use of more energy efficient equipment and heat recovery systems.

Moreover, there are opportunities for using more environmentally benign sources of energy such as replacing fuel oil or coal with natural gas or co-generation of heat and power on the site. In the long run, the initial capital investment will often be profitable because it normally embodies a general upgrade of the energy system which triggers lower energy demand. Final savings of course depend on the price ratio traditional vs. cleaner fuels.

By-products

As was already mentioned in the section on effluent treatment techniques any polluting substances should be prevented from entering the effluent stream. Valuable by-products such as whey or blood can be used as commodity, if economic markets can be found. This of course depends on the scale of the operation and the proximity to potential markets.

4.1.4 Conclusions on the likely impact of environmental regulation on competitiveness in the dairy and meat industry in Germany

The above analysis has shown that productivity levels in both the German dairy and meat industry tend to be among the highest in the sample countries. This will be a helpful factor in adjusting to environmental regulation because high productivity plants are assumed to be in a good position to absorb environmental compliance costs. Germany might also benefit from a sophisticated environment industry when it comes to the implementation of water and energy saving technology as sketched in paragraph 4.1.3 above. In chapter seven it will be shown that environmental regulations are just one of several cost increasing factors for the German dairy and meat industry and that they are not among the most important ones. Other factors like the already mentioned continued concentration and rationalisation of the German food industry as well as lack of international marketing efforts were estimated to be more critical for Germany.

4.2 Structural background of the supply chain: packaging, food and re tail industry in Germany, the Republic of Ireland and the UK

This section presents the key structural features of the supply chain in the packaging, food and retail industry in Germany, Ireland and the UK. Again the purpose is to give an overview over the essential points relating to the relationship between competitiveness and environmental issues for the sectors under concern. The focus is on the inclusion of activities at all stages along the production chain of food packaging, the dairy and drinks industry as well as the retail industry. The finding is that only small impacts of packaging regulation on competitiveness in the German sector are to be expected. Interviews for the empirical part of that case study were undertaken in 1997 and 1998. Therefore the data on structural issues mainly concentrate on the late 1990ies.

4.2.1 Structural background of food packaging materials

Not only do the sample countries Germany, Republic of Ireland and United Kingdom have varying packaging standards as shown in chapter 3.3, but Table 4.2-1 also illustrates the relative size in terms of employment of the sample sectors and hence the importance of considering the environmental/competitiveness relationships in these sectors. For the Republic of Ireland it is worth noting that packaging is less important in terms of employment. In the EU paper-based, plastics and light metals packaging are the dominant forms of packaging measured by value and are included as principle types of packaging in this study. Since glass is also an important material for food packaging, we include it in our study as well.

Concerning the comparative labour productivity the available data of the Statistical Office of the European Communities is often aggregated to a higher level than ideally wished. For example, the manufacture of glass containers is contained within the entire glass manufacturing industry; metal containers are represented within all metal goods (i.e. non-electrical and non-furniture); paper and board packaging is just one part of all paper and board products; and plastic packaging is represented by the whole plastics industry. Taking the relative productivities, in most cases the rankings are: West Germany first, with the UK second and Republic of Ireland (most of the time) at the bottom (see Table 4.2-2).

The competitive position of each material type which is important for food packaging will now be briefly considered in terms of industry structure, main players, and prospects for growth. Emphasis is not only put on the EU as a whole, but also on the relevant detail for the sample countries Germany, the Republic of Ireland and the United Kingdom. The influence of environmental trends are shown as well.

Table 4.2-1 Relative importance of the selected packaging branches, employment, 1986, 1991, 1995 and 1998

Branch (NACE classification) and country	**1986**	**1991**	**1995**	**1998**
Finished metal goods (e.g. cans)				
West Germany	288 865	349 116	299 347*	137 043*
United Kingdom	132 148	130 863	123 971	61 195
Republic of Ireland	3 194	4 215*	2 206	4 258
Glass				
West Germany	65 185	71 553	365 773*	62 862*
United Kingdom	35 719	38 174	207 618	31 668
Republic of Ireland	4 506	3 791*	9 595	3 994
Conversion of paper and board				
West Germany	99 061	124 949	122 445*	103 155*
United Kingdom	108 743	101 035	46 081	79 815
Republic of Ireland	3 184	2 923[a)]	4 473	4 764
Plastics**				
West Germany	213 980	290 285	365 773*	271 773*
United Kingdom	119 504	145 391	207 618	170 033
Republic of Ireland	4 743	6 059*	9 594	9 765

Notes: [a)] 1990.
* Data for 1995 and 1998 are shown for Germany, not only West Germany.
** Data for 1995 include rubber sector as well.
Data for 1998 were available in units only instead of '000.

Source: Statistical Office of the European Communities (1990, 1995a, 1999, 2000).

Table 4.2-2 Gross value added at market prices per employee, 1990, 1991, 1995 and 1998 (compared using PPPs, (West)Germany = 100)

	Glass	Metal goods	Paper and board	Plastics**
1990				
West Germany	100	100	100	100
United Kingdom	93	83	96	94
Republic of Ireland	71	69	114	78
1991				
West Germany	100	100	100	100
United Kingdom	84	73	90	84
Republic of Ireland	–	–	–	–
1995				
Germany*	100	100	100	100
United Kingdom	89	82	69	69
Republic of Ireland	74	55	77	69
1998				
Germany*	100	100	100	100
United Kingdom	95	113	102	81
Republic of Ireland	76	–	96	72

Note: All sectors defined according to the NACE. A single consumer expenditure category PPP, judged to be 'representative' of the value added of a given industry, was applied to each sector.

* Data for 1995 and 1998 are shown for Germany, not only West Germany.

** Data for 1995 include rubber sector as well.

Sources: Statistical Office of the European Communities (1995a, 1995b, 1999, 2000); own calculations using PPPs.

4.2.1.1 Light metal packaging made of aluminium and tinplate

For industrial classification purposes this is defined as the manufacture of containers with walls less than 0.49 mm thick or a maximum capacity of 40 litres. It includes, for example, metal cans for carbonated drinks which are included in this study.

There is some degree of economies of scale within drinks can production (the setting up of a single production line requires a fixed investment of at least Euro 40 million). Although the market was in the early nineties still fairly fragmented, there were already five major groups; Pechiney (France), Carnaud-Metalbox (France–UK), VIAG Continental Can Europe (Germany), Crown Cork CIE (USA) and PLM (Sweden). The largest of these in 1993 still had only 6 per cent of the European market (admittedly, at the global level Carnaud-Metalbox was recently taken over by Crown Cork and Seal CIE of USA).

Concerning the exclusive production of tinplate cans Rasselstein Hoesch GmbH (RHG) is the market leader not only in Germany, but also in Europe[19]. World-wide this company takes the second place after Nippon steel. Market shares of RHG in 1996 were 70 per cent in Germany, 18 per cent in Europe and 7 per cent world-wide (see Breitenacher et al., 1997).

Concerning the consumption of beverage cans the UK and Germany show the highest shares in Europe (see Table 4.2-3). The data refer to the mid-nineties when this study was undertaken.

Table 4.2-3 Consumption of beverage cans in the EU, 1994

Country	No. (million)	Percentage share
UK	8 300	30
Germany	5 800	21
EU	27 700	100

Source: The Canmaker, June 1995, p. 33.

There is a wide regional distribution of plants given the relatively low value/volume ratio of the finished products, i.e. it is not usually profitable to transport light metal packaging more than 300 km from the point of manufacture. Therefore, traditionally there has not been much trade in this activity. Table 4.2-4 presents a distribution of beverage can plants in Europe.

Concerning growth, for example, during the first half of the 1990s beverage can consumption increased by 20 per cent (the metal share of all drinks containers reached 15 per cent and is forecast to rise further). The following trends which are important both from an economic and environmental point of view are likely to continue: decreased thickness (at least for tinplate) and increased quality of metal substrate, improved quality of protective coatings, use of laser welding, stricter recovery and hygiene standards and more sophistication in labelling and other decorative and branding features (e.g. shaped cans).

[19] While in the EU only 8 % of all light metal packaging is made from tinplate (i.e. steel with a zinc layer), in Germany about 30% of all tinplate packaging is used for the production of drink cans (Statistical Office of the European Communities, 2000). The reason for this lies in the longstanding tradition of the use of steel extrusion technology in Germany (personal communication with Dr. Holländer, 9.10.1997). Aluminium cans are produced in only one plant in Germany, mainly for exports. However, aluminium is used for the production of lids. According to a spokesman of PLM the size of aluminium lids has been reduced over time, not at least due to customer pressure who have to pay relatively high DSD fees for the lids made of aluminium (personal communication with PLM, 12.2.1998). The separation of tinplate from other metal waste is relatively easy since iron is magnetic.

Table 4.2-4 Producers of beverage cans in Europe, 1995

Country	Producer	Location	No. of Lines	Material[a)]
Austria	PLM	Enzesfeld	2	A
Belgium	CMB Bevcan	Mechelen	2	T
France	CMB Bevcan	Custines	2	T
	Continental Can Europe	Dunkirk	3	T
	Nacanco	Dunkirk	2	T
	PLM	La Ciotat	2	A
Germany	Continental Can Europe	Weissenthurm	3	T
	Continental Can Europe	Hassloch	3	T
	Continental Can Europe	Hermsdorf	1	T
	Nacanco	Gelsenkirchen	3	T
	PLM	Berlin	3	T
	PLM	Recklinghausen	3	A
Greece	Alucan	Patras	2	A
	Hellas Can (CMB)	Corinth	2	A
Italy	CMB Bevcan	Parma	2	A
	Nacanco	Nogara	2	A
	Nacanco	San Martino	1	A
	Tubettificio	Anzio	1	A
	Tubettificio	Lecco	1	A
Netherlands	Continental Can Europe	Oss	2	T
Poland	Continental Can Europe	Radomsko	1	A
	Can Pack	Brezesko	1	A
Spain	CMB Bevcan	Agoncilto	2	T
	National Can Iberica	Madrid	2	T
	National Can Iberica	La Selva	2	A
Sweden	PLM	Malmö	4	A
Turkey	Nacanco	Izmir	1	T
	Nacanco	Izmir	1	A
	CMB Bevcan	Izmit	1	A
UK	CMB Bevcan	Braunstone	1	T
	CMB Bevcan	Braunstone	1	A
	CMB Bevcan	Carlisle	3	T
	CMB Bevcan	Carlisle	2	A
	Continental Can Europe	Rugby	3	A
	Continental Can Europe	Wrexham	4	A
	Nacanco	Milton Keynes	3	A
	Nacanco	Runcorn	2	T
	Nacanco	Runcorn	1	A
	Nacanco	Wakefield	3	A
[a)] A = Aluminium; T = Tinplate.				

Source: The Canmaker magazine, June 1995.

4.2.1.2 Plastics

Plastics processing is a highly fragmented activity (up to 27.000 producers across the EU with only about 12 companies employing more than 1.000 people) placed in between highly concentrated sectors (i.e. the big petrochemicals enterprises which supply the polymers, and the automobile, electrical and retail firms to which most plastic products are sold). In 1998 a third or Euro 35.6 billion of the European production value took place in Germany where plants are on average the largest in the EU. The second largest producer is the UK with a production value of Euro 17.4 billion.

Packaging and plastic types

In 1998 plastic packaging was with about 40 per cent the largest single component of total European demand for plastics. Over recent decades plastics have increasingly substituted for glass. This is particularly the case in food, distribution and beverage activities (the packaging proportion of total plastics consumption was 21 per cent for Germany and 36 per cent for the UK).

Despite the impact of environmental regulations, the plastics industry in the EU shows particular dynamic growth. E.g. refillable PET bottles are now firmly established in eight of the 16 Western European countries with a growing trend, e.g. also in Germany at the late 1990ies a switch from galls to PET has started. In 1994, PET (which was just beginning to be used for water) accounted for 23.5 per cent of volume compared with 24.3 per cent held by PVC (Food Technology International Europe, 4 May 1995). Since 1995 PET has been considered to be the standard pack for carbonated soft drinks rather than glass. It is expected that PET demand will grow at more than 10 per cent per year until the end of the century (Packaging Week, 24 August 1995). The 'pool' of refillable PET carbonates bottles is becoming established in eight countries, leading to a decreased demand for new bottles. Most growth will come from the introduction of new size (Food Technology International Europe, 4 May 1995).

Output growth prospects for plastic packaging have been for a long time reckoned to be fair to good (Statistical Office of the European Communities, 1995 and 1997, Commission of the European Communities, 1999). For the coming years, growth rates will probably be about 4–5 per cent annually. Since there are overcapacities for some types of plastics, e.g. for PP, producers aim at increasing their margins through a lower labour intensity. Between 1993 and 1998 Germany has reduced the number of employees by 50 per cent down to 7.6 employees per million Euro of production value.

4.2.1.3 Glass containers

In 1998 producers of glass and glass goods achieved a production value of about 30 billion Euro (see Table 4.2-5). Glass was one of the first materials for which an efficient recycling system was installed. Since 1993 the production of glass and glass products follows an upwards trend. In the UK and Germany growth rates of 4.8 and 5.2 per cent were achieved between 1993 and 1998. Value added levels are highest in Germany (see Table 4.2-6).

Table 4.2-5 Production of glass (million Euro), 1998

Country	1998
Germany	8 044
United Kingdom	3 256
EU15	30 000

Source: Statistical Office of the European Communities (2000).

Table 4.2-6 Levels of value added in glass production, 1993, 1994 and 1998

Country	Value added (Euro m)		
	1993	1994	1998
Germany	2 963	3 008	3 291
Republic of Ireland	112	130	160
United Kingdom	1 114	1 041	1 572

Source: Statistical Office of the European Communities (1995, 1997, 2000a).

Glass Packaging

Despite the increasing competition from tinplate, plastics (PET) and paper cartons, the apparent consumption of glass packaging has increased between 1993 and 1998 by 14.2 per cent and amounted to 17 million tonnes. There were overcapacities in the production of bottles and other containers so that producers looked for market niches. This process has been successful and imports of glass containers between 1993 and 1997 declined by 6.9 per cent.

There is growing low-cost supply from third countries with low wages and lower environmental expenditures than within the EU. Companies have answered this with more innovation, higher product quality and service as well as investments in energy saving equipment. But this restructuring has also led to a steady decline in employment between 1993 and 1997.

Industrial structure

In terms of container glass production the following players are significant: Saint Gobain in six countries, BSN (France), Gerresheimer Glas (Germany), Ruhrglas (part of VIAG, Germany), Hermann Heye (Germany), Nienburger Glas (Germany), AVIR (Italy), PLM (Sweden), Rockwater (UK) and United Glass (UK).

The container glass industry has good prospects for future output growth subject to the impact of increased competition from Eastern Europe, substitution towards PET in the drinks bottle market and relatively high environmental costs. Niche markets will continue to play a role.

4.2.1.4 Paper packaging

In the EU, in volume terms, packaging paper grades account for 40 per cent of total paper production, graphic paper grades account for 50 per cent and hygiene and speciality papers for 10 per cent. The largest production capacities for graphic and packaging paper are found in Finland, France, Germany, Italy, Sweden and the United Kingdom. Table 4.2-7 illustrates the relative levels of value added in paper and board converting and Table 4.2-8 provides some data on output, consumption and trade.

Table 4.2-7 Value added in paper and board converting, 1993, 1994 and 1998

Country	Value added (Euro m.)		
	1993	1994	1998
Germany	4 747	4 919	4 808
Republic of Ireland	–	–	228
United Kingdom	3 316	3 408	3 584

Source: Statistical Office of the European Communities (1995, 1997, 2000a).

Table 4.2-8 Output, consumption and trade of EU paper and board converting (Euro '000s, current prices)

Economic variables	1988	1992	1995
Consumption	38 400	49 894	53 855
Production value (Euro m)	39 617	51 298	56 050
Exports	1 792	2 299	3 427
Imports	574	895	1 232
Employment ('000s)	428.5	446.3	393.3

Source: Statistical Office of the European Communities (2000) - Panorama of European Enterprises 1999.

Industrial structure of paper packaging

The following packaging companies, originating from the sample countries, were included in the EU's ten largest: Arjo Wiggins Appleton (UK), PWA (Germany, though acquired by Sweden's SCA in 1995), Jefferson Smurfit (Republic of Ireland), Haindl (German), David S. Smith (UK), Saint-Gobain Paper-Wood (France, though acquired by Jefferson Smurfit in 1994) and Cartiere Burgo (Italy).

The Single Market has led to increased merger and acquisition activity. Thus, between 1989 and 1993, there was a fall in the number of pulp and paper mills in the EU as the industry became more concentrated and the average output and productivity of mills increased. Major merging activities took place in Germany, Finland and Sweden where the largest mills are located in Europe. Many of the European mills are non-integrated and a further concentration process is expected. However, this might not affect the scale of the remaining operations. Many niche products require small flexible production units close to their markets.

Concerning international competitiveness the EU paper industry has in general a good position, in particular scoring well on labour productivity. The larger European paper groups have developed economies of scale for core product grades which are sold across national frontiers rather than owning groups of small factories which sell to national markets and emphasise on a high degree of product diversification.

The industry has been target of several hard-hitting environmental campaigns which focused on four issues– the quantity of forest wood used in paper production; the high and growing consumption of paper; the proportion of paper which is recycled; and the use of chlorine and its compounds for bleaching pulp and paper to high levels of brightness.

In response to this several measures were undertaken. In 1995 the international study group for the certification of Sustainable Forest Management was established and in 1996 its members (among them EU Member States) committed themselves to ISO 14000. Moreover, recycling rates were steadily increased. The industry has also phased out elemental chlorine in favour of chlorine dioxide with the goal of totally chlorinefree production. Environmental aspects will continue to be of great importance and have already reached the status of a competitive marketing element.

4.2.1.5 Summary of conclusions regarding the packaging materials with special emphasis on the situation in Germany

Light metal packaging

1. Though metal containers make up only a few per cent of total household waste (by weight) they do represent a large share of total packaging (16 per cent) by value.

2. The relative ease of recycling some metals is notable (particularly tinplate and aluminium). For historic reasons drinks cans in Germany are mainly made from tinplate. At the time of the empirical study Germany already used a levy to promote recycling.

3. The demand prospects for this industry are good.

4. Further light weighting possibilities of tinplate constitutes a particular competitive advantage for the German industry, both from an economic and environmental perspective.

Plastics

1. Like metal containers, these represent a large proportion of all packaging (30 per cent).

2. There is a wide range of plastic types (sometimes distinguished by economy of material usage) and there have been substitution effects between these.

3. The prospects for continued consumer demand growth are fair to good.

4. As for the other materials, at the time of the empirical study, Germany already uses a levy ("green dot fee") to promote recycling.

Glass

1. Glass containers (i.e. mainly bottles) represent a less important share of packaging than either metal cans or plastic containers. Nevertheless, their share is still substantial.

2. Glass bottles remain under strong competition from aluminium, PET plastic and paper cartons.

3. Glass has one of the highest rates of recycling of any industrial activity. Germany has been for a long time at the forefront in the EU in terms of pushing up the rate of recycling (up to about three-quarters of consumption), which reflects the impact of the German DSD system, while the UK and Republic of Ireland lag behind (still less than one-third). As the EU Directive on Packaging and Waste Packaging is implemented it will presumably force the lagging countries to increase their recycling rates.

Paper and board

1. There is a strong and ongoing concentration process in the industry. Already before 1995 most paper mills in the EU were subject to non-EU ownership (i.e. Sweden and Finland).

2. Fairly high rates of recycling have already been achieved (the EU Directive would push these up still further).

3. Some of the methods of paper processing have been the subject of environmental concerns, e.g. chlorine bleaching.

4. Environmental aspects have reached the status of a competitive marketing element in Germany and particularly in the Nordic countries.

Table 4.2-9 summarises the technical characteristics of the described packaging materials and also gives economic and environmental details of the German packaging market.

Table 4.2-9 Packaging Materials in Germany

Material	Economic Characteristics*	Environmental Characteristics	Technical Characteristics
Packaging Paper	Paper industry holds the top position within the whole packaging industry: - 38.7 % of all packaging material, - 15.2 % of total industry turnover.	1 t of used paper helps to save $5m^3$ of wood; price must cover costs of collecting, sorting, transport, but must also be competitive! required recycling quota in the Packaging Ordinance (since 1.7.1995): 64 % (actual recycling in 1993: 55 %)	**low quality paper**: no protection against moisture or heat, not very watertight; **high quality paper:** watertight, fat resistant; **coated paper:** (by wax or paper): improves quality
Cartons Fibreboard			see above; coatings improve techn. characteristics depends on weight and quality of paper
Plastics	Position 2 within the packaging industry: - 28.7 % of all packaging material, - 11.2 % of total industry turnover.	plastics must be cleaned and sorted before recycling; required recycling quota: 64 % (actual recycling in 1993: 29 %)	variety of different plastics; main common characteristics: light, easily malleable, resistant, good use for compounds
Aluminium	Metal industry holds position 3 within the packaging industry: - 22.7 % of all packaging, - 8.9 % of turnover; aluminium is used much less in Europe than in US, economic advantages of tinplate: (r) stable steel price relative to aluminium, (r) tinplate can be produced in very thin layers.	recycling needs only 5% of the energy needed for **aluminium** production from bauxite; 100% recycling possible; high scrap value of aluminium has made secondary aluminium production as large as primary one; since **tinplate** is magnetic, it can relatively easily be separated from other metal waste; but: downcycling; required recycling quota for aluminium and tinplate: 72 % (actual recycling of aluminium in 1993: 7 % through DSD; a lot of aluminium is also collected outside the DSD systems e.g. when fillers collect lids; recycling of tinplate in 1993: 35 %)	main disadvantage: cannot be produced as thinly as tinplate

To be continued

Material	Economic Characteristics*	Environmental Characteristics	Technical Characteristics
Tinplate			tight, easily malleable, can easily be printed on
Glass	Position 4 within the packaging industry: • 6.6 % of all packaging, • 2.6 % of total industry turnover, • innovation: light glass bottle.	glass recycling saves costs and energy; secures market position against competitors; required recycling quota: 72 % (actual recycling in 1993: 62 %)	resistant against almost all materials, neutral in taste and odour, transparent, tight
Wood	Position 5 within packaging industry: • 3.2 % of all packaging, • 1.2 % of total industry turnover.	recycling of valuable resources; no recycling quota	packaging mainly made from conifers; quality of packaging depends on moisture, weight, stability of raw material

Note: *Data refer to 1990; only West Germany. Sources: Bojkow (1989), Breitenacher et al (1997), Klöckner (1992).

4.2.2 Structural background of the food industry: soft drinks/mineral water and dairy industry

This section analyses the competitive situation of the soft drinks and mineral water as example users of the packaging types described above. For an analysis of the dairy industry see already chapter 4.1.1.

The soft drinks and mineral water sector comprises soft drinks that are prepared with water, sugar, fruit juices, fruit extracts and/or other flavourings in addition to mineral water, both still and sparkling. In the EU, between 1987 and 1997 the consumption of soft drinks has increased by more than 40 per cent and mineral water consumption even rose by 50 per cent.

In terms of sectoral size (measured by value added within the various EU member states) the German sector is the largest followed by those in the UK and France (concerning solely mineral water Italy and Germany are the most important producers). In terms of employment the rank order is Germany, Spain, Italy and the UK. Taking soft drinks and mineral waters separately, production capacity in both is dependent on domestic consumption levels and generally sectoral size is positively related to population size and domestic per capita consumption levels of soft drinks and mineral water.

Germany, for instance, had in 1997 an annual per capita consumption level of soft drinks of over 80 litres, resulting in high levels of productive capacity and employment in both sectors. However, in comparison to 1990 per capita consumption has decreased by 1.5 litres in Germany. In the UK consumption is concentrated mainly within soft drinks, and again production and employment is consistent with those domestic consumption levels, whereas mineral production is much smaller and is more export dependent. Total consumption of soft drinks in the UK has risen from 73 litres to 86.1 litres per head between 1990 and 1997 and in the Republic of Ireland even 99.1 litres per head were reached in 1997 in comparison to 67.9 litres in 1990. Notwithstanding rising consumption levels, EU production managed to cover domestic demand and generate a surplus for export.

Within the soft drinks and mineral water sector, it is difficult to make direct comparisons at the international level in relation to the competitiveness of different productive systems, because of the importance of transport costs on retail prices which does to some extent protect each domestic market from other European producers. Such costs also limit the level of extra-EU exports. Relative labour cost levels shown in Table 4.2-10 for the early 1990ies are consistent with a Germany–UK rank order; firm size differences do not appear to provide an obvious explanation. For the Republic of Ireland GVA per head was not available, but in terms of physical productivity (million litres per employee) the Republic of Ireland produces at a lower level than both Germany and the UK.

Table 4.2-10 Soft drinks and mineral water combined labour productivity, 1990, 1991 and 1998 (gross value added per head at market prices, PPPs and other indicators, West Germany = 100)

	GVA per head	Relative labour costs per head (compared using average market exchange rate)	Average enterprise size (turnover) (compared using average market exchange rate)
1990			
West Germany	100	100	100
United Kingdom	90	71	163
Rep. of Ireland	–	–	–
1991			
West Germany	100	100	100
United Kingdom	88	74	155
Rep. of Ireland	–	–	–
1998			
Germany	100	100	100
United Kingdom	95	69	–
Rep. of Ireland	–	98	–

Note: Sector defined according to the NACE, i.e. pre-revisions 4280, then NACE Rev. 1 15.9. A single PPP, judged to be 'representative', was applied, i.e. 'Non-alcoholic beverages'. All data relate to only those enterprises with 20 or more employees.

Source: Statistical Office of the European Communities (1994, 1995a, 2000).

Competitive pressures in the German soft drinks industry

Overcapacities, pressure on price from the retailer side, polarisation of consumer preferences towards low cost products on the one hand and premium products on the other as well as high cost burdens, not least those caused by investments in the packaging area, have weakened the profitability of the beverage industry. Packaging should be easily recyclable and also fulfil promotional demands. Product innovations, new marketing and distribution tools are seen as key factors for strengthening the competitiveness of the industry (Breitenacher, 1997).

4.2.3 Structural background on the food retailing sector

Retailing is defined as the distribution of goods to the final consumer. Recent years have seen the distribution industry within the EU exhibit increasing levels of concentration in terms of a number of large operators and a greater degree of vertical and horizontal integration. The concentration phenomenon is most evident in northern Member States while there remains a high number of independent enterprises in the south, although future reduction in their numbers is almost certain (Statistical Office of the European Communities, 1994a). These developments are also reflected in the number of super and hypermarkets in the individual countries (see Table 4.2-11).

Table 4.2-11 Number of Super- and Hypermarkets > 2500 m², 1st of January each year

Country	1990	1994	1997	1998
Germany	996	1207	1392	1402
Italy	99	192	247	268
Spain	–	74	77	–
UK	580	862	1052	1110

Source: Statistical Office of the European Communities, 2000.

Growing use of franchising and technological developments (e.g. electronic point of sale) is changing the sector. Table 4.2-12 indicates that in productivity terms, at least as approximated by turnover per head measured using consumer PPP, the UK retail sector was in the early 1990ies higher than that in West Germany and the Republic of Ireland. In the Republic of Ireland, however, turnover per head was in 1996 higher than in the UK (Statistical Office of the European Communities, 2000).

Table 4.2-12 Retailing: enterprise numbers, turnover and employment, 1996

Economic variables	Germany*	Republic of Ireland	United Kingdom
Number of enterprises ('000)	422.6	20.1	207.0
Turnover (Euro bn)	259.5	11.4	237.4
Employment ('000)	2 194	120.7	2 485.6
Turnover per head[a] (Euro '000)	118	94.4	79.9
Turnover per head, using PPP[b] (Germany=100), 1991 all countries	100	96	107

Notes:
*Germany includes East Germany. Data refer to 1991.
[a] Converted to ECU using the average market exchange rate.
[b] Converted to ECU using the Purchasing Power Parity (PPP) rate of conversion for final national consumption.

Source: Statistical Office of the European Communities (1994a, 1997, 2000).

The internationalisation of retailing groups within the EU is admittedly much slower than that for other industrial groupings, but the degree of internationalisation is likely to rise with the emergence of EU-wide shopping centres. Employment within the sector has been growing in most member states and increases of between 2 per cent and 25 per cent were experienced between 1988 and 1991 in the main EU countries. However, more recent estimates indicate that employment within the sector has experienced some decline.

Further consideration of retailing with particular reference to food retailing in Germany

It is notable that Germany has several extremely large food retailing chains and in general the German food companies are larger than the main grocery companies in the UK. In Northern Ireland and the Republic of Ireland even the largest firms are still small by European standards. Such broad differences between the sample areas could be highly significant. The companies in Germany and Great Britain are much more likely to be characterised by whatever economies of scale operate within retailing. There may be a general tendency for larger companies to achieve higher standards of environmental performance because they can potentially devote more financial and human resources to this task. Retailers in Germany and Great Britain are much more likely to have monopsony positions relative to suppliers (e.g. food processors and packagers) than their counterparts in Northern Ireland and the Republic of Ireland.

Prospects for EU food retailing

Like the rest of the retailing sector, food based retailing is being affected by the liberalisation of transportation and telecommunications within the Single Market. There is also the impact of EU legislation on packaging as well as harmonisation with respect to safety and hygiene. Rates of indirect taxation are also to be brought closer together. Although there has been some movement towards internationalisation and increased concentration, cultural differences are likely to ensure that national grocery markets remain differentiated.

4.2.4 Likely impact of packaging regulation on competitiveness in the supply chain of the packaging, food and retail industry

This section analyses the likely impact of packaging regulation on competitiveness in the supply chain of the packaging, food and retail industry. It is found for Germany that the Packaging Ordinance exerts environmental pressure along the entire supply chain with the fillers carrying the largest burden. Still, this impact is not large enough to cause damage to competitiveness.

4.2.4.1 Likely impact of packaging regulation on packaging materials

Packaging regulation both on an EU and a national level sets recycling targets for various packaging materials. In the following the likely impact of this regulation on the competitiveness of the selected case study packaging materials will be outlined.

Light metal packaging

The EU Packaging and Packaging Waste Directive sets targets for a reduction in the volume of packaging and an increase in the extent of recycling for light metal packaging. Germany already has relatively stringent regulations (the Töpfer Packaging Ordinance; the DSD; see chapter 3.3).

In the light metal packaging activity recycling is usually straightforward. In the case of tinplate the energy requirements are relatively low. Steel based materials also have the advantage that the recovery from scrap can be facilitated by use of magnets. This is not true for aluminium but in that case the critical feature is the high value of the material. Also, the energy requirements to recycle from aluminium scrap are only 5 per cent of those needed for primary smelting. At the same time, the industry, especially in Germany, has seen its environmental record come under attack. Concern is usually expressed about the activities of the primary part of the industry, i.e. the deforestation that results from bauxite mining. According to some sources the price of aluminium sheet is set to increase by 40 per cent over the next five years (Statistcal Office of the European Communities, 1997). This is likely to encourage some substitution towards tinplate. At the time of the study, in the mid 1990ies, one deterrent to increased recycling of tinplate was that its scrap value just covered collection costs. Nevertheless the EU average recycling rate is 50 per cent. In 1995 35 per cent of the metal used in European aluminium drinks cans had been recycled (and only 16 per cent in the UK) compared to rates of 62 and 66 per cent respectively in the US and Japan.

Another important implication of environmental pressure stemming from legislation is a movement towards light weighting. Tin plate cans may soon be reduced in weight from 25 g to 18 g. This constitutes a special competitive advantage for Germany where the drinks cans industry uses almost exclusively tinplate and not aluminium.

Plastics

Packaging regulation will also impact on the recycling of plastics. Total EU production of recycled plastic in the mid-1990s was growing at 20 per cent per annum. Usage of recycled plastic grew from 1.5 million tonnes in 1993 to 2.4 million tonnes in 1996 (i.e. 17 per cent of the total 14 million tonnes of plastic waste). The imperative of the German Packaging Ordinance to increase the extent of recycling is likely to change the design of plastic packaging (e.g. to reduce the need to dismantle into components). However, as experiences in Germany show, plastics recycling is expensive and allowances for energy recovery from plastics are increasingly made.

Glass

In terms of the re-use of scrap material, glass manufacturing probably now has one of the highest recovery rates of any manufacturing activity (Germany about 76 per cent, Republic of Ireland about 37 per cent and 24 per cent in the UK in 1997, Statistical Office of the European Communities, 2000). It can be expected that the EU Directive on Packaging and Packaging Waste will in the long run lead to higher levels of recycling in these EU countries which lag behind the Directive's standards.

Paper and board based packaging

Waste paper recovery in 1993 represented 41 per cent of supplies to production of paper and board. EU Member States vary according to the stage they have reached with respect to their waste management strategies. The four leading European countries in terms of 1994 recovery rates were: Austria (66 per cent), the Netherlands (63 per cent), Germany (60 per cent) and Sweden (60 per cent). The same rank order, though at slightly higher levels, is true for 1997. From the sample countries Germany reaches the highest recycling levels (see Table 4.2-13).

Table 4.2-13 Recycling rates for paper and board (%)

Country	1975	1980	1985	1990	1997
West Germany[1]	35	33	40	40	76
Republic of Ireland	22	4	–	3	–
United Kingdom	28	29	27	31	24

[1] Data in 1997 refer to Germany.

Note: Rates are defined as percentages of apparent national consumption, i.e. domestic production plus imports but minus exports, and are likely to be somewhat approximate and perhaps unreliable. It is not clear whether the apparent spectacular drop in the Republic of Ireland rate was a real phenomenon.

Source: Statistical Office of the European Communities (1995 and 2000).

One implication of the general EU pressure against landfill as a method of waste disposal will be increasing pressure not to use laminated board. There has also been some tendency for food to be packed in plastic bags as opposed to cardboard boxes so to allow the more economical use of space in domestic refrigerators. However, convenience aspects can push back environmental issues and reinforce the use of composite packaging.

4.2.4.2 Likely impact of packaging regulation on competitiveness in the food industry: Soft drinks/mineral waters and dairy industry

Packaging regulation also influences the packaging mix of the fillers. In the following possible consequences of regulation on the packaging decisions in the soft drinks/mineral waters and dairy industry are shown.

Soft drinks/mineral waters

In the early 1990ies, when the Packaging Ordinance was designed, the most heavily used packaging type for soft drinks in the EU was non-returnable plastic. This was especially the case in Italy (64.7 per cent) but also in other countries, e.g. Republic of Ireland (56 per cent), the UK (51 per cent), Spain (43.9 per cent) and Belgium (43.4 per cent). In Germany and Portugal, returnable glass/plastic was preferred (respectively 76.7 per cent and 53.1 per cent). Metal packaging accounts for a modest share and was most popular in the UK (30.4 per cent). Other types of packaging (cardboard, non-returnable glass and pre-/post-mix) did not account for more than 10 per cent of the packaging used by the Member States.

The trend towards the replacement of glass with plastic materials has been very strong within the entire drinks sector; however, the pace of replacement has varied across national industry segments. Compare, for example, the very strong use of plastic refillable bottles in Germany (though still behind the rate of use of around 90 per cent in the Scandinavian economies) to the continuing predominance of non-returnable plastic in Italy and the Republic of Ireland. In soft drinks and mineral water the swing towards first PVC and then PET has been strong. While the share occupied by metal cans has remained steady in the soft drinks sector, the use of glass has dropped steeply. There has been a swing towards the use of PET in both industry segments. Packaging regulation presses on the introduction of reusable PET; this has been increasingly the case in Germany.

Dairy industry

As far as the dairy processing industry is concerned, some of the biggest problems stem from water usage (see food case study), but also from the disposal of packaging and recycling of used containers whilst keeping in line with national and/or EU-level required standards of practice. Current European packaging practise for dairy products like milk represents a wide variety of types: plastic, composites, board and glass. However, recycling targets and associated fees could have the potential to change the packaging mix as will be illustrated for Germany (see below).

In 1995 the German dairy industry contributed with DM 750 million almost a fifth to the licence fees of the Duales System Deutschland AG. Hülsemeyer (1997) argues that this reduces competitiveness of the industry, since 1 kg of raw milk is on average burdened by about 4 Pf. in Germany in contrast to less than 2 Pf. in Belgium and Austria and no payments, as of yet, for recycling in Denmark, Spain, the Netherlands, Italy, UK and France. In addition the current fee structure disadvantages some dairies over others according to their assortment. For example, producers of yoghurts filled in plastic pots are at a cost disadvantage relative to producers of UHT milk filled in cartons (the DSD fees for board are lower than for plastics) and these in return have to pay more disposal fees than the producers of butter which is wrapped in composite foil made of aluminium and paper (Meier, 1997). With respect to polystyrene yoghurt pots the average fee per pot is 1.8 Pf. and the fee for a paper pot which is coated inside and outside with PE is only 1.3 Pf.. The relatively favourable total energy consumption of the production and disposal of a paper pot makes it preferable from an environmental point of view. In addition, the DSD fees are lower as shown above. However, technical problems are more frequent in the use of a paper pot than of a plastic pot. This fact and simultaneously the improved printing technology for plastic pots has led to a substitution back to plastics (Langowski, 1997). Moreover, apart from environmental regulation also convenience aspects will play an important role in packaging of dairy products.

4.2.4.3 Likely impact of packaging legislation in the retail sector

Legislative pressure on the retail industry can trigger strong influences throughout the supply chain because of the powerful gatekeeper position of the retail sector.

If both producers and retailers comprehend environmental management as a chance to gain competitive advantages, this can often trigger innovation throughout the vertical chain. Subsequently co-operation will be established with suppliers, the packaging industry, consumer and environmental associations.

In regard to their product assortment retailers can influence producers in several ways ("ecology-pull-strategy") (Meffert/Kirchgeorg, 1992):

— conscious selection of suppliers under environmental aspects,
— influence on suppliers in regard to product design and characteristics e.g. via directives on the selection of material use.

In the light of an increasing concentration and co-operation process in the retail sector there is a vertical integration which is connected with a growing influence on the production side. The effectiveness of an ecology-pull-strategy is largely determined by the relative power of retailers - retail demand power vs. strong trade marks of producers. Important decisions are taken in the negotiations between the big retailers and producers. The possibilities of influence for SMEs are limited. The increasing orientation of the big producers towards the European market strengthens the problem.

Retailers can also influence consumers by using an "ecology-push-strategy", e.g.:

— environmentally friendly product assortment,

— information services for consumers,

— deposit systems.

Since retailers are in direct contact with final consumers, not only legislative pressure, but also consumer concern for the environment can be a driver for environmental management. There are three basic strategies by which retailers can take environmental aspects into consideration (Meffert/Kirchgeorg (1992)):

1. A first group of retailers, in particular discounters, are defensive with regard to environmental protection measures. Their basic concern is cost-orientation.

2. A second group of retailers could be described as undertaking environmental actions on a case-by-case basis; i.e. firms do not try to establish an individual environmental profile.

3. Other retailers follow an active environmental basic strategy and thrive to gain competitive advantages by integrating environmental aspects into their management.

4.2.5 Conclusions on the likely impact of environmental regulation on competitiveness in the supply chain of the packaging, food and retail industry in Germany

The above has shown that German packaging legislation is likely to trigger adjustments not only concerning the achievement of recycling targets, but already with respect to the production of packaging in order to avoid and/or reduce waste packaging. Thus, as a consequence of recycling targets and related licence fees, the entire supply chain of the packaging, food and retail industry will be under environmental pressure.

It is already a fact that the Packaging Ordinance of 1991 made German retailers use their demand power in order to pass on the legislative pressure to their suppliers. E.g. in Germany it was the retail sector which urged their suppliers in the packaging and food industry to reduce transport packaging and there was also an induced trend for monomaterials. Also, as a result of negotiations with the fillers, packaging manufacturers undertook innovative activities like light weighting (e.g. of drinks cans made of tinplate) and introduced new designs (plastics packaging) in order to reduce both packaging waste and DSD fees for their clients. Since the fillers (in our case the soft drinks/mineral water and dairy industry) have to pay the main part of the DSD licence fees, any impacts on competitiveness arising from the Packaging Ordinance can be assumed to be largest for them. The extent of a possible competitiveness impact depends on

— the degree of packaging substitution from materials with relatively high licence fees (e.g. plastics) to those with lower fees;
— the ability to absorb additional costs (in turn depending on productivity) and
— the possibilities to compensate for the additional costs elsewhere.

However, not only legislative pressure, but also convenience aspects play an important role in food packaging. This can explain why currently some of the environmental trends of the mid nineties tend to disappear in Germany. In chapter eight of this study the detailed cost and competitiveness impacts of the packaging initiatives triggered by legislation will be examined for all industries of the supply chain sample. It will be shown that competitiveness impacts are indeed largest for the fillers; however, these impacts are not big enough to cause damage to competitiveness.

4.3 Structural background of the cement industry in Germany, Spain and the UK

The purpose of this section is to outline the competitive situation of the cement industry in the EU, Germany, Spain and the UK. Special attention is paid to relevant impacts of environmental requirements on the factors which characterise competitiveness within the industry (section 4.3.1). In section 4.3.2 likely economic impacts of clean air regulation for the cement industry in the sample countries are tackled. The interviews for the empirical part of the study were undertaken in 1999 and 2000. Most of the background data therefore refer to the late 1990ies.

4.3.1 Competitive situation of the cement industry in the EU, in Germany, Spain and the UK

This section gives details on the structural background of the cement industry in the sample countries Germany, Spain and the UK. It starts off with some general points and an overview of the industry at EU level.

Product, demand, economies of scale and capacity utilisation

Cement is a binding material which when mixed with water takes on the property of hardness and becomes water resistant. The two major categories are pure Portland cement[20] and composite Portland cement where different amounts of main constituents are added, e.g. slag, silica fume, pozzolana, fly ash, burnt shale, limestone. The choice of process is to a large extent determined by the state of the raw materials (dry or wet). In Europe, more than 75 per cent of production is based on dry processes due to the availability of dry raw materials. Wet processes are more energy intensive, and thus more expensive. Plants using semi-dry processes are likely to change to dry technologies whenever expansion or major improvement is required. Plants using wet or semi-wet processes normally only have access to moist raw materials. This is the case in Denmark and Belgium, and to some extent in the UK.

Cement is a construction material and demand therefore is determined by the business cycles of the construction industry. Cement demand is inelastic to price: If construction slows, then a reduction in cement prices will not boost construction. On the other hand, if cement prices rise, demand will not decrease because there is no real substitute for cement in the short run. In 1973 the EU industry represented 27 per cent of world demand and has gone down to just 12 per cent in 1997. However, due to higher European prices this represents 20 per cent of world cement value (The Cement Industry, 8. Dec. 1998). Currently, within the EU, as regards cement demand there are two different trends. Whilst

[20] Modern-day cement was patented in 1824 by an Englishman named Joseph Aspdin who called his invention Portland cement because it resembled a natural limestone quarried on the Isle of Portland, in England (Dumez and Jeunemaître, 2000).

consumption in Northern Europe followed a decreasing path, Southern Europe has been experiencing growth until around 1992 when growth started to slow down. In 1997 the largest cement consumers in Europe were Italy (22 per cent), Germany (19 per cent) and Spain (15 per cent).

Cement production requires massive investment which is written off over several decades. Over the last 30 years technology of kilns has changed to the dry technology with cyclone preheaters. This has gone along with an increase in capacity and greatly improved energy efficiency. Creating new capacity by increasing plant size and thereby lowering unit costs means that economies of scale are achieved. These economies of scale relate only to individual plants; the acquisition of additional plants does not trigger a cost advantage with regard to unit production costs. Today, the minimum optimal size (MOS) for a kiln is considered to be 3,000 tonnes per day (Wagner and Vassilopoulos, 2000).

Up to this size the unit costs decrease, if capacity is fully utilised. Since fixed costs are high in relation to variable costs, the break-even point is high. That implies that profits in the industry are very sensitive to the level of production capacity. Important cash flow is generated when production increases beyond the break-even point. Conversely, when activity is low, losses are felt both quickly and severely.

Structure of the cement industry at European level

The EU cement industry consists of some 250 plants operated by 64 groups. It produced in 1995 1420 million tonnes, employing just under 60,000 persons. Already since the 1970ies the European industry has been consolidating through merger and acquisitions (Wagner, 1979). Today the largest companies are multinational enterprises with plants in the EU, US and CEEC. Among these companies are Heidelberger (Germany), Holderbank (Switzerland), Italcementi (Italy) and Lafarge (France). They control most of the French, Belgium and Moroccan market (see Table 4.3-1 below). In the last years they all have invested in cement plants in Eastern Europe as well. The turnover of each of them exceeded Euros 3 billion in 1997.

Table 4.3-1 Market shares of the largest cement companies in 1998

	France	Belgium	Spain	Italy	Turkey	Morocco	North America
Italcementi	33%	18%	8%	32%	9%	17%	6%
Lafarge	35%	n.a.	13%	3%	19%	41%	14%
Holderbank	15%	35%	7%	7%	n.a.	22%	14%
Heidelberger	14%	38%	n.a.	n.a.	15%	n.a.	11%

Source: http://www.nettradecenter.se/ntcframe.html

Although a lot of merger activities has taken place, the competitive situation in the European countries varies. Looking at the market share of the largest three manufacturers they account for around 50 per cent in Germany, Italy, Spain and Poland. Highly concentrated ownership can be found in the UK and France where the largest three producers control more than 80 per cent of the market (see Table 4.3-2).

Table 4.3-2 Market share and number of plants of the three largest producers in 1996

	France*	Germany**	Italy***	Spain****	UK*****	Poland******
Market share of largest 3 manufacturers	82%	48%	55%	56%	94%	57%
No. of their owned plants	32	20	43	27	20	7
Total no. of plants		66	86	37	22	20

* Lafarge, Ciments Francais, Vicat
** Heidelberger, Dyckerhoff, Schwenk in West Germany; for East Germany: 50% by Lafarge, RMC, Dyckerhoff
*** Italcementi, Buzzi/Unicem, Cementir
**** Cemex, Cementos Portland, Lafarge
***** Blue Circle (over 50%), Castle, Rugby
******Lafarge, Heidelberg/CBR, CRH

Source: Dresdner-Kleinwort-Benson, April 1998; The Global Cement Report 1998.

Ownership structure of companies differs among the countries. In the UK the three major companies are owned by institutional shareholders, whereas the majority of Holderbank and Italcementi is family owned (by Schmiedheiny resp. Pesenti). In Ireland there is a monopoly situation where the company is owned by Irish and UK institutions.

The decrease in cement demand over the last decades in Western Europe has led to a decrease in capacity utilisation. Given the high capital costs in the cement industry this has strong effects on unit costs (see section on production costs below).

Competitive situation of the cement industry in Germany, Spain and the UK

In the following the competitive situation of the cement industry in the sample countries Germany, Spain and the UK will be analysed. The section starts with an overview and then examines the individual factors which influence competitiveness (e.g. market structure, production costs, etc.).

Overview

Within the countries under consideration the German cement industry is the largest producer in terms of number of plants, employees and total production. Spain takes the top ranking with respect to labour productivity measured as annual output per employee. The main variable responsible for this is her extremely low labour force in comparison to both Germany and the UK. Germany has the lowest energy consumption per head and a middle position concerning electricity consumptions. Spain is the only country which exports more than it imports (see Table 4.3-3).

Table 4.3-3 Comparison of German, Spanish and UK cement industry

	Germany	Spain	UK
Most recent year available	1999	1997	1998
No. of plants	66	43	22
Employees	11 372	5 464	5 000
Production cement in 1000 tonnes	36 000	27 933	12 409
Tonnes of cement/employee	3 105	5 112	2 482
Energy consumption (kcal/kg clinker	715	844	1 000
Electricity kWh/t cement	108	106	112
Import of cement (1000t)	4 466	3 044	1 300
Export of cement (1000t)	2 929	5 572	600
Import – Export	1 537	-2 528	700

Sources: Bundesverband der Deutschen Zementindustrie e.V., British Cement Association and information of Oficemen.

Market structure in Germany, Spain and the UK

The German industry has undergone considerable restructuring already before 1985. This was due to constantly shrinking cement consumption until 1985, when it was little over 50 per cent of the 1972 cement sales. During this period 24 works and 22 companies were decommissioned and an intense concentration took place. Simultaneously a modernisation process was started which triggered environmental improvements early onwards. While in 1970 only 29 per cent of industry's capacity came from plants with cyclone preheaters, these covered 84 per cent of the capacity in 1996 (see Table 4.3-4 below). Furthermore, rationalisation meant that smaller and/or inefficient kilns were shut down. The total number of kilns halved over this period from 168 to 82 while output dropped only by 15 per cent.

Table 4.3-4 Numbers, types and capacity of kilns in the years 1970 and 1996 in Germany

Type	1970		1996	
	Number of kilns	Capacity in %	Number of kilns	Capacity in %
With cyclone preheater	48	29	51	84
With grate preheater	53	31	21	14
Drying kiln	7	4	1	1
Shaft kiln	42	25	8	1
Total	168	100	82	100

Source: Bundesverband der Deutschen Zementindustrie e.V., different years.

The reduction of kilns by 50 per cent at a similar output shows that economies of scale have been improved. The number of German plants with less than 200,000 tonnes of yearly production, for example, have been decreased from 45 per cent of the industry's capacity to just 7 per cent from 1963 to 1996, while the plants with more than 1,000,000 tonnes have increased their share to 25 per cent (see Table 4.3-5). Indeed, from 1972 to 1990 average kiln sizes have increased by around 60 per cent in Italy, 40 per cent in Germany, 80 per cent in France and the UK, and 200 per cent in Belgium.

Table 4.3-5 Economies of scale in the German cement industry

Plants with a production of	1963		1974		1982		1991		1996	
	No.	%	No.	%	No.	%	No.	%	No.	%
0 - 200,000t	44	45	23	23	25	34	22	8	18	7
201- 500,000t	35	36	35	35	42	36	17	19	25	27
501- 1,000,000t	16	17	17	17	20	22	24	59	17	41
over 1,000,000t	2	2	2	8	10	8	3	14	6	25
Total no. of plants	97		83		74		66		66	

Source: Bundesverband der Deutschen Zementindustrie e.V., different years.

In 1998 the largest three companies controlled approximately 60 per cent of the German kiln capacity with 17 plants, while the three next largest owned another 20 per cent of the market with six plants. The remaining 20 per cent of the market are taken up by a number of smaller companies (see Table 4.3-6 below). To tal clinker capacity is almost 40 million tonnes of which 98 per cent is produced by dry kilns with preheaters and lepol kilns.

Table 4.3-6 German cement producers, number of clinker works and market share in 1997

Company	No. of clinker works	% of market
Dyckerhoff	6	22
Heidelberger	7	21
Schwenk	4	16
Lafarge	2	6.5
RMC	2	7.0
Holderbank	2	5.5
Total clinker capacity in thousand tonnes	39,760	
2.5 M tonnes of special cements and 2 M tonnes of mothballed/closed capacity have been deducted.		

Source: Dresdner Kleinwort Benson Research (1998).

As in Germany a concentration process has taken place in Spain during the last 20 years. Whereas in 1979 almost half of the then 50 plants were of small size, only about 12 per cent of the reduced number of 43 plants was small in 1997 (see Table 4.3-7).

Table 4.3-7 Structure of Spanish cement industry by cement capacity

	Number of Plants	
Plants with capacity of	1979	1997
Up to 500,000 tonnes	24	5
501-1,000,000 tonnes	15	18
1,000,001-2,000,000 tonnes	11	15
> 2,000,000 tonnes		5
Total number of plants	50	43

Source: Cemento-Hormigon, La Industria del Cemento en España, Portugal y Latinoamérica, (Barcelona, 1997). Memoria del Cemento, 1979.

In comparison to the German and also the UK cement market the Spanish market is relatively fragmented with many European players represented. Cemex, a Mexican multinational, is the clear market leader with 26 per cent, followed by Cementos Portland (owned by the Spanish conglomerate Portland Valderrivas) with 17 per cent and four other companies sharing another 34 per cent. Only one of these four companies, Uniland Cementera, is privately owned by two families. Production is relatively concentrated geographically in Madrid, Catalonia and Andalusia and accounts in these regions for more than 50 per cent of the total Spanish production (Dresdner Kleinwort Benson Research, 1998).

The UK market is the most concentrated one from all sample countries with only three main cement manufacturers and two small ones (see Table 4.3-8). Within the last ten years a strong restructuring has taken place with a reduction in the number of plants from 48 to 22. Ongoing investments and an implementation of a new plant by each of the three large companies characterise the present development.

Table 4.3-8 UK cement producers and capacity share in 1998

Company	% of capacity
Blue Circle	52
Castle Cement	23
Rugby	20
Sean Quinn	3
Buxton Lime	2
Total clinker capacity in 1000 t	13 902

Source: Information provided by the British Cement Association.

Blue Circle industries are the market leader in the UK accounting for about 52 per cent of the total market capacity, followed by Castle Cement (now belonging to Heidelberger Zement) with a capacity share of about 23 per cent of the market. Rugby Cement (now taken over by Readymix) operates six plants with a capacity of about 2.7 million tonnes and a market share of 21 per cent. All three companies have invested in new plants and/or major modernisation of existing plants, replacing small old plants or wet kilns, bringing both significant environmental and economic benefits.

Comparison of employment and productivity in Germany, Spain and the UK

In contrast to the use of capital and energy the use of labour is only of minor importance for the production of cement. Consequently labour productivity is very high. Growth of productivity has always been greater than growth in production. E.g. in Germany in 1995 the productivity per working hour was 1.6 tonnes, whereas it was only 0.5 tonnes in 1960 (see Table 4.3-9). This reflects an increased cement production (plus 36 per cent) and a decreased labour volume (minus 60 per cent), but a less than proportional reduction of employees (minus 44 per cent) during that period. Box 4.3-1 gives details on labour productivity in the highly modern East German cement plants.

In the short term, with an unchanged capital stock, labour productivity can hardly be changed. The greatest influence on labour productivity therefore stems from investment in new kilns which increases clinker capacity per kiln and leads to a concentration process which enhances labour productivity.

Table 4.3-9 Employment and productivity in the German cement industry

Year	Number of employees	Labour volume in 1,000 hours	Labour productivity in tonnes per employee	Labour productivity in tonnes of cement per working hour
1960	22 326	50 912	1 100	0.482
1965	22 551	50 028	1 486	0.670
1970	17 176	37 818	2 182	0.991
1975	15 000	27 290	2 198	1.208
1980	13 126	23 720	2 524	1.397
1985	11 844	19 607	1 937	1.170
1990	10 148	17 045	2 731	1.626
1993[1]	13 292	21 999	2 444	1.477
1994	12 797	21 131	2 823	1.719
1995	12 437	20 577	2 678	1.618

[1] Since 1993 including the new states.

Source: Buttermann, 1997.

Box 4.3-1 Labour productivity in East Germany

Cement works in the former GDR were bought by West German, French and British investors. All seven plants were modernised with an investment of several billion Euros to satisfy the most modern European standard. In 1998 the East German capacity amounted to 8.5 million tonnes and the works supplied about 20 per cent of German production. Average rotary kiln capacity has been 1,894 tonnes/day in West Germany and 2,400 tonnes/day in East Germany. This illustrates the much larger capacity and the higher potential to utilise economies of scale in East German production locations. The average consumption per capita in the West is about 420 kg, while it is nearly 580 kg in the East which can be explained by the building boom in Berlin. However, the sales volume which had been attained in the 70ies has not been reached again.

The Spanish cement industry has drastically reduced its labour force by 40 per cent between 1988 and 1997. This and a parallel increase in output has led to a productivity increase per employee of 89 per cent. In 1995 Spanish labour productivity was with 5,432 tonnes per employee almost double as high as the German level of labour productivity. Towards the end of the 1990iese the gap becomes less wide (see Table 4.3-10). In the UK clinker and cement production is about equal because extenders are generally sold to the ready-mix producers who tend to do the blending themselves[21]. Production output has dropped by 25 per cent between 1989 and 1998. Due to the restructuring of the production works capacity utilisation dropped only by 9 percentage points. Having about

[21] About 54% of cement consumption is going into ready-mix concrete products, 26% in precast products and about 20% into bagged cement (International Cement Review, October 1998).

5,000 employees in 1998 labour productivity reached about 2,482 tonnes per head. This is similar to the German level (see Table 4.3-11).

Table 4.3-10 Employment and productivity in the Spanish cement industry

Year	Number of employees	Labour volume in 1,000 hours	Labour productivity in tonnes per employee
1988	8 995	n.a.	2 709
1992	7 869	8 940	3 331
1993	6 593	7 643	3 629
1994	5 901	6 938	4 518
1995	5 980	7 334	4 764
1996	5 545	6 917	5 012
1997	5 464	6 967	5 432

Source: Oficemen, informe anual 1992, Anuario de 1996, Anuario 1997, unpublished data provided by Oficemen, own calculations.

Table 4.3-11 Major indicators of the UK cement industry

Year	1989	1995	1998*
No. of plants	48	22	22
Employees	n.a.	n.a.	5 000
Production of clinker in 1000 t	15 234	11 371	12 372
Production of cement in 1000 t	16 849	11 805	12 409
Tonnes of cement/employee	n.a.	n.a.	2 482[22]
Capacity utilisation in %	97	96	89
*most recent year available			

Source: British Cement Association, different years.

Production costs in the cement industry

Given the large investment costs necessary to set up a cement plant, fixed costs are very high and significant relative to variable costs. The Table 4.3-12 below shows an exemplary calculation of different cost factors. Total unit costs can be broken down in about 56 per cent of fixed costs and in about 44 per cent of variable costs. Depreciation and financing charges as part of fixed costs can make up for about 40 per cent of total unit costs - depending on the age of equipment and type of technology used. The next biggest cost block are fuel and electricity costs with a share of 16 per cent in total unit costs. Maintenance and labour costs each add another 8 per cent to unit costs.

[22] Productivity in terms of clinker per employee is slightly above the figure for Germany.

In an international perspective production costs vary with respect to a large range of factors: for example, labour costs, quality of raw materials and type of additives, specific consumption of energy etc. As about 70 per cent of variable costs are incurred by energy, this factor receives greatest attention for cost reduction. This is also important from an environmental perspective since CO_2 emissions can be reduced.

Table 4.3-12 Rough estimate of factors contributing to unit costs

Fixed Costs	**% of total unit costs**	**% of fixed costs**
Depreciation & financing charges	40%	70%
Maintenance	8%	13%
Personnel, rents, rates	8%	15%
Admin. & Overheads	1%	2%
Subtotal	56%	100%
Variable Costs	% of total unit costs	% of variable costs
Fuel (thermal)	16%	36%
Electricity	15%	35%
Raw Materials	11%	25%
Paper bags, refractories, wear parts, lubricants, royalties	2%	4%
Subtotal	44%	100%

Source: Shenoy, Chako, McGovern, 1997.

Example: Production cost structure in Germany

The average cost structure for the German cement industry is presented in Table 4.3-13 below. It needs to be kept in mind that the census data is based on enterprises and not on plants. Depending on the degree of diversification the actual data can be different from the data shown below.

The data allow an interesting observation concerning energy costs. Although energy costs account for a large percentage of variable costs, they have decreased in real terms between 1976 and 1994 by 3.2 per cent because energy efficiency has increased (see below).

Table 4.3-13 Cost structure of the German cement industry, in DM/tonne (in prices of 1995)

Year	1976	1981	1990	1991	1992	1993[1]	1994
Fixed costs	18.46	16.98	16.13	13.22	11.62	12.96	12.65
Variable costs, of which	87.79	107.53	91.09	86.21	83.68	95.42	87.58
Energy	42.96	50.82	32.24	32.46	30.91	31.02	28.98
Raw material	18.67	27.20	28.69	25.87	25.57	31.07	28.62
Labour	26.16	29.52	30.17	27.88	27.20	33.33	29.98
Total costs	106.25	124.51	107.22	99.43	95.30	108.38	100.23
Mark-up	36.58	21.76	19.09	12.79	12.02	18.24	15.69
Turnover	142.82	146.27	126.30	112.21	107.322	126.61	115.93

[1] Since 1993 including the new states.

Sources: Buttermann, based on census data, 1997.

Energy efficiency in the sample countries

Optimisation of process engineering for the rotary kiln plants has induced significant reductions in the specific fuel energy consumption for the European cement production. Modernisation and building of new plants brought about a decline of about 35 per cent since 1960 with the rate of decrease diminishing over the last years (Cembureau, 1997). A reduction of about 3 per cent in the coming years is forecasted; after that no further major potential of improvement is expected.

Differences in the specific energy consumption with respect to technology can be considerably. Depending on raw materials and the type of process, the fuel consumption reaches about 3,200 to 5,500 MJ/t clinker (Cembureau, 1997). In general, countries which still operate a number of wet processes show a higher energy consumption. But changes to other processes are under way, e.g. in the UK a number of kilns in the wet process will be replaced by a modern semi-wet process. Besides taking into account the required environmental standards the new production process and equipment will reduce fuel usage by one-third.

In comparison to Spain and the UK Germany has the lowest energy consumption per head (see Table 4.3-14).

Table 4.3-14 Energy and electricity consumption in the German, Spanish and UK cement industry

Country	Germany	Spain	UK
Most recent year available	1999	1997	1998
Energy consumption (kcal/kg clinker	715	844	1000
Electricity kWh/t cement	108	106	112

Sources: Bundesverband der Deutschen Zementindustrie e.V., British Cement Association and information of Oficemen.

Because of its high-energy costs, the cement industry has been searching for alternative fuels and waste. The technologies have been introduced more than a decade ago and are now well established. Waste consumption nowadays represents 12 per cent of the industry's fuel mix and is used in 12 of the 15 EU Member States. This represents for the industry a vast contribution to cost competitiveness since energy costs amount up to 36 % of variable costs. Novel technologies allow some EU companies to use up to 50 per cent waste as fuel.

Existing EU legislation covers already some of the hazardous waste and cement kilns do have to comply with those levels. However, the current Directive in force does not cover NO_x emissions. The forthcoming Waste Directive shall be comprehensive in that all waste shall be covered. The new Directive shall also include limits on NO_x emissions (see also Appendix A.3).

Electricity consumption

Electricity is mainly used for grinding of stone and cement and to date there are no breakthrough technologies to reduce the vast amount of electricity used. Electricity prices in turn vary widely across countries. Effectively, large discounts may exist for large consumers.

While the energy consumption for the kiln has been reduced continuously in the last 40 years the electricity consumption has been increasing from 80 kWh/tonne cement in 1960 to 110kWh /tonne cement in 1990 in West Germany. Among the reasons for this are a higher use of electricity for environmental equipment, a higher use of coal instead of oil in modern mills, finer grinding of cement particularly of composite cement and a more automatic process.

Since cement milling requires the largest part of electricity, special efforts are taken to improve the cement mills. Ball mills are simple to operate, relatively easy to maintain, but because of their great specific power consumption more efficient grinding processes have been developed[23]. With vertical roller mills (Vertikal-Wälzmühlen), high-pressure grinding rolls (Gutbet-Walzenmühlen) and horizontal roller mills (Horizontal-Wälzmühlen) the energy requirement can be reduced substantially. If ball mills are run in combination with other types of mills, considerable energy savings can be achieved. The costs of servicing and maintenance is generally low with ball mills and highest with high-pressure grinding rollers. If high-pressure grinding rollers and horizontal rollers mills are used, separate dryers might be necessary for moist mill feed.

Concerning electricity consumption there is not a lot of variation within the sample countries. Germany occupies a middle position (see Table 4.3-14 above). In the UK electricity consumption per tonne dropped during 1989 and 1998 by 11 per cent. In Spain electricity consumption per tonne of clinker fell by 13 per cent until 1995 but increased until 1997 by 5 per cent again. This increase could be due to higher electricity consumption due to finer cement milling and/or to more environmental measures.

Transportation costs

A further point which influences competitiveness is the importance of transportation costs relative to production costs. Cement is a heavy low unit price product so that transport costs rapidly become a very important factor in defining any producer's customer base. The costs of transportation vary depending on whether it is done by road, rail or sea with road transportation being the most expensive, followed by rail and sea transport being cheapest. In Europe the majority of cement is delivered by road. In the EU road transport costs usually limit the delivery radius to about 200 km (Fiederer et al., 1994). The European Commission estimates a delivery distance of 120 km only.

Rail freight cannot transport cement competitively to distances exceeding 200-400 km. In order to use the railway system usually the plant and the customer must have railway track connections. This is one of the reasons why the transport by rail is hardly used.

Trade

Interregional trade is not very extensive. One important reason for this are the existence of business routines and relations with competitors. These routines

[23] Ball mills also have the greatest noise emission. Therefore they have to be operated inside closed buildings with special ventilation and air-extraction equipment.

stabilise the behaviour of cement factories and prevent a significant alteration of the natural markets determined by the relation between production and transportation costs because one would have to expect a retaliation of competitors (a negative example are the so-called Westfalian price wars during which quite many cement factories had to close in Germany (Spenner, 1995)).

In international terms, because of the high transportation costs, exports and imports of cement are also small, not exceeding 6-7 per cent of world production. On average EU exports amount to almost 15 per cent of total production. Of all EU countries Belgium (35 per cent), Denmark (51 per cent), Greece (50 per cent), Luxembourg 58 per cent), Spain (22 per cent) and Sweden (48 per cent) export more than average (Bundesverband der Deutschen Zementindustrie e.V., 1998). High export rates from countries like Belgium and Luxembourg can be explained by their relatively small size so that deliveries into the neighbouring country fall easily into the 200 km radius. The majority of Spanish exports uses the relatively cheap waterway and has Africa and the US as major destination. Greece also uses the waterway, has the highest intra-EU trade share and exports about half of its production mainly to Italy, Spain, US, Asia and Africa. (Cembureau, 1997).

Imports into the EU from extra-EU countries have been increasing in the latest years, mainly from Poland and the Czech Republic into East Germany and Austria (e.g. in 1997 15.3 per cent of whole German consumption was covered by imports (Bundesverband der Deutschen Zementindustrie e.V., 1998). Romania, Tunisia and Turkey export heavily to Spain. Germany and Spain have filed anti-dumping complaints against the relevant CEEC, but the cases were closed without the imposition of duties. Exports to the EU from Turkey and CEEC have risen from 2.5Mt in 1985 to 9Mt in 1993.[24]

Quantity wise imports from South East Asia to the EU still play a minor role. Exports from Eastern European non-adjacent countries also have been quite small. For instance total exports from Russia have just reached 800,000 tonnes in 1999.[25] In quantity terms these imports are relatively insignificant, but they disturb the prevailing prices and can induce considerable price discounts by the

[24] Panorama of EU Industry, 1997

[25] Imports from South East Asia, Saudi Arabia, Turkey and Lebanon to Belgium and the Netherlands in the first 10 months of 2000 has been 600,000 tonnes cement. Prices were about 20% below market price, information by Cembureau.

national cement manufacturers. The prices for these cements were said to be about 10 Euro cheaper than the prices at national ports[26].

With respect to the sample countries cement imports are of varying importance and depend on unique events or cyclical trends of the construction industry. In detail the following can be said:

German reunification opened the cement market to East Germany and Eastern Europe. Cement imports increased rapidly from 1.9 million tonnes in 1990 to 6.9 million tonnes in 1995 (7.1 million tonnes in 1994) and then again dropped to 4.2 million tonnes in 1998. With 1.9 million tonnes Poland was the main importer in 1998. In 1996 imports from the Czech Republic, Poland and the Slowak Republic accounted for 75 per cent of the imports to Germany (Bundesverband der Deutschen Zementindustrie e.V., 1998). German production of cement in this decade reached a top in 1994 with 36 million tonnes but then dropped again. Cement exports totalled 2.9 million tonnes in 1998 of which over 80 per cent were shipped to EU countries.

In Spain the high rate of kiln utilisation driven by the expansion of infrastructure where concrete was used instead of asphalt and a relatively high price of Spanish cement (e.g. $ 75 or Pta. 9500 in 1996 in central Madrid) has attracted imports since 1995. However, Spain exports more than it imports.

Imports to the UK accounted for about 1.3 million tonnes of cement in 1998 which originate mainly from Greece, Spain and the Baltic region (International Cement Review, Jan. 1998). UK exports have been low with 0.6 million tonnes.

Cement prices in the sample countries

The price of cement differs widely. In West Germany it has been estimated to be $ 79.8 in 1996 (about 60 Euros) and remained constant until 1999 (Dresdner Kleinwort Benson) while the price in East Germany was only at $ 66.5 in 1997 (50 Euros) as a consequence of East European, in particular Polish imports. Also companies operating on both sides of the borders (i.e. Lafarge, CBR and German companies) have been trading cheap cement originating from Poland. In 1999 the East German price increased only slightly to 52 Euros.

In order to keep imports low, cement prices along the Spanish coastal regions have been kept low. For example, Southern coastal prices stayed in 1998 at

[26] Interviews in November and December 2000 with executives of participating cement companies.

about 48.8 $ (Pta 8,300) (see Dresdner Kleinwort Benson Research, 1998). Consequently imports levelled out to about 3 million tonnes in 1998 and 1999. For 1998 the Global Cement Report quotes ex-works prices for bulk cement to be Pta. 9,600 in the Basque Country, Pta. 8,400 in Andalusia and Pta.9,900 in Madrid for 1998. Average Spanish cement prices were expected to be Pta. 8,756 in 1999 (Dresdner Kleinwort Benson Research, 1998).

In the UK in 1997 average prices for bagged products varied between 85 and 93 Euros per t.

4.3.2 Likely impact of clean air regulation on competitiveness in the cement industry

The likely impact of clean air regulation on competitiveness of the cement industry depends on the nature of the corresponding environmental initiatives, i.e. whether they are end-of-pipe measures which increase costs or whether they are clean technology solutions which can decrease both emissions and costs. An increase in costs through end-of-pipe measures may be compensated elsewhere in the plant so that the overall effect may be neutral in the end. During the interviews with plant managers it was generally stated that environmental technologies cause additional costs unless energy or dust is reduced. This, however, clearly depends on the specific circumstances in each plant.

In contrast to the food and packaging case study there is a clearly defined list of abatement measures for the cement industry. This list of currently commercially available techniques for the prevention or reduction of emissions and waste in the cement industry was recently published in a background reference document as part of the work of the European IPPC Bureau in Seville (EIPPC Cement BREF).

In the following all abatement measures for NO_x, SO_x and dust reduction currently known in the cement industry are briefly described and their theoretical impacts are shown as quoted in the literature or told by the suppliers of these technologies (interview with Krupp Polysius). A summarising table of expected environmental and economic effects for most of the measures is shown at the end of this section (see Table 4.3-17). As far as possible, the current use of abatement techniques in the cement industry in our sample countries is presented. Special attention is given to the German cement industry since, it is the environmentally most advanced industry among the selected sample countries. The question arises which impacts follows for competitiveness of the German cement industries (see empirical findings in chapter nine).

General measures

Many general measures can be used to both optimise the process and improve the environment. The general measures mostly yield an economic benefit.

For instance, neuronal networks and other expert systems are used to optimise the process. Correct and even fuel quantities help to reduce NO_x by 10-25 per cent. Other general measures can also achieve SO_2 reduction, e.g. the correct mixture of raw materials. But usually additional techniques are required in order to comply with standards required e.g. by the German TA Luft. Large plants have their own engineers, whereas small plants co-operate with machine suppliers.

Measures for dust reduction: Electrostatic precipitators and bag filters

The most commonly used techniques for dust reduction are electrostatic precipitators and bag filters.

In Germany traditionally electrostatic precipitators (EPs) are used and fabric filters only in individual instances (Schneider et al., 2000). In an EP the dust becomes electrostatic between a wire and a shield. The wires can get worse with time, the dust must not be too dry. The more chambers there are in an electrostatic precipitator, the greater are the reserves for dust collection. All dust abatement measures require a disproportionately high amount of electrical energy with decreasing dust content in the clean gas. Electrostatic precipitators have the advantageous characteristic of a low pressure drop. At normal clean dust levels they need less electrical energy and can therefore be operated at lower costs than fabric filters. However, electrostatic precipitators have to be switched off during peaks in CO concentrations and this leads to a temporary increase in dust emissions. In general there is big heat stream with 150 m^3/s in an EP. If there is a disturbance, the EP looks like a balloon or can even explode. To date operational optimisation of the majority of German cement works has reduced the need for shut downs.

In cement works fabric filters are mainly used after cement mills, coal grinding plants and in many small sources of dust. If they are used after the kiln, then there are no increased emissions due to operationally-induced peaks in CO concentrations. But fabric filters need more maintenance and also a high level of attention during operation because they are sensitive to short-time temperature peaks. Moreover the wear of the filter bags shortens the service life and the operational reliability. As a consequence operating costs are higher for fabric filters than for electrostatic precipitators. German companies reported that they use bag filters at smaller machines only because the temperature is lower and danger of fire is not so high.

BAT measures for lowering SO_2 emissions

The levels of SO_2 in the exhaust gas from rotary kilns are essentially related to the content of volatile sulphur compounds in the raw materials.

Three different BAT techniques are currently suggested for SO_2 abatement dependent on the initial level of SO_2 concentration. The addition of hydrated lime (absorbent addition) to the kiln meal, or raw gas, and the use of a dry or a wet scrubber. A fourth, but very rarely used techniques is the activated carbon technique. About 20 kiln installations of the total of 252 rotary kilns in operation in Europe are equipped with secondary SO_2 abatement techniques (see Table 4.3-15 below). In the following more details on the currently suggested SO_2 abatement techniques are given:

a) Absorbent addition
If initial SO_2 emissions are not extreme, it is sufficient to use additives like slaked lime or quicklime in the combustion process. The investment costs are quite low, and the necessary changes in equipment are made by the plants themselves. Absorbent addition is with a total cost of 0.13-0.45 Euros/t of clinker the cheapest SO_2 abatement measure.

b) Dry scrubber
The dry scrubber is the most expensive measure for lowering SO_2 emissions with a total cost of 3.15 to 3.45 Euros/t clinker. The plants already using or planning a dry scrubber have very high SO_2 emissions up to 2,000 mg/m^3. A certain level of SO_2 is necessary, otherwise cement becomes rigid at a very early point. If raw materials contain pyrite or SO_2, then high SO_2 emissions are to be expected. The raw materials in East Germany and also in the Beckum area in West Germany contain pyrite. On average a third to a quarter of all German plants have this problem.

c) Wet scrubber
The wet scrubber is a further method for lowering very high SO_2 emissions. It is installed as an end-of-pipe technology after the dedusting equipment. Slurry is sprayed which leads to absorption of SO_x. The resulting gypsum is reused in the cement mill. It is highly efficient in terms of SO_2 reduction, but it requires a lot of energy and maintenance. Investment costs are 7.5 million Euros. The use of the wet scrubber produces a lot of fog because the gas streams have to be cooled down to zero degrees. Some machine suppliers work on an alternative process

integrated solution with lower investment costs. Total costs for a wet scrubber vary between 1.45 and 2.6 Euros/t clinker.

Table 4.3-15 SO_2 reduction in the European cement industry

Country	Absorbent addition	Wet scrubber	Dry scrubber	Activated Carbon
Austria		1		
Belgium	2			
Denmark		2		
Germany	11			
Italy	1			
Sweden		1		
Switzerland			1	1
UK		1		
Total	14	5	1	1

Source: Schneider et al. (2000) and own additions.

d) Activated carbon
The activated carbon technique was considered as very expensive by the machine supplier of the cement industry. Only one Swiss plant has installed it so far. The filter cannot only absorb SO_2, but many other pollutants. The Swiss government has subsidised that investment.

Measures for lowering NO_x emissions

Concerning NO_x reduction there are primary and secondary measures. Often a combination of both is needed.

Primary NO_x measures are flame cooling (water injection about the main burner) and low-NO_x burners. Also the addition of mineralisers to improve the burnablity of the raw meal is possible. Achievable emission levels with these measures are estimated to be above 400 mg/m^3. Secondary measures for a further emission reduction are staged combustion and SNCR. If a value of lower than 200 mg NO_x/m^3 should be achieved, then SCR must be used. The distribution of NO_x abatement measures in the European cement industry is illustrated in Table 4.3-16 below.

In the following more details on NO_x abatement techniques are presented:

a) Flame cooling
In Germany flame cooling is not used because it can lead to higher energy consumption and to technical burning problems. A certain reduction of NO_x can be achieved by regulating the flame. Long flames produce less NO_x, but this can be

done only within limits. Otherwise the quality of the product is damaged (colour, strength, easiness to mill it).

b) Low NO_x burners

These have been on the market for a long time and decrease NO_x and energy consumption simultaneously. Investment costs are moderate.

c) Multi staged combustion

With multi staged combustion (MSC) 500 mg NO_x/m^3 can be met in most cases. There are no additional operating costs. However, for economic reasons retrofitting of kilns with staged combustion and a calciner (which is the precondition for the use of staged combustion) just for NO_x abatement is hardly undertaken (there was one example in the sample).

Table 4.3-16 NO_x abatement measures in the European cement industry

Country	flame cooling	use of mineralisers	staged combustion	SNCR	SCR
Austria	3	1	1		
Denmark		1			
France	10	1			
Germany		4	7	15	1
Italy			3		
Sweden			1	2	
Switzerland				1	
Total	13	7	12	18	1

Source: Schneider et al. (2000)

MSC is undertaken in the area of the calcinator. Air is added at different levels in order to improve the combustion. It was specified that investment costs are about 1.5 2.5 million Euros relative to total investment costs of 50-75 million Euros. This process can only be implemented in plants with a calcinator which is a rarely used technology in Germany. In order to use this technology in plants without calcinator it is always necessary to install a precalcinator. This increases the capacity of the kiln by 30-40 per cent. This in turn requires to adapt the size of the preheater and cooler. Investment costs would be 8-10 million Euros. If all this is not economical, then selective non-catalytic reaction (SNCR) is required. An advantage of MSC is that there are almost no operating costs. There are only normal maintenance costs except for cleaning with air pressure. The computer operators also get some training. Training normally requires about 1 week and is offered by suppliers.

Different countries prefer different technologies. For instance in France calcinators with preheaters are preferred. The efficiency of MSC depends very much on the type of fuel material used. Plastics, brown coal, contaminated oils and other materials with a fugitive character are secondary fuel materials which are adequate. Petcoke is less suitable.

d) SNCR (selective non-catalytic reduction)

With SNCR normally emission levels between 500 and 800 mg/m^3 are achieved, but depending on the design of the plant also levels of 200 mg NO_x/m^3 can be reached. So far there is only one plant in Sweden where with SNCR 200 mg NO_x $/m^3$ are reached (Pott, 1999). This plant has a calciner which makes the long residence time for the necessary reactions during SNCR available. The reduction of NO_x emissions by 85 per cent from 1200 mg to 200 mg was also possible because a very high molar ratio of 1. to 1.2 was used and the unconverted ammonia created at this high molar ratio is collected in a downstream SO_2 scrubber. The technology is based on the injection of ammoniac water. Investment costs are between 0.4 and 0.75 million Euros. This is due to the obligation to construct high quality tanks. Costs vary according to capacity. NO_x reduction lies between 30 and 50 per cent. Ammoniac emissions depend very much on the characteristics of the raw material. Also the proportion between ammoniac and NO is decisive. A Mol factor of more than 1.3 means that ammoniac is not completely used and ammoniac emissions are increasing. Operating costs are low except for the ammoniac water. Additional skill needs are low, some safety instructions are required. A skilled mechanic could do the job.

e) SCR

SCR is currently used only in one German plant where a demonstration project is started during the year 2000. Costs are quite high and from Table 4.3-17 it can be seen that primary measures and SNCR are cheaper than SCR.

It is difficult to say whether SCR will come on the market. SCR leads to a SO_2 reduction of 80-90 per cent, but investment costs are 3-3.5 million Euros for a 2,000 t/a plant. Up to now it has not been tested in industrial plants, test were just done in pilot tests in Sweden, Austria, and Germany. One does not know how long the catalyst will last. One German plant has now applied at the Umweltbundesamt for a subsidy for industrial use.

Table 4.3-17 Overview about pollution reduction techniques for the cement industry and its environmental and economic effects

Emission Reduction Techniques	Kiln systems Applicability	Reduction Efficiency	Reported emissions mg/m³ [1]	kg/tonne[2]	Reported costs [3,7] Investment	Operating
NO_x Reduction Techniques						
Flame cooling	All	0-50 %			0.0 -0.2	0.0-0.5
Low-NO_x burner	All	0-30 %	400-	0.8-	0.15-0.8	0
Staged combustion	Precalciner	10-50 %	<500-1000	<1.0-2.0	0.1-2	0
	Preheater				1-4	0
SNCR	Preheater and Precalciner	10-85 %	200-800	0.4-1.6	0.5-1.5	0.3-0.5
SCR – data from pilot plants only	Possibly all	85-95 %	100-200	0.2-0.4	ca. 2.5 [4] 3.5-4.5 [5]	0.2-0.4 [4] No info. [5]
SO_2 reduction techniques						
Absorbent addition	All	60-80%	400	0.8	0.2-0.3	0.1-0.4
Dry scrubber	Dry	up to 90%	<400	<0.8	11	1.4-1.6
Wet scrubber	All	>90%	<200	<0.4	6-10	0.5-1
Activated carbon	Dry	up to 95%	<50	<0.1	15 [6]	no info.
Dust reduction techniques						
Electrostatic precipitators	All kiln systems		5-50	0.01-0.1	2.1-4.6	0.1-0.2
	Clinker coolers		5-50	0.01-0.1	0.8-1.2	0.09-0.18
	Cement mills		5-50	0.01-0.1	0.8-1.2	0.09-0.18
Fabric filters	All kiln systems		5-50	0.01-0.1	2.1-4.3	0.15-0.35
	Clinker coolers		5-50	0.01-0.1	1.0-1.4	0.1-0.15
	Cement mills		5-50	0.01-0.1	0.3-0.5	0.03-0.04
Fugitive dust abatement	All plants		–	–	–	–

1) normally referring to daily averages, dry gas, 273 K, 101.3 kPa and 10% O_2
2) kg/tonne clinker: based on 2000 m³/tonne of clinker
3) for NO_x and SO_2: investment cost in 10^6 Euros and operating cost in Euros/tonne of clinker, normally referring to a kiln capacity of 3000 tonne clinker/day and initial emission up to 2000 mg NO_x/m³
4) costs estimated by Ökopol for a full scale installation (kiln capacities from 1000 to 5000 tonne clinker/day and initial emissions from 1300 to 2000 mg NO_x/m³), operating costs ca. 25% lower than for SNCR
5) costs estimated by Cembureau for a full scale installation
6) this cost also includes an SNCR process, referring to a kiln capacity of 2000 tonne clinker/day and initial emission of 50-600 mg SO_2/m³
7) for dust: investment cost in 10^6 Euros and operating cost in Euros per tonne of clinker for reducing the emission to 10-50 mg/m³, normally referring to a kiln capacity of 3000 tonne clinker per day and initial emission up to 500 g dust/m³

Source: EIPPC Cement BREF (2000)

Change in fuel materials

This is not a recognised pollution abatement technique, but it is widely used in the industry so that its effects shall be illustrated here.

The change from primary to secondary (alternative) fugitive materials is relatively easy. One only needs a measuring device and respective containers. However for tyres a higher investment is needed. They must be stored in-house, have to be separated, weighted and fed into the system. For an equipment which can feed 2 tonnes per hour, costs are about 1 million Euros. If the German waste market is such that a company uses tyres, it receives about 30 Euros per tonne from the supplier. Moreover it saves 75 Euros per tonne of coal which would have been needed instead of the tyres. This leads to a saving of 1.25 million Euros per year. Some German companies planned to substitute up to 75 per cent of its fuel by secondary materials until the year 2001.

4.3.3 Conclusions on the likely impact of environmental regulation on competitiveness in the German cement industry

The above analysis has shown that cement is a commodity product and that cost competitiveness is decisive for business success. Therefore the impact of additional costs caused by environmental regulation is an important issue for the industry, especially for Germany where regulation is more stringent than in the other sample countries. It could be shown that the German cement industry already uses costly pollution abatement techniques which are not frequently used elsewhere. Simultaneously, within the sample countries the German cement industry is a forerunner concerning energy efficiency; this will be a helpful factor for keeping production costs at the lowest possible level. However, with the exception of the new East German plants, labour productivity in Germany is relatively low, e.g. Spain reaches a higher level. Since the level of productivity is hypothesised to be associated with the ability to absorb compliance costs, this could imply a negative aspect for competitiveness in the German cement industry. However, the empirical results in chapter nine show that there are only a few instances where German sample plants are disadvantaged by stringent environmental regulation. It will be shown that there are sources for cost compensation which are large enough to prevent a loss in competitiveness caused by environmental regulation.

4.4 Conclusions on the likely impact of environmental regulation on competitiveness in the Germany sample industries

On a national average, both in the food case study and the packaging supply chain example Germany shows the highest productivity levels as well as the highest degree of strictness of regulation.

According to our principle hypothesis that high productivity plants are also in a position to absorb environmental compliance costs, it can be expected that German plants in the food industry and the packaging supply chain case are not put at a major economic disadvantage caused by environmental legislation. In the food industry, Germany may benefit from a strong environmental industry which supplies waste water technology. In the packaging supply chain case competitiveness impacts are assumed to be stronger for the fillers due to the inherent characteristics of the DSD system.

Only in the cement branch German productivity lays behind the other sample countries, but the industry again shows the highest degree of stringent legislation in the sample countries. Simultaneously, it could be shown that the German cement industry operates costly environmental equipment which is not used in the other sample countries. Thus, prima facie, the German cement industry could be put at a cost disadvantage in this case.

The chapter also showed that environmental pressure is just one of many competitive pressures in each case study industry.

5. Hypotheses, measurement of variables and research method

The main focus of the study is to examine the impact of stringent environmental regulation on competitiveness of German firms relative to other Member States of the EU where environmental regulation is currently less stringent. For this purpose three industry studies have been selected in different environmental problem areas where German legislation was notably more stringent than the regulation in selected comparator countries. Firms in Germany, the Republic of Ireland, the UK and Spain were sampled in the food industry, the packaging sector (including a supply chain) and the cement industry to represent contrasts in environmental legislation and enforcement as well as in economic performance (mainly productivity), skill and R&D input and age of capital stock.

The first section of this chapter outlines the underlying hypotheses used in the research, the inherent key variables and their measurement. The second section examines the method of sampling and the history of this research technique.

5.1 Hypotheses and measurement of key variables

The sample data was analysed according to common hypotheses concerning the effects of adjustment to environmental pressure. The basic consideration in the hypotheses was that the sources responsible for international differences in productivity levels could be causally associated with the effects of the solutions to the environmental pressures: On the one hand, varying productivity levels could influence the ability to solve environmental problems, on the other hand they could also have an impact on the direction of any feedback on productivity levels through environmental measures. In this context the main question of the impact of stringent environmental regulation on competitiveness of German firms relative to other countries will be examined. With respect to the above basic consideration one could a priori think that German industry is particularly apt to comply with stringent legislation since it is a high productivity economy. The following hypotheses further differentiate that thought and add aspects like the small firm size in Germany which could make the country more vulnerable in the face of environmental policy. Furthermore the key variables for the testing of the hypotheses are defined. Also the measurement of the key variables is illustrated.

Hypotheses:

Productivity and cost of compliance: The study sought to measure the cost of environmental compliance and find a relationship with the economic performance characteristics of the sample plants.

It was hypothesised that the proportional cost of environmental compliance relative to turnover incurred by the firms is likely to be a negative function of the productivity level (Hitchens et al., 1998, 1999, 2000, 2001). This would be because firms with the capabilities to achieve high productivity will also find it easiest to react to environmental pressures without the penalty of reduced output and employment. It implies for international comparisons that countries/regions with relatively high productivity (levels of competitiveness) will have relatively low compliance costs. Also, within a country/region those firms with above average productivity could be expected to have relatively low compliance costs. According to this hypothesis compliance costs in Germany and relative to the other sample countries could a priori be expected to be low because the country is a high productivity economy.

Competitiveness and environmental initiatives including environmental awareness: The study also sought to link productivity and compliance costs to the number of environmental initiatives undertaken and the underlying environmental awareness.

A favourable environmental performance is hypothesised to be linked to characteristics of high productivity (best practise) plants. Just as these high productivity plants would be more aware of any gains in efficiency than the average, their management and workforce can be expected to undertake a higher number of environmental initiatives because they show a higher environmental awareness. Still, these plants are hypothesised to be able to keep compliance costs low for a number of reasons also related to best practise firms:

— *Type of environmental initiative and compliance costs*: In general, compliance costs could be assumed to increase with the number of environmental initiatives. High productivity plants, however, will not only react more actively to environmental pressure with a higher number of environmental initiatives than other plants with lower productivity, but also prefer clean technologies to end-of-pipe technologies. This is because the latter is normally associated with additional costs and losses in productivity, whereas clean technology can improve productivity, reduce compliance costs and simultaneously bring emissions down (Coenen et al., 1995).

— *Skills and R&D*: The type of environmental activity is also anticipated to be connected with the skill and R&D endowment of plants. Plants with a well trained labour force and a strong R&D basis will also integrate environmental activities in their work routines and come up with a higher number of individual, clean technology solutions and are likely to be more successful in reducing environmental costs than their counterparts lacking these skills (interview with Bundesinstitut für berufliche Bildung, 2000; Jenkins, 1999). Regulation could lead firms to increase their R&D efforts, thus leading to a decrease in production costs (Oates et al., 1993) and stimulating productivity growth (Dufour et al., 1998; see Kennedy (1994) for a formal argument). In comparison to the UK Germany is said to have a better skill base and the R&D activity is also higher (Daly et al., 1985; Prais, S.J., 1981; Steedman and Wagner 1987 and 1989; Mason et al., 1994). Therefore Germany might have relatively good preconditions for the use of clean technology.

— *Plant age*: The age of a plant and its machinery is likely to have an impact on its environmental performance and on compliance costs (Alpay, 2001). Modern plants with younger machines are more productive and less polluting than plants using older machines. They can reach better environmental standards because the newest technology also embodies best environmental technology and are allowed to achieve environmental goals at a lower cost than with old equipment which would need expensive retrofitting.

Influence of plant size on environmental performance and compliance costs: Small and medium sized enterprises (SMEs), i.e. those employing fewer than 500 persons, are potentially vulnerable in the face of environmental policy (Jackson, 1994). Their problems frequently relate to lack of access to financial resources, limited international experience, skill deficiencies because of limited ability to provide benefits and competitive wages to employees and dependence on large firms to which small firms subcontract (Ettlinger, 1997). However, many sectors of the German manufacturing industry consist of a large percentage of SMEs. Therefore the study sought to analyse the influence of plant size on environmental performance and compliance costs.

Plant size is assumed to have an influence on the adjustment to regulation and related environmental pressure (Henriques and Sadorsky, 1996). Large plants are hypothesised to undertake more environmental initiatives because they often have more financial and human resources for the solution of environmental problems than small and medium sized plants and have also better opportunities to research cost efficient environmental measures. Large multiplant firms could possess certain economies of scope across several markets in complying with a regulation; that is large diversified companies may be able to spread compliance costs across several markets (Arnold, 1995). Thus the proportional cost of compliance relative to turnover is likely to be a negative function of the size of plants/firms.

History of regulation: The research aimed at analysing the influence of the history of regulation on the ability of plants to develop solutions to environmental pressures.

It was hypothesised that plants in countries with a history of stringent environmental legislation would be in a better position to develop or gain access to environmentally beneficial solutions than their counterparts in other countries with a history of less stringent regulation. This would be because their supplier network is more experienced with solutions to environmental problems and can offer a more differentiated supply of environmental goods. Germany has a very long history of environmental regulation and a very developed supply of environmental goods of which it could potentially benefit when adjusting to legislation (Wackerbauer, 2002).

Green market pressure in the supply chain relations: Not only regulatory pressure forces firms to react, but also environmental pressure coming from the market place.

Not only regulatory pressure forces firms to react, but also environmental pressure coming from the market place (Hitchens et al., 2000; Dobilas and Macpherson, 1997; Henriques and Sadorsky, 1996). Relations within the supply/production chain are likely to be both an influence upon, as well as being influenced by, the level of environmental standards, e.g. a manufacturer may find it easier to increase the environmental standards of its products if it has a reliable and competent base of suppliers to draw on. A manufacturer may be forced to upgrade product and process environmental standards by pressure coming from the customers of plants in the three sectors under study.

Green consumer pressure: The study aimed at measuring the influence of green consumer pressure on plant competitiveness.

Plants which operate in a market where final demand for clean products and processes is large and consumers are prepared to pay a premium, will be more involved in environmental measures than other firms which do not have similar customer groups and are not rewarded for their environmental activities (Sprenger, 1996). Currently green consumer pressure can be expected to be higher for Germany than for Spain or the UK where the general level of environmental awareness among the population is just beginning to grow. The demand for environmentally sound products is expected to grow in the future because environmental awareness of consumers will increase (Steger, 1993; Welford, 1995).

Market structure (packaging and cement study only): The industries researched in this study (food, packaging and cement) operate in different market structures. This will influence a firm's strategy towards the environment.

In a relatively competitive market like the food industry, constant striving for new cost efficient environmental solutions is required because chances to pass on costs to customers are small. In an oligopoly like the cement industry individual incentives for renewal and updating always depend on what other members of the oligopoly will undertake. On the one hand, adjustment to new environmental requirements may therefore be delayed in an oligopolistic market. On the other hand, if an industry achieves super-normal profits, it is to a certain extent protected against additional costs arising from environmental compliance and the pursuit of wider management goals which includes environmental protection is possible (Hettige et al., 1996). Thus, market power will allow to pass on environmental cost increases (Alanen, 1996). Also, where a firm is acting in a protected local market, it can pass on any higher cost arising from environmental regulation than if it was active in a global market with competition from countries with laxer regulation (Levêque, 1993). Monopsony power in the retail sector (packaging supply chain study) could be used to exert environmental pressure and force suppliers to adopt environmental initiatives.

Ownership: The industries under concern in this study are to varying degrees composed of multinational enterprises and the study sought to find out the influence of ownership on adjustment to environmental standards and its competitive implications.

Ownership is likely to be of influence on the environmental response a plant or a firm gives to environmental regulation. The industries under concern in this study are to varying degrees composed of multinational enterprises. Multinational companies are likely to show better environmental performance than other enterprises because they export both the environmental and the efficiency standard of their home country (at least in their new production facilities, see Epstein and Roy, 1997 and 1998). Therefore, where multinational branch plants are sampled in those parts of the EU with lower environmental standards/enforcement, they are expected to undertake a higher number of environmental initiatives and are in a better position to reduce environmental costs than indigenous companies producing the same products.

Location: The research also wanted to find out whether plant location had any impact on compliance costs and competitiveness.

Location can affect the cost of compliance. The assumption that environmental regulation has differential impacts between regions and countries would be related to differences in competitiveness between Member States of the EU which are based on the underlying differences in productivity and skills/R&D (Barras et al., 1997). These factors are related to the efficient adjustment to regulation. Therefore plants in peripheral locations are likely to be less advanced on environmental matters, to undertake less environmental initiatives and to be potentially subject to higher environmental costs because of lack of access to suppliers, language barriers, higher transport costs etc. than plants in centrally located regions.

Employment impact: The study aimed at measuring the employment impact of adjustment of environmental legislation.

The direction of the employment implication of adjustment to stringent environmental regulation is not clear a priori. The available macro-econometric evidence has shown that the overall employment effects of environmental policies are likely to be small.

High standards and strict enforcement: Overall, the research aimed at analysing the influence of the intensity of regulation on the competitiveness of plants.

Given the above hypotheses, although high environmental standards and strict enforcement may represent a short term cost and burden to the firm, they could in the medium and longer term push firms on to a higher growth path by forcing them to make product and process changes which yield higher competitiveness. If this happened it would represent part of the so-called "double dividend", i.e. gains in environmental performance would also be accompanied by increased economic performance.

Key variables and their measurement:

The study aimed at measuring the costs of environmental compliance, their systematic relationship with plant performance and their overall influence on plant competitiveness.

Environmental measures and cost of compliance: The research sought to measure the cost of compliance with environmental regulation and focused on environmental initiatives in specific areas (waste water, solid waste and clean air). Every action undertaken by the company as a consequence of environmental pressures was itemised. Hereby it was attempted to take account of all related costs and environmental as well as economic effects (capital and running cost for the chosen abatement techniques, emission reduction (if any); training and skill requirements; maintenance costs; effects on process efficiency; capacity; price; output; market share; employment and profitability as far as possible).[27]

Cost of compliance and relationship with plant performance: The research also aimed to identify whether, and if so, how the costs of compliance are systematically influenced by plant and other supply side factors. A main input was the productivity and value added performance of the plants. Interviews sought to measure comparative value added per head and partition this between physical and quality elements. Data were also collected on the R&D activity of a plant, plant age, skills, plant size, ownership, location, export performance, firm growth, etc. It was recorded whether any of these characteristics are likely to give rise to a competitive advantage or disadvantage by country, region or industry structure. Thus, competitive impacts of regulation could be analysed within a country and between sample countries.

Impact of compliance costs on plant competitiveness: The importance of the net cost of compliance relative to other factors which influence competitiveness was examined. Companies and plants were interviewed on the nature of demand incl. supply chain pressure for environmentally friendly processes and products, the degree of competitiveness they faced in the market and on other key factors influencing the competitiveness of the industries e.g. labour costs, exchange rates etc.

5.2 Research method

For a robust testing of the above hypotheses the need for a detailed production of empirical data was recognised. Therefore the matched plant comparison was selected as method of in-depth analysis of the problem. The matched plant tech-

---

[27] It was not possible to develop weights for each environmental initiative within or across sample plants because the corresponding capital and running cost data were not always available in the necessary breadth and depth. However, for each case study a classification list of all possible environmental initiatives was available (see Box 6.1-1 for the food study, Box 6.2-1 for the packaging study and Box 6.3-1 for the cement study). Therefore a precise count of numbers of environmental initiatives was possible and broad differences in the numbers of initiatives signalled differences of environmental performance of individual sample plants. To a certain extent environmental performance data was also directly available. However, no relationship between environmental performance and numbers of environmental initiatives could be established, probably because of the weighting problem and the lack of knowledge about the intensity of use of initiatives.

nique is an interview-based sample survey technique which is comparable to a benchmarking exercise and which has the advantage of generating detailed micro-level data (Daly, Hitchens and Wagner, 1985). While no formal model is used for the specification of a production function, the technique has yielded robust measures of the importance of a range of factors influencing relative competitiveness in a variety of industries across the EU, among them Britain, Germany, the Czech Republic and Hungary (e.g. Hitchens, Wagner and Birnie, 1990; Hitchens, Wagner and Birnie, 1993; Mason, van Ark, Wagner, 1994; Hitchens et al., 1995). Links were traced between product quality and innovation, efficiency and process innovation and connected to that it was shown that the presence of intermediate and higher level skills (such as Meister in Germany) attributed to higher productivity levels. Lately the method was extended for the exploration of the relationship between productivity and environmental performance across the EU (see Hitchens et al., 1998, 2000 and 2001).

The method provides valuable insights for both top management and policy makers through a systematic comparison of supply-side features of the firm after controlling for size, ownership and product type. It then allows for an analysis of strengths and weaknesses of similar firms. Focus is put on the explanatory factors for these differences in performance.

The research focused on a sample of plants showing characteristics along a spectrum from low to best practice firms defined both by productivity and environmental performance in each industry and country under study. As far as possible a typology of several classes of environmental performers (e.g. compliance, compliance plus and excellence; see also Table 5.2-1 for a further example of classification) in the manufacturing industry was developed and put in relation to the productivity classification.

Table 5.2-1 Example of a Classification of Environmental Management Technologies

Type of Environmental Management Technologies	Classification
Use of conventional technologies that meet compliance standards	Reactive
Adoption of state-of-the-art technologies for pollution reduction	Proactive
Use of proven, reliable technology for risk minimisation; adoption of "fail-safe" systems	Crisis Preventive
Adoption of state-of-the art "clean" technology for pollution prevention and waste elimination	Strategic

Source: Florida (1996), p. 115.

The review of the literature on the relationship between environmental regulation and competitiveness also revealed a number of methodological difficulties which the current matched plant approach seeks to avoid. The interview-based in–depth approach of the matched plant technique allows access to sometimes confidential data on environmental costs and economic performance. This is particularly important since the focus of the study was on the effects of clean technology solutions which are not covered in the census data. Clean technology is important because it is an element of the mechanisms of innovation inherent to the Porter type argument. Within the framework of the matched plant technique both quantitative and qualitative data can be collected. This allows for statistical testing of data according to hypotheses and adds analytical rigour derived from analysis of qualitative data. With the co-operation from management of industrial enterprises net costs of compliance can be calculated and their importance relative to other factors of competitiveness can be determined. Moreover, concerning the testing of the effects of environmental initiatives detailed micro-level data can be collected.

Although the case studies included a sufficiently large number of cases for some statistical analysis, the samples used for the case studies are still small. As was shown in the literature review, larger samples could be analysed through alternative methods like econometric modelling of panel data (as far as this exists) e.g. with a distance function approach (where best practice plants are captured and efficiency differences between them are analysed, similar to the matched plant technique where plants along a spectrum of low to best practice plants are selected); the measurement of total factor productivity and regression analysis; analysis of imports and exports (revealed comparative advantage used in trade studies) or locational studies (focus on foreign direct investment); anecdotal evidence from very small case studies (e.g. extensive reporting on individual plants without further statistical testing). The advantage of matched plant technique lies in the collection of more detailed data at the plant level which in this case (data on the use of clean technology) would not have been included by the census data.

6. Description of total sample for each country and industry

This chapter examines the major economic and environmental characteristics of each case study sample. A breakdown is given by size, ownership, product, location and a number of performance characteristics like growth, productivity, skills, environmental initiatives, and so on. These variables were important for matching firms between countries (discussed later on in each section of the respective industry study). In the cement study, it was also possible to record emission levels. This makes the section on the cement sample longer than the illustration of the other two samples.

Sampling was in each industry undertaken in the light of three considerations. Firstly, the sample should be broadly representative of the industry in each country under concern; secondly, the sample should include the principal firms in that industries (this usually implied interviewing large producers) and the third aspect was to include firms which could be matched internationally in terms of size and product.

6.1 Dairy and meat sample

6.1.1 Target sample size, directories, response rates and data collection

10 plants in each the dairy and meat industry in the sample countries Germany and Republic of Ireland/Northern Ireland were targeted so that a total of about 40 interviews was envisaged. Standard industry directories were used for sampling and companies were selected at random except within the required sampling framework. Response rates were considerably higher in the Irish economies than in Germany. In East Germany face to face interviews were conducted in Thuringia, Saxony and Saxony-Anhalt and in West Germany firm interviews took place in Bavaria and Baden-Württemberg. In Northern Ireland firms interviewed were located in all counties except Londonderry, while in the Republic of Ireland sample companies were drawn from counties Cavan, Cork, Monaghan, Sligo and Leitrim. All interviews took place between summer 1995 and spring 1996.

6.1.2 Sample size

Eventually a total of 47 firms was interviewed with more interviews in the Republic of Ireland and Northern Ireland in order to facilitate appropriate matching with German firms. In Germany an approximately equal number of interviews were conducted in East and West Germany (see Table 6.1-1).

Table 6.1-1 Numbers of dairy and meat firms sampled in each region

Industry	East Germany	West Germany	Northern Ireland	Republic of Ireland
Dairy	5	5	6	9
Meat	5	6	6	5
Total	10	11	12	14

6.1.3 Size distribution of sample plants and firms

Tables 6.1-2 (a) and (b) show the size distribution of plants sampled. About a quarter of dairy and meat plants are of very small size, employing fewer than 50 persons. Another quarter of dairy plants and 18 per cent of meat plants are of medium size with between 50 and 100 persons employed, while more than 40 per cent of dairy and 50 per cent of meat plants are large. Only a handful of these large plants employ over 500 persons. Thus, by and large the sample is characterised by small and medium sized plants which matches with the industry structure in each country.

Table 6.1-2 Size distribution of dairy and meat sample firms

Number of dairy plants according to size band

Employees	East Germany	West Germany	Northern Ireland	Republic of Ireland	Total
< 20	1	1	1		3
20-49	1	1	1	1	4
50-99	3		2	2	7
100-199		2		2	4
200-499			2	2	4
500-999		1		2	3
Total	5	5	6	9	25

Number of meat plants according to size band

Employees	East Germany	West Germany	Northern Ireland	Republic of Ireland	Total
< 20		1		1	2
20-49		3			3
50-99		1	1	2	4
100-199	3		2	1	6
200-499	2	1	2	1	6
500-999			1		1
Total	5	6	6	5	22

6.1.4 Economic characteristics of sample firms

Ownership

Tables 6.1-3 (a) and (b) show that the dominant legal form of business organisation sampled was the limited company. In dairying more than a quarter of sample plants were organised as co-operatives and in the meat industry, mainly in slaughtering, several plants were municipally owned. In dairying PLCs were the least represented legal form of business, in the meat sample the share of PLCs was equal to that of municipally owned plants.

Table 6.1-3 (a) Number of dairy plants according to ownership

Ownership	East Germany	West Germany	Northern Ireland	Republic of Ireland	Total
Co-op	1	3	3		7
Ltd.	4	2	2	5	13
PLC			1	4	5
Total	5	5	6	9	25

Table 6.1-3 (b) Number of meat plants according to ownership

Ownership	East Germany	West Germany	Northern Ireland	Republic of Ireland	Total
Municipal	1	3	1		5
Ltd.	4	3	3	2	12
PLC			2	3	5
Total	5	6	6	5	22

Products sampled

Tables 6.1-4 (a) and (b) show the range of products sampled. The main dairy products included liquid milk and associated milk products like cheese, butter, butter milk, yoghurt, cream and powdered milk. One plant in the Republic of Ireland also produced alcohol and cream liqueur. Beef and pork dominated the meat products in sample plants, less important was the processing of lamb and chicken. One plant in Germany also produced salami.

Table 6.1-4 Products produced by dairy and meat sample plants

(a) Number of products sampled in the dairy industry

Product	East Germany	West Germany	Northern Ireland	Republic of Ireland
Liquid milk	1	2	4	3
Cheese	3	4	3	3
Powdered milk	3	1	2	4
Butter	2	4	2	2
Butter milk	1	2	1	3
Yoghurt	1	1	1	1
Cream	1	1	1	3
Low fat spreads			1	1
Whey powders	1			2
Ice cream				1
Cream Liqueur				1
UHT				1
Casein		1		1
Alcohol				1

(b) Number of products sampled in the meat industry

Product	East Germany	West Germany	Northern Ireland	Republic of Ireland
Beef	3	5	3	2
Pork (incl. Bacon, ham)	4	6	3	3
Lamb	2	4	1	1
Chicken	1		1	
Turkey		1		
Salami		1		

Skills and qualifications/ research and development

Skills and qualifications were documented because it was hypothesised that firms with an abundance of skills and research and development activities will undertake more environmental initiatives to reduce waste water.

Table 6.1-5 shows the average percentages of certified skills and qualifications found in sample plants. The share of the labour force included holds academic diplomas and degrees e.g. in dairy, food or agricultural science as well as technical qualifications or apprenticeship (craft) training e.g. mechanical and electrical fitters. Excluded is the share of employees trained on the job. On average Irish plants have the lowest percentage of qualified or skilled personnel compared with their German counterparts. The East German average in the dairy industry is risen by one very high level of skills (94 per cent) in one company. This company indicated that it had an overhang of highly qualified staff from GDR times.

Table 6.1-5 Average percentages of certified qualifications in the dairy and meat sample

Industry	East Germany	West Germany	Northern Ireland	Republic of Ireland
Dairy	49	39	10	9
Meat	17	19	12	9

Table 6.1-6 shows employment in research and development as a percentage of total firm employment. In dairying the Irish economies are at the forefront, in the meat industry Germany employs slightly more persons for research and development than Ireland.

Table 6.1-6 Employment in research and development as average percentage of total firm employment in the dairy and meat sample

Country	Dairy	Meat
Germany	1.1	0.5
Ireland	1.8	0.4

Machinery

The age of machinery is a relevant variable because it was hypothesised that plants with modern machinery and equipment will have relatively good environmental performance. This would be because up-to-date machinery usually embodies best available technology. Age of plant was measured according to the percentage of machinery in each plant which was younger than five years old, between five and 10 years old as well as over ten years old, in physical terms, where each unit of plant was weighted according to its importance in the final output.

Table 6.1-7 shows that Irish sample plants were found to have on average the oldest machinery. However, also West German dairy plants had a fair share of machinery being older than 10 years. East German plants are relatively young indicating the degree of modernisation having taken place after the reunification.

Table 6.1-7 Machine age by country and region in the dairy and meat sample (% of total machinery in age group categories)

Age of machinery in years	East Germany	West Germany	Northern Ireland	Republic of Ireland
Dairy:				
Under 5	53	36	22	20
6-10	32	14	24	28
Over 10	15	48	54	52
Meat:				
Under 5	100	48	37	36
6-10	0	32	38	46
Over 10	0	17	26	18

Exports

The extent to which plants export can be considered as an indicator of competitiveness which later on in the analytical part of this case study will be linked – among other things - to environmental costs in order to identify the relationship between competitiveness and the cost of regulation. Table 6.1-8 shows the average export ratios measured as a percentage of turnover. For Irish plants exports were especially important, mainly due to the small size of their home market.

Table 6.1-8 Average export ratios in dairy and meat sample firms as % of turnover

Export ratio	Germany	Ireland
Dairy	19.6	58.0
Meat	6.6	44.7

Growth

As a further indicator of competitiveness growth was measured in terms of employment change over the period 1990-1995. Companies have been ranked according to their rate of growth (see Tables 6.1-9 (a) and (b)). The strongest employment decline was recorded in East German dairying, still as a consequence of restructuring after reunification. Meat plants in the Republic of Ireland showed the overall strongest employment growth with more than 20 per cent during 1990 and 1995.

Table 6.1-9 Employment change in dairy and meat sample plants, 1990-1995

Dairy (number and %)

Employment growth	East Germany	West Germany	Northern Ireland	Republic of Ireland
40 % +		1		1
5 to 39%			2	1
+/– 0 to 5 %		3	1	
Clear decline	5	1	3	5
Av. Increase in %	–	+43	+16	+180
Av. Decline in %	–56	–29	–15.6	–25.8
Overall change in %	–56	+22	–15	+25.7

Meat (number and %)

Employment growth	East Germany	West Germany	Northern Ireland	Republic of Ireland
40 % +	1	1	2	2
5 to 39%	2	1	1	1
+/– 0 to 5 %		1	1	1
Clear decline	2	3	1	1
Av. Increase in %	+34	+39	+41.3	+40.3
Av. Decline in %	–27.5	–15.7	–14	–15
Overall change in %	+5.4	+6.7	+18.3	+21.2

6.1.5 Waste water initiatives undertaken by sample firms

There are a number of methods available for reducing water use and the resulting effluent production, e.g. water recycling, steam segregation and water recovery through waste water and by-product separation. The frequency and type of these waste water initiatives were taken into account. Box 6.1-1 presents the list of waste water activities which was found among sample plants.

Waste water initiatives were similar for both meat and dairying, with the exception of methods and techniques required for the collection of blood. The number of different initiatives undertaken by sample plants is shown in Table 6.1-10 (a) and (b). German plants undertook the greatest number of initiatives with West German plants having a slightly higher number in place than their East German counterparts.

Box 6.1-1 Overview of waste water initiatives in sample plants

(1)	design of plant so as to minimise waste water (e.g. update drains)
(2)	reduction of effluent produced by cleaning in place incl. energy saving measures for production of hot wash water
(3)	recycling of rinses/water monitoring
(4)	separation of waste water
(5a)	collection of blood (meat industry only)
(5b)	recovery of valuable wastes from waste water
(6)	waste disposal technology (e.g. compressor)
(7)	effluent plant

Table 6.1-10 Number of waste water initiatives undertaken by dairy and meat sample plants

(a) Dairy

No. of initiatives	East Germany	West Germany	Northern Ireland	Republic of Ireland
0				1
1			1	
2			1	1
3	2		1	5
4	2	2	3	
5	1	2		2
6		1		
Sample size	5	5	6	9

(b) Meat

No. of initiatives	East Germany	West Germany*	Northern	Republic of Ireland
0				
1				
2			1	1
3	1		3	3
4	4	2	2	
5		3		1
6				
Sample size	5	6	6	5

*In one plant of the overall 6 sample plants no definite data on the number of environmental initiatives were obtained.

Tables 6.1-11 (a) and (b) give an overview about the kind of initiatives which were most frequently undertaken. In dairying the most important initiatives were cleaning in place, recycling of rinses, recovery of valuable waste and installation of effluent plants. In the meat industry the most important activities were recycling of rinses, separation of waste water, blood collection, cleaning in place and also use of effluent plants.

Table 6.1-11 Types and numbers of waste water initiatives undertaken in dairy and meat sample plants

(a) Dairy

Type of initiatives	East Germany	West Germany	Northern Ireland	Republic of Ireland
Design of plant	3	2	2	4
Reduction of effluent by cleaning in place	3	5	4	6
Recycling of rinses/monitoring	5	5	4	3
Separation of waste water	0	2	0	3
Recovery of valuable waste	3	5	0	2
Waste disposal technology	0	0	4	3
Effluent plant	4	1	5	8
Sample size	5	5	6	9

(b) Meat

Type of initiatives	East Germany	West Germany*	Northern Ireland	Republic of Ireland
Design of plant		1	2	1
Reduction of effluent by cleaning in place	1	2	4	1
Recycling of rinses/monitoring	2	5	1	1
Separation of waste water	3	5	4	5
Blood collection	3	5	6	4
Recovery of valuable waste	0	2	1	1
Waste disposal technology	0	0	0	2
Effluent plant	2	1	6	4
Sample size	5	6	6	5

*In one plant of the overall 6 sample plants no definite data on the number of environmental initiatives were obtained.

Sample plants had also undertaken environmental awareness initiatives. They were recorded in terms of certified environmental policy schemes. Table 6.1-12 shows that only a few plants had a written environmental policy in place with Irish dairies being the leaders.

Table 6.1-12 Number of dairy and meat sample plants with written environmental policy

Industry	Germany	Ireland
Dairy	3	5
Meat	1	1

Only one of the German dairy companies had a written environmental policy. The manager of this small West German company had developed a "green philosophy" for from the beginning onwards. The company had specialised on ecologically produced food products. The largest two dairies in West Germany were about to introduce an eco-auditing process. In one company this was mainly due to the environmentally friendly attitude of the manager, the manager of the other firm said that the eco-audit was required by market pressure.

In the meat industry both in Germany and Ireland only one plant each had introduced a written environmental policy; the plant in Germany was the same small plant specialised on ecologically produced food products already mentioned above in dairying.

6.1.6 Matched plants and firms in the dairy and meat sample

The Tables shown in the preceding sections have illustrated that sample plants can be broadly matched across the chosen countries, especially concerning size distribution, legal form of business and products (see Table 6.1-13).

Table 6.1-13 Numbers of plants broadly matched in the dairy and meat sample

	Dairy	Meat
Ireland to Germany	14	10

However, more careful matching is required in each industry in order to measure productivity and environmental performance across countries. This is considered in chapter seven where the empirical results on the impact of stringent waste water legislation on competitiveness are presented.

6.2 Description of sample in the supply chain packaging industry, food and retail industry

6.2.1 Target sample size, directories, response rates and data collection

A minimum of 10 retailers, 15 dairy processors and soft drinks producers and 25 packaging manufacturers and machine suppliers were targeted in order to have a sample of about 100 plants in the two sample clusters Germany and Republic of Ireland/United Kingdom. In packaging, firms manufacturing the main types of material producers relevant for the packaging of soft drinks, mineral water and dairy products were considered. These included paper and board, glass, metal cans and plastics.

Lists of firms were obtained from standard industrial directories. In Germany these included *Seibt* (1997 and 1998) and *Europäisches Handelsinstitut*. In the Republic of Ireland/UK they included *Kompass*, *Northern Ireland Trade Directory*, *Kelly's Business Directory* and *The Cream of Irish Industry*. Interviews with relevant trade associations and regulators and also material obtained at industrial exhibitions provided background information on firms. Firms were selected randomly within the sampling frame, including market leaders, and the selection was a representation of that industry in the country. Sample firms were matched by size and product type with firms in the other sample countries.

Sample companies were first contacted by telephone, then detailed information was faxed or posted to them. In general, firms in all countries were willing to co-operate. Response rates were in the Republic of Ireland/UK higher than in Germany (Table 6.2-1).

Table 6.2-1 Sectoral and country response rates (%)

Sector	Republic of Ireland/UK	Germany
Retailing	83	56
Dairy/drinks	89	53
Packaging	79	45

The micro data on which this study are based is derived from the completion of detailed questionnaires during face to face interviews with senior managers in 1997–8. These interviews lasted up to half a day and were followed by a visit to the shop floor where technical points that arose during the interview were illustrated. All data refer to the plant in dairy/drinks and packaging industries (though where services were provided from another site or office, e.g. R&D, such relevant data were included). Head offices were occasionally visited to improve or add to the plant level data. In the retail sector the primary source of information was the head office. In several cases further data were made available by fax and telephone following the interview.

6.2.2 Sample size

Tables 6.2-2 (a) and (b) show the number of companies visited in each country, region and industry. Fifty or more plants were sampled in each country.

Germany

While a total of 57 interviews were made in Germany, sufficiently detailed responses for later analyses were available only from 54 firms and plants. Interviews across the three sectors (retailing, processing and packaging) predominated in West Germany. In East Germany face to face interviews were conducted in Thuringia, Saxony, Saxony-Anhalt and Brandenburg. Since no indigenous retail chain is left in East Germany, only West German headquarters of the large supermarket chains were visited. Two smaller independent East German retailers did not wish to participate in the study. Concerning producers of packaging machinery, the main suppliers are likewise situated in West Germany. This is also true for producers of plastic packaging for the food industry. Interviews with firms from all sectors in West Germany took place in Baden-Württemberg, Bavaria, Berlin, Hamburg, Lower Saxony, North-Rhine Westphalia and Rhineland Palatinate.

Table 6.2-2 (a) Number and geographical distribution of plants sampled in Germany

No. of plants in each sector	West Germany	East Germany	Total
Total	41	13	54
Dairy	6	4	10
Soft drinks	1	4	5
Retail	10	–	10
Total packaging	24	5	29
Plastics	9	–	9
Metals	3	1	4
Glass	2	1	3
Composites	1	–	1
Machinery	4	–	4
Paper & board	5	3	8

Republic of Ireland/UK

Fifteen firms were sampled in Northern Ireland and 20 in the Republic of Ireland. The majority of processing firms was located in the two Irish economies while packaging suppliers and retail customers were located in both parts of Ireland and Great Britain. Hence the 22 firms sampled in Great Britain represented, in the main, the major UK supermarket retail customers and packaging suppliers used by those processing firms.

In Northern Ireland seven visits were made to plants located in County Antrim, four in County Armagh, one in County Down and three in County Tyrone. In the Republic of Ireland ten firms were located in the Greater Dublin area and the remainder in the south-west of Ireland in counties Cork, Limerick and Kerry. Visits took place to firms located in all the UK Standard Regions with the exception of the North East and the South, one company interviewed was located in South Wales. Most visits (11) were to firms in the Midlands and Yorkshire and the South East (4).

Table 6.2-2 (b) Number and geographical distribution of plants sampled in Northern Ireland, the Republic of Ireland and Great Britain

No. of plants in each sector	Northern Ireland	Republic of Ireland	Great Britain	Total
Total	15	20	22	57
Dairy	3	6	1	10
Soft drinks	3	1	2	6
Retail	4	3	4	11
Total packaging	5	10	15	30
Plastics	4	1	3	8
Metals	–	–	5	5
Glass	–	1	2	3
Composites	–	–	1	1
Machinery	–	–	3	3
Paper & board	1	8	1	10

6.2.3 Size distribution of sample plants and firms

Tables 6.2-3 (a) and (b) show the size distribution of firms and plants sampled by country.

In the case of Germany 44 plants, 29 in packaging and 15 in dairy/drinks were visited. The bottom half of Table 6.2-3 (a) shows that this represents 15 and 27, respectively, different dairy/drinks and packaging companies. In two cases in the packaging sector (metal cans and plastics), two interviews each were made with plants owned by the same firm. Only head offices of German retailers were visited (unlike the case of the Republic of Ireland/UK where shops were visited as well).

On average the spread of sizes of packaging plants visited in Germany was slightly wider (with more large plants represented) than was the case in the Republic of Ireland/UK. Average dairy/drinks and packaging firm size was slightly larger in Republic of Ireland/UK. The number of large retail firms visited was similar in all sample countries. A smaller subsample of firms and plants was also matched by size. These are considered more fully at the end of this chapter.

Table 6.2-3 (a) Size distribution of sample plants and firms, Germany

No. of plants in each sector	1–99	100–499	500–999	1 000 and over	Total
Total plants	14	19	5	6	44
Dairy and soft drinks	4	7	1	3	15
Retail	–	–	–	–	–
Packaging	10	12	4	3	29
Total firms	14	21	5	12	52
Dairy and soft drinks	4	7	1	3	15
Retail	–	2	–	8	10
Packaging	10	12	4	1	27

In the Republic of Ireland/UK sample, while a total of 57 visits were made, these were to 53 distinct companies. In five cases (three in packaging and two in retailing) interviews also took place with subsidiary companies and in the case of retailing, branches of chains with headquarters in Great Britain but located in Northern Ireland.

Table 6.2-3 (b) Size distribution of sample plants and firms, Republic of Ireland/UK

No. of plants in each sector	1–99	100–499	500–999	1 000 and over	Total
Total plants	16	31	3	–	50
Dairy and soft drinks	6	8	2	–	16
Retail	3	1	–	–	4
Packaging	7	22	1	–	30
Total firms	4	14	6	29	53
Dairy and soft drinks	2	3	2	9	16
Retail	–	–	2	8	10
Packaging	2	11	2	12	27

6.2.4 Economic characteristics of sample firms

Ownership of firms

Tables 6.2-4 (a) and (b) show a breakdown of sample firms by ownership type. The Republic of Ireland/UK sample was more dominated by PLCs and particularly multinational enterprises (MNEs). Firms forming part of a larger group were also important in Germany.

Table 6.2-4 (a) Sample firms by ownership type, Germany

No. of plants	Packaging	Dairy/drinks	Retailing	Total
PLCs	4	–	5	9
including MNEs	4	–	–	4
Ltd	25	12	4	41
including those in groups	16	4	–	20
Co-operative	–	3	1	4

Table 6.2-4 (b) Sample firms by ownership type, Republic of Ireland/UK

No. of plants	Packaging	Dairy/drinks	Retailing	Total
PLCs	16	10	7	33
including MNEs	9	5	–	14
Ltd	13	4	2	19
Co-operative	1	2	–	3

In the Irish economies 62 per cent of visits were to PLCs of which 44 per cent were members of a larger group; 44 per cent of firms were limited companies and three Irish firms were co-operatively owned. In Germany the majority (76 per cent) of companies were limited and half were also members of a larger group.

Products sampled

Table 6.2-5 (a) shows the number of firms manufacturing glass, metal, paper and board and plastic packaging and packaging machinery that were sampled in each country. These products were important for the packaging of milk and other dairy products and for mineral water and soft drinks (including especially plastic pots and containers, glass bottles and glass jars, metal cans, cardboard products and so on). Packaging machinery firms are included because they supply the technology, machinery and packaging lines for bottling, canning, crate washing, PET bottle pre-forms, cardboard packaging and so on. The total of 59 packaging firms sampled produced 134 products. In only five cases were products sampled in one country and not in another.

Table 6.2-5 (b) gives the same breakdown for dairy and drink products. The 31 food firms sampled produced 87 products. Liquid and powdered milk (including milk beverages), cheese, butter, yoghurt, cream and ice cream were the main dairy products produced. Table 6.2-5 (b) also shows that, in addition to soft drinks, the production of mineral water was important, especially in Germany (soft drinks based on mineral water were also produced). Only three products sampled are represented in one country and not the others. Alcohol, concentrate and animal feed is neither produced in Germany nor the UK, but is an important product in the Republic of Ireland.

Table 6.2-5 (a) Packaging: number of products sampled, by country

Products	Northern Ireland	Republic of Ireland	Great Britain	West Germany	East Germany	Total
Total bottles:	1	1	10	3	2	17
Glass	–	1	4	3	2	10
PET	1	–	2	–	–	3
HDPE	–	–	3	–	–	3
PVC	–	–	1	–	–	1
Total metal cans:	2	–	5	3	1	11
Tinplate	–	–	3	1	–	4
Aluminium	–	–	2	1	1	4
Total aluminium:	1	–	–	1	–	2
Lids	–	–	–	1	–	1
Bags	1	–	–	–	–	1
Total paper&board:	1	7	3	8	6	25
Sacks/bags	–	1	–	–	1	2
Corrugated board	1	1	2	2	–	6
Cardboard boxes, folding cartons	–	4	–	4	3	11
Cartons	–	–	1	1	–	2
Cutsize board	–	1	–	3	–	4
Total plastic-various	5	1	1	12	–	19
Plastic packaging	–	–	3	4	1	8

Table 6.2-5 (b) Dairy and drinks industry: number of products sampled, by country

Products	Northern Ireland	Republic of Ireland	Great Britain	West Germany	East Germany	Total
Liquid milk	1	3	1	1	1	7
Skimmed milk	–	–	–	–	2	2
Cheese	–	5	–	2	2	9
Powdered milk	2	4	–	3	3	12
Butter	1	3	–	3	2	9
Yoghurt	–	1	1	3	2	7
Cream	2	–	–	2	2	6
Low fat spread	–	1	–	–	–	1
Whey powder	–	1	–	1	1	3
Ice cream	1	–	–	1	–	2
UHT	1	–	–	1	–	2
Casein	–	1	–	1	–	2
Fruit Juice	–	1	1	1	–	3
Alcohol	–	1	–	–	–	1
Concentrate	–	1	–	–	–	1
Others	1	–	–	1	1	3
Animal feed	–	1	–	–	–	1
Soft drinks	3	–	2	1	–	6
Mineral water	2	–	–	6	2	10

Skills and qualifications

Table 6.2-6 shows the percentage of the workforce with skills. These include all formal qualifications (degrees, technical qualifications including apprenticeships). The data show that in all industries German firms engage more skilled persons than their Republic of Ireland/UK counterparts. In the Republic of Ireland/UK there were significantly more qualified persons employed in the packaging industries (mainly in paper, glass and metal manufacturing) than in the dairy/drinks industries. The regional breakdown within countries shows little variation.

Table 6.2-6 Average percentage of firm employment with formal qualifications

Sector	Great Britain	Republic of Ireland	Northern Ireland	All	Median
Dairy	– (1)	19.0 (5)	8.7 (2)	16.1 (7)	8.0
Drinks	19.0 (2)	5.0 (1)	5.1 (3)	9.7 (6)	5.5
Plastic	45.0 (1)	– (1)	6.8 (2)	19.5 (3)	7.0
Paper	32.0 (1)	28.9 (4)	30.0 (1)	29.6 (6)	27.5
Glass	17.5 (2)	– (1)	–	17.5 (2)	17.5
Metal	51.5 (2)	–	–	51.5 (2)	51.5
Machinery	–	–	–	–	–
All packaging	35.8	28.9	14.5	28.8	20.0
		East Germany	**West Germany**	**All**	**Median**
Dairy		64.3 (3)	55.7 (7)	58.3 (10)	55
Drinks		95.0 (1)	54.0 (4)	62.2 (5)	65
Plastic		–	69.5 (9)	69.5 (9)	61
Paper		53.6 (3)	40.0 (6)	45.2 (9)	37
Glass		100.0 (1)	48.0 (2)	65.3 (3)	39
Metal		100.0 (1)	100.0 (3)	100.0 (4)	100
Machinery		–	83.5 (4)	83.5 (4)	87
All packaging		72.2	66.7	68.0	50
Note: Number of observations in brackets.					

Machinery

Table 6.2-7 shows the percentage of machinery less than 5 years old in the various industries. The data show that in the Republic of Ireland/UK the age of dairy/drinks machinery at the median plant is older than in Germany. In the packaging industries machinery age is similar in the Republic of Ireland/UK and Germany. Comparison of the mean age of plant indicates that German equipment is in all industries but the metal packaging sector younger than that in the Republic of Ireland/UK.

Table 6.2-7 Average percentage of machinery less than 5 years old

Sector	Great Britain	Republic of Ireland	Northern Ireland	All	Median
Dairy	– (1)	23.3 (3)	15.0 (3)	19.2 (6)	20.0
Drinks	37.5 (2)	50.0 (1)	59.0 (2)	48.6 (5)	50.0
Retail	–	–	–	–	–
Plastic	41.7 (3)	– (1)	71.0 (4)	51.1 (8)	77.5
Paper	20.0 (1)	34.3 (7)	95.0 (1)	39.4 (9)	30.0
Glass	41.5 (2)	– (1)	–	41.5 (2)	41.5
Metal	25.0 (1)	–	–	25.0 (1)	75.0
Machinery	– (3)	–	–	–	–
All packaging	36.2	34.3	75.8	43.6	50.0
		East Germany	**West Germany**	**All**	**Median**
Dairy		51.7 (3)	54.3 (7)	48.3(6)	50
Drinks		– (3)	72.5 (7)	49.5 (10)	70
Retail		–	73.3 (3)	73.3(3)	–
Plastic		–	69.1 (9)	69.1(9)	75
Paper		75.0 (3)	25.8 (6)	42.2(9)	25
Glass		100.0 (1)	25.0 (2)	50.0(3)	50
Metal		33.0 (1)	45.3 (3)	42.3(3)	33
Machinery		–	48.8 (4)	48.8(4)	50
All packaging		71.6	48.2	52.6	50
Note: Number of observations in brackets.					

R&D

Table 6.2-8 shows the percentage of the labour force engaged in R&D, broken down by industry and country. In the dairy industry, firms in the Irish economies have the highest R&D; in the drinks industry German firms record the highest level. The numbers of persons engaged in R&D in packaging, by country, varies between industry, but was highest in the West German mechanical engineering sample (design of packaging machinery).

Table 6.2-8 Percentage of the labour force engaged in R&D

Sector	Great Britain	Republic of Ireland	Northern Ireland	All	Median
Dairy	– (1)	5.5 (5)	1.2 (1)	4.6 (6)	2.4
Drinks	1.4 (2)	– (1)	1.0 (1)	1.3 (3)	1.1
Plastic	8.0 (1)	– (1)	1.4 (4)	2.7 (5)	1.4
Paper	– (2)	2.0 (3)	0.2 (1)	1.5 (4)	1.0
Glass	3.0 (1)	– (1)	–	3.0 (1)	3.0
Metal	1.3 (1)	–	–	1.3 (1)	1.3
Machinery	– (3)	–	–	–	–
All packaging	4.11	2.00	1.16	2.17	1.35

Table 6.2-8 (continued)

Sector	East Germany	West Germany	All	Median
Dairy	1.2 (2)	1.6 (5)	1.5(7)	0.9
Drinks	0.6 (1)	2.7 (4)	2.3(5)	0.6
Plastic	–	2.4 (9)	2.4(9)	2.0
Paper	– (3)	2.2 (4)	2.2(4)	0.0
Glass	1.4 (1)	0.6 (2)	0.9(3)	0.7
Metal	5.4 (1)	1.9 (3)	2.8(4)	2.9
Machinery	–	8.1 (4)	8.1(4)	7.5
All packaging	3.4	3.2	3.2	1.0
Note: Number of observations in brackets.				

Exports

Table 6.2-9 shows the number of firms exporting and exports as a percentage of sales. Dairy and drinks exports from the Republic of Ireland are much higher than from any other region shown. Exports from the packaging industries as a whole are on average greater from the UK than from Germany though there is a wide variation by sector and region, as shown for plastics where the Republic of Ireland is a leading exporter. Four of the ten German retailers export, i.e. have shop chains abroad, compared with none of their Republic of Ireland/UK counterparts.

Table 6.2-9 Numbers of firms exporting in each region and industry and exports as average percentage of sales

Industry	EG	WG	UK/NI	ROI
Dairy (no.)	2/4	6/6	2/4	6/6
Exports/Sales (%)	4.75	25.2	28.25	76.25
Drinks (no.)	0/1	–	3/5	1/1
Exports/Sales (%)	0	11.3	5.2	95
Retail (no.)	–	4/10	0	0
Exports/Sales (%)	–	5.58	0	0
Packaging (no.)	2/4	24/25	16/19	4/10
Exports/Sales (%)	2.8	27.42	31.28	18.5
Paper and board (no.)	0/3	3/3	2/3	2/8
Exports/Sales (%)	0	31.67	5.2	17.75
Plastics (no.)	–	9/9	6/7	1/1
Exports/Sales (%)	–	17.3	28.93	60

Growth

Table 6.2-10 (a) and (b) shows employment change at sample plants and firms during the period 1992–6. The Table shows a wide diversity of experience between regions, countries and industries. Employment growth in the median plant in Germany was strongest in drinks and metal manufacturing. This was influenced by strong growth in a Greenfield mineral water plant in Germany and by an indigenous, but modernised and flourishing metal cans factory in East Germany. Employment decrease was strongest in the German glass industry. In the Republic of Ireland/UK employment growth was strongest in the drinks industry and plastics packaging. There was a slight decrease in metal manufacturing.

Table 6.2-10 (a) Average employment growth, 1992–7 in sample plants located in Great Britain, the Republic of Ireland and Northern Ireland

Sector	Great Britain	Republic of Ireland	Northern Ireland	All	Median
Dairy	– (1)	73.3 (6)	–8.7 (3)	46.0 (9)	8.0
Drinks	19(1)	10.0 (1)	19.0 (3)	17.2 (5)	19.0
Retail	6(1)	334.3 (3)	– (4)	252.3 (4)	–
Plastic	10(1)	13.0 (1)	18.3 (4)	16.0 (6)	18.0
Paper	– (2)	11.0 (7)	0 (1)	9.6 (8)	0
Glass	66(1)	– (1)	–	66.0 (1)	66.0
Metal	–1(3)	–	–	–1.0 (3)	–1.5
Machinery	– (3)	–	–	–	–
All packaging	13.4	11.25	18.4	12.8	10.0

Table 6.2-10 (b) Average employment growth, 1992–7 in sample plants located in Germany

Sector	East Germany	West Germany	All	Median
Dairy	–4(2)	1.7 (7)	0.4(9)	0
Drinks	125(1)	13.3 (4)	35.6(5)	30.0
Retail	–	11.9 (7)	11.9(7)	2.1
Plastic	–	41.1 (9)	41.1(9)	0
Paper	43(3)	40.8 (6)	41.5(9)	10.0
Glass	–51(1)	–21.5 (2)	–31.3(3)	–21.0
Metal	47(1)	21.2 (3)	29.8(3)	44.35
Machinery	–	8.0 (4)	8.0(4)	0
All packaging	25	27.8	27.5	0

Note: Number of observations in brackets.

6.2.5 Solid waste and environmental awareness initiatives undertaken by sample firms

Interviews focused mainly on initiatives to reduce solid waste, as these were most directly related to packaging, and in particular the hierarchy of treatment of packaging incorporated in the EU Packaging Directive. This section describes the sample according to the type and number of solid waste initiatives undertaken in each country. It also includes indicators of environmental awareness of management in plants and firms.

Solid waste initiatives

When it comes to the production of packaging and its disposal the EU packaging hierarchy stresses the superiority of packaging reduction, reuse, recycling and recovery above that of landfill. The research has classified the solid waste initiatives found in the sample plants under broadly the same headings and differentiates a bit more (see Box 6.2-1 for a classification).

Different from the broad headings used in the Packaging Directive packaging reduction has been further broken down into light-weighting where the packaging material is changed and into initiatives like pack sizing and innovative design where the packaging material itself remains unchanged; packaging reuse involves a number of trips being made by the same package; and recycling covers its post-life transformation into a new primary or secondary market product. Packaging recovery is often used to mean the conversion of used packaging into some secondary product or energy conversion as opposed to recycling, which refers to the production of a 'new' substitute product similar to the original. We have strayed from this convention and any recycling, whether producing a substitute or not, is classified under the broad heading of recycling. Recovery on the other hand has been used to describe the active collection of waste to recycle, e.g. collection of scrap material belonging to others or the placing of bottle/can banks on premises.

Box 6.2-1 Classification of solid waste initiatives found in the sample plants

1. Reuse materials
   a) Pallets
   b) Bottles
   c) Packaging
   d) Trays
   e) Bins
   f) Fixtures
   g) Displays
   h) Steel containers
   i) Paper sacks/cardboard boxes
   j) Transit packaging
   k) Plastic bags/containers
   l) Food scrap
2. Recycle
   a) Foil
   b) Plastics
   c) Process scrap (PET etc.)
   d) Paper waste or board
   e) Outer/Transit packaging
   f) Plant for recycling
   g) Inks used in production
   h) Aluminium
   i) Plastic film wrap
   j) All materials
   k) Glass
   l) Stell cans
   m) Wood
   n) Cork
   o) Batteries
3. Specially designed packaging to allow for easy recycling
4. Packaging recovery
   a) General recovery/segregation
   b) Bottles – banks
   c) Cans – collections
   d) Recovery of plastics
   e) Associated recovery company
5. Material switching
6. Light-weighting
   a) Plastics
   b) Down-gauging: $\Downarrow$ can container thickness/diameter
   c) Down-speccing paper/board thickness – $\Downarrow$ Quality
   d) Transit packaging
   e) Glass
   f) Sale of unpackaged produce
   g) Development of new material
   h) Variable thickness boxes (loadbearing framework)
   i) High density polyethylene bags
   j) Compound (increase)
7. Packaging reduction without changes to material used
   a) Reducing label sizes/packaging volumes
   b) New designs to remove need for outer packaging
   c) Larger packs
   d) Eliminate excess packaging
8. Waste reduction
   a) Compactor
   b) Bailer
9. Waste segregation
10. Waste targeting
   a) Reduce food waste
   b) Waste targets
   c) Consumer orientated waste avoidance programme
11. Recycling of process waste

In addition to these areas we have also included some measures that may reduce landfill or aid alternative methods of disposal. These include the segregation of waste (thus easing recycling), the setting of waste targets and the reduction of waste volume through the use of compactors and bailers.

Table 6.2-11 (a) and (b) summarise the percentage of firms in each country undertaking different solid waste initiatives with detail shown for each sector. Overall in Germany 480 initiatives or 8.9 per firm were identified and in the Republic of Ireland/UK sample 256 solid waste initiatives or 4.7 per firm with the most important initiatives being packaging reuse and recycling in each sample country.

In comparison with the Republic of Ireland/UK, a greater percentage of German firms are engaged in waste initiatives for seven of the ten types of initiative shown. The exceptions are waste targeting and waste reduction (through the use of bailers and compactors) as well as packaging reduction where the engagement of firms in all sample countries was at 24 per cent on an equal level; moreover, the level of activity is broadly similar for light-weighting and waste reduction. Viewed on a sectoral basis more German firms undertake initiatives in the majority of subsections given. In retailing for eight of the ten solid waste initiatives more German firms are active; it is also the case for nine out of ten initiatives in the packaging industry, but only for five initiatives in the dairy/drinks sector.

The larger average number of initiatives undertaken by German firms arises from a greater percentage of firms engaged in each individual waste activity. It is noteworthy that in four areas, especially, there is more activity undertaken by German firms. This is in packaging design (particularly German retail and packaging companies are more active than their Irish/UK counterparts); material switching where Irish/UK packaging companies lag behind; packaging recovery (otherwise the high recycling targets given by the German regulation could not be reached) and waste segregation which both are particularly important across all sectors compared with the two Irish economies (see Box 6.2-2 for details of German initiatives in these areas). In the German packaging industry these activities are strongly driven by regulation, in dairy/drinks and retail the driver is costs as a consequence of more stringent regulation, but also market forces, i.e. some of the developments happen anyway. Table A.6-1 in the Appendix A.6 to this chapter gives further details on other important types of solid waste initiatives used in Germany.

On average, in all sample countries retail firms undertake the highest number of initiatives and are followed by dairy/drinks and finally packaging companies.

Table 6.2-11 (a) Solid waste initiatives: number of initiatives and percentage of companies by sector in Germany

Initiative description	All companies		Dairy/Drinks	
	Number	%	Number	%
Reuse materials	113	89	33	93
Recycle materials	134	96	27	93
Special design	11	20	2	13
Material switching	15	28	5	33
Packaging recovery	63	98	14	93
Lightweighting	37	59	6	40
Packaging reduction	33	24	6	33
Waste reduction	25	46	5	33
Waste segregation	44	81	12	80
Waste targeting	5	9	2	13
TOTAL	480		112	
Initiatives per firm	8.9		7.5	
	Retail		**Packaging**	
	Number	%	Number	%
Reuse materials	31	100	49	83
Recycle materials	34	100	73	97
Special design	6	60	3	10
Material switching	3	30	7	24
Packaging recovery	20	100	29	100
Lightweighting	3	30	28	79
Packaging reduction	21	60	6	7
Waste reduction	7	70	13	45
Waste segregation	8	80	24	83
Waste targeting	3	30	0	0
TOTAL	136		232	
Initiatives per firm	13.6		8.0	

Table 6.2-11 (b) Solid waste initiatives: number of initiatives and percentage of companies by sector in the Republic of Ireland/UK

Initiative description	All companies		Dairy/Drinks	
	Number	%	Number	%
Reuse materials	57	75	23	94
Recycle materials	59	71	17	69
Special design	3	5	3	19
Material switching	7	13	4	25
Packaging recovery	19	20	1	6
Lightweighting	34	58	14	81
Packaging reduction	17	24	10	38
Waste reduction	31	47	10	50
Waste segregation	11	20	6	38
Waste targeting	18	33	8	50
TOTAL	256		96	
Initiatives per firm	4.7		6.0	
	Retail		**Packaging**	
	Number	%	Number	%
Reuse materials	15	89	19	60
Recycle materials	10	67	32	73
Special design	0	0	0	0
Material switching	2	22	1	3
Packaging recovery	15	78	3	10
Lightweighting	0	0	20	63
Packaging reduction	4	56	3	7
Waste reduction	8	89	13	33
Waste segregation	2	22	3	10
Waste targeting	5	56	5	17
TOTAL	61		99	
Initiatives per firm	6.8		3.3	

Box 6.2-2 Details on solid waste initiatives undertaken by German firms

Reuse initiatives in Germany are partly distinguished from those applied in Northern Ireland and also the Republic of Ireland due to differences in packaging legislation. Within the sample this difference was visible with respect to reuse of bottles in the German dairy and drinks industry. The German Packaging Ordinance prescribes that a quota of 72 per cent of all soft drinks, mineral water and beer must be bottled in reusable containers. The quota for milk filled in reusable bottles is 17 per cent. Two of the three German dairies which bottle fresh milk use reusable glass bottles, and four of the five mineral and soft drinks producers fill the beverages in reusable glass bottles; one company also introduced reusable bottles made from PET for the bottling of soft drinks. All retailers offered reusable bottles; some had separate zones in their shops where only beverages are sold and the deposit is refunded. In terms of material switch a relatively low share of all companies, only 28 per cent, were active. In **dairy/drinks** the five following substitutions took place (driver is shown in brackets):

from gabletop to glass bottle for UHT milk (marketing reasons, but failed);
from plastic wrapping for ice-cream to paper wrapping (costs);
from aluminium butter foil to paper foil (law/costs);
from plastic wrapping for grated parmesan to reusable jars (market);
from reusable glass bottles to reusable PET bottles for soft drinks (market).

In the retail industry three explicit cases of material switch were discovered:

from a variety of plastic materials for service packaging down to five different, easily recyclable materials (costs/law);
from several materials used for a single type of packaging to monomaterials (costs);
support away from gabletops for milk to plastic pouches (costs/consumers).

Seven examples were given in the packaging industry:

from PVC to PP or PET (was mentioned three times in the plastics industry, driver was law);
from styropor to paper flakes as filling material (in a plastics company, for cost reasons);
from grey board to corrugated board (East German paper processing plant, had quality reasons);
from aluminium coated paper trays for the fish industry to printed paper trays (paper processing plant, customer driven);
from flat pouches to a different type of pouch thus saving packing material (new machinery to do so has been developed in the packaging machinery industry because of legislation).

The German companies were also very active in waste segregation (81 per cent) into different fractions (mostly paper/board, plastics and glass) and waste reduction through compactors (46 per cent). With respect to products, 79 per cent of the firms visited in the packaging industry undertook light-weighting efforts. Retailers light-weighted the sales packaging directly supplied to the customers in the shops, for example plastic pouches, paper wrappings, etc. used in fresh food departments. Excess packaging has been removed as far as possible as a first consequence of the Packaging Ordinance, with retailers strongly pushing removal of excess transit packaging. Other efforts like special design for less packaging or reduction of label sizes have taken place only to a lesser extent.

Environmental awareness initiatives

Box 6.2-3 shows the classification of environmental awareness variables found in the sample. Environmental policies, audits, designated environmental managers and meetings were taken as indicators of environmental awareness/efficiency. Evidence also included companies with an environmental management scheme (either ISO 14000 or EMAS). Some companies provided environmental information for their employees and others provided information to the public via an environmental report. Staff training was a further consideration and may involve employees being sent on environmental courses.

Box 6.2-3 Classification of environmental awareness variables found in sample plants

1. General environmental measures
   - Audits
   - Environmental manager
   - Team/committee
   - Environmental policy
   - Environmental information service/staff training
   - Environmental sponsorship
   - Eco- balances
2. Packaging guidelines
3. Improved working conditions

Table 6.2-12 shows the percentage of companies in which there was evidence of environmental awareness and management. The Table shows clearly that German firms undertook more than twice as many environmental awareness initiatives as their counterparts in the Republic of Ireland/UK (see Box 6.2-4 for details on environmental awareness measures undertaken in Germany and Table A.6-2 in the Appendix A.6 to this chapter for a description of all awareness initiatives in Germany). Again German retailers were the most active companies.

Table 6.2-12 Environmental awareness initiatives: number of initiatives and percentage of companies by sector

Country/Sector	Number	%	Average per firm
Republic of Ireland/UK			
All companies	49	56	0.86
Dairy/Drinks	16	63	1.00
Retail	11	78	1.22
Packaging	12	43	0.40
Germany			
All companies	111	81	2.16
Dairy/Drinks	34	87	2.26
Retail	34	100	3.90
Packaging	43	72	1.51

Box 6.2-4 Details on German environmental awareness initiatives

More than 90 per cent of all 54 companies interviewed in Germany had implemented environmental awareness initiatives such as environmental audits, policy, training etc., but also measures to standardise packaging and improve working conditions. All retailers undertook environmental awareness initiatives, followed by 81 per cent of the companies in the dairy/drinks industry and 72 per cent in the packaging industry.
Concerning environmental culture in the ten East German firms visited (five in dairy/drinks and another five in packaging), the most striking element was that at the time of the visits none of the firms had implemented an eco-audit; only one firm planned to finish an audit in 1998, another one thought about starting one in 1998. This lack of awareness is also reflected in the on average lower number of initiatives undertaken in East German firms. Some of the firms mentioned that they still lag behind in terms of quality management and that the introduction of a quality management system has priority over an environmental management system.

6.2.6 Matched plants and firms in the supply chain sample

Table 6.2-13 shows the number of plants which could be matched with counterparts in the other country. Some firms and plants could be matched twice. For example, eight dairy firms in both Irish economies were matched with nine in Germany (i.e. one was matched with two firms).

Analysis of data on the basis of matched comparisons is undertaken to show differences between the samples on key variables in the analysis. Importantly, it is also used to compare the impact on productivity performance of differences in waste costs and the use of environmental initiatives between matched pairs of firms in the samples.

Table 6.2-13 Number of firms and plants matched by sector and country

Sector	ROI/UK : Germany
Dairy	8: 9
Drinks	5: 5
Retail	7: 8
Plastics	4 : 4
Paper	7 : 6
Glass	2 : 2
Metal	2 : 2
Machinery	3 : 2
Total	38 : 36

6.3 Cement sample

6.3.1 Target sample size, directories, response rates and data collection

The very detailed nature of the work inevitably implied that sample sizes were small and a number of 10 cement plants respectively in Germany, Spain and the UK was targeted. Since the cement industry is largely owned by international companies, in addition a series of interviews at headquarters was foreseen. These interviews aimed at investigating the various environmental strategies and their impacts on competitiveness of the companies.

In order to gain access to the plants, mainly Cembureau's latest World Cement Directory of 1996 was used. Moreover, cement associations in Germany, Spain and the UK provided helpful advice and information on which plants to include in the study. Response rates were 92 per cent in Germany, only 55 per cent in Spain and 90 per cent in the UK.

The in-depth interviews at plant level were carried out between May 1999 and April 2000 using a structured questionnaire and lasted up to half a day. Normally interviewees were the plant managers assisted by an environmental manager and an accountant. After the interviews the plants were visited. Interviews at headquarters took place in November and December 2000 in order to test the initial analysis of the data with the top managers of international companies responsible for environmental strategies.

6.3.2 Sample size

At present, about 78 per cent of European cement production is undertaken in dry process kilns. Thus, the study focused on this most commonly used process route. In total, 18 interviews could be undertaken in dry process plants in the three main sample countries Germany, Spain and the UK (see Table 6.3-1 below). Most interviews have taken place in Germany mainly because access to plants was very difficult in Spain and three interviews in UK semi-dry plants were not included in the final analysis. In addition to the main sample countries two interviews took place in Austria, and one each in Sweden and Norway. These visits have been made to get more information about adjustment to the prevalent stringent legislation in these countries.

From the eight dry process plants visited in Germany six were located in West Germany and two in East Germany. The latter plants have been visited in order to reflect the special situation in East Germany where after the German reunification cement plants have been rapidly modernised with high capacity dry kilns. From the five Spanish plants three were located in Andalusia, one close to Madrid and one in Catalunia. This reflects a mixture of environmental stringency on a national level in Spain. In the UK visits included one visit to an independent company and four visits to plants of the three main players. The plants were located throughout the UK.

Table 6.3-1 Number of sample plants

Germany	Spain	UK
11	5	5*
*One plant had both old wet kilns and a new dry kiln which eventually will replace the capacity of the wet kilns.		

6.3.3 Size distribution of sample plants and representativeness of sample

As stated above the sample plants should not only be of comparable size bands across the sample countries, but their size distribution should also be representative of the industry in each country. To control for this factor the size distribution of the sample was compared with the national size distribution. For Germany and Spain the national distribution was published, for the UK the national distribution was calculated from data in the World Cement Directory (Cembureau, 1996).

Table 6.3-2 shows two important aspects of the sample. Firstly, it reveals a fairly close correspondence between the size distribution of the sample plants and the national industry average for each country. Small plants in Germany, the country with the largest share of small plants producing up to 500,000 tonnes of cement per year, are very well represented with a share of 62 per cent in the sample vis-à-vis 60 per cent on the national level. Spain has the lowest share of small plants with only 12 per cent in the total industry, which is approximately matched by the sample distribution with 20 per cent. In the large size class with above 1 million tonnes Spanish plants have the largest share (46 per cent) which is also reflected closely in the Spanish sample (40 per cent). The UK sample of the small and medium sized plants represents the national distribution fairly well. But the large size class is slightly over-represented. However, limited access to plants in all sample countries has made a closer matching of sample with the national size distribution impossible.

Secondly, the Table also demonstrates that a matching of plants among the same size groups of all sample countries is possible. Due to the varying number of plants between the size groups some plants were matched more than one time with a counterpart of the same size.

Table 6.3-2 Size distribution in the sample (figures in brackets show national level)

Annual output of:	Germany		Spain		UK	
	No.	%	No.	%	No.	%
0-500,000 tonnes (small)	5 (40)	62 (60)	1 (5)	20 (12)	1 (6)	20 (33)
501-1,000,000 tonnes (medium)	1 (17)	12 (26)	2 (18)	40 (42)	1 (7)	20 (40)
> 1,000,000 tonnes (large)	2 (9)	24 (14)	2 (20)	40 (46)	3 (5)	60 (27)
Total	8 (66)	100	5 (43)	100	5 (18)	100

Source: Own data collection in 1998/99.
Germany: Bundesverband der Deutschen Zementindustrie, Zahlen und Daten, 1999/2000.
Spain: Cemento-Hormigon, 1997.
UK: Cembureau, World Cement Directory, 1996.

6.3.4 Economic characteristics of sample plants

Kiln capacity, plant capacity and capacity utilisation

Average kiln capacity is largest in Germany because the recently modernised East German plants have modern and large kilns. Spain occupies a medium position, kiln capacity is smallest in the UK (see Table 6.3-3 below). German plants also have the highest average plant capacity. However, their rate of capacity utilisation is with 70.5 per cent the lowest in the sample. This reflects the situation of excess capacity where several plants do not use all their kilns installed at the plant. In the UK and Spain average plant capacity and also the rate of capacity utilisation are equally high.

Table 6.3-3 Capacity of the sample plants*

Capacity	Germany	Spain	UK
Av. capacity of kilns in tonnes/day*	2 300	2 000	1 600
Av. capacity of plant in tonnes/day*	3 700	2 900	2 900
Av. capacity utilisation in % of running kilns	70.5	86.8	91.7
Av. production of plants in 1998 in '000* tonnes	950	1 000	950
Sample size	8	5	5

*Figures are rounded to the next hundred.

Ownership

Two of the eight German plants were independent, the other six were owned by internationally operating companies. All Spanish plants belonged to multinationals and the majority of UK plants as well (see Table 6.3-4). This ownership structure also reflects the national level where from all sample countries only the German cement industry is known to consist of a considerable number of family owned plants.

Table 6.3-4 Ownership structure of sample plants

Type of ownership	Germany	Spain	UK
Independent companies	2	0	1
Multinational companies	6	5	4
Total	8	5	5

Kiln age, other technical characteristics and average investment

On average the year of kiln installation is earliest in Germany with 1976. For Spain and the UK it is 1980 (see Table 6.3-5). The years refer only to the date of installation and do not contain information about any modernisation which might have been done in the meantime. Cement manufacturers stressed that the age of construction of a kiln is not very indicative. With each repair and reconstruction of equipment an adaptation to modern technology takes place. This process starts with small things like better weigh feeders to control the fuel feed and becomes very obvious with the implementation of modern process controls and fuzzy logic to optimise processes which also triggers a positive environmental effect.

As a proxy for this modernisation activity average yearly investment measured over the last five years as percentage of sales was used. Investments into new kilns have not been included in this calculation as this would distort the data. For this reason a large share of the expenditure for East German kilns and mills has not been included. The average yearly investment during the last five years as a percentage of sales is highest in the UK sample showing the upgrading measures having taken place there recently. The investment share is a bit lower in the German sample. German cement plants stressed their continuous investment strategy for replacing and maintaining a good shape. The works contained in the Spanish sample have on average invested half the amount of the UK sample average. However, about half of the Spanish plants had already started a permanent strategy for continuous updating.

Table 6.3-5 Kiln age, other technical characteristics and average investment in the sample plants

Kiln age, other technical characteristics and average investment	Germany	Spain	UK
Sample size	8	5	5
Av. year of kiln installation	1976	1980	1980
Av. investment in % of sales	4.9*	3	6
Precalciner in % of plants	40	0	60
Expert system in % of plants	80	60	60

*two large and completely modernised plants in East Germany have not been included for calculation of the average yearly investment.

Further indicators of how modern a plant is are the use of precalciners and expert systems. Precalciners and the use of expert systems are considered to be part of the best available techniques for the production of clinker in new or updated cement plants (EIPPC Cement BREF, 2000). They were used at about half of the kilns in the UK (see Table 6.3-5). In Germany about forty per cent of plants had precalciners and in the UK sixty per cent, whereas in Spain none of the plants used them. Expert systems have been installed in many works in recent years to run the plants more efficiently. They were more frequently used in Germany than in the UK and Spain.

Skills and qualifications

Plants stressed that the position of the kiln controllers was among the most demanding and most important jobs in a cement factory. In Germany the Verein Deutscher Zementwerke e.V. offers special training courses for kiln controllers where also environmental issues play an important role. In UK plants a trend could be found to hire and train engineering graduates as kiln controllers. In Spain it was stressed that new and younger employees would be hired for more demanding tasks.

Concerning general operator skills in the kiln and cement area there was a wide difference between the countries which reflects the role of national formal qualification schemes more than anything else. In Germany 70 per cent of operators were formally trained as electricians and mechanics. In East German plants almost all operators had gone through extensive vocational training. Only one of the German plants had a low percentage of skilled workers. Most plants in the other countries show a very low percentage of formally trained workers. The advantage of formal skills is seen in a higher flexibility of the workforce to move between different jobs as well as in a better mastering and more intensive control of the processes. This is hypothesised to improve not only productivity, but also to affect energy consumption as well as emissions of gases and dust (for the results see analytical part in chapter nine).

Labour productivity

Labour productivity was calculated as the output of cement divided by the employment in the kiln and cement area including the maintenance personnel. Maintenance was included as in some of the plants the workers in the kiln area were also involved in maintenance and vice versa. Outsourced maintenance was included by adding numbers of outsourced personnel in full-time equivalents.

Table 6.3-6 Average labour productivity* (measured as annual cement tonnes per employee in cement, kiln and maintenance departments) and hours worked per week

Average labour productivity*	Germany	Spain	UK*
Productivity	8,550	9,700	10,000
Hours worked per week	38	40	40

*Remark: For labour productivity measure it was not possible to distinguish between different technologies of kilns within the same plant. Therefore some of the output and employment of the wet technologies are included.

Labour productivity was highest in UK and Spanish plants. Labour productivity in German plants is 15 per cent lower than in the British plants, although the two large high productivity East German plants are included. Given similar technical equipment in German cement plants as in the other EU-countries the low productivity figure is at a first glance surprising. However, for an interpretation of the productivity figures mainly two considerations are of importance:

— Working time in Germany has been 38 hours per week, whereas in the other countries it is 40 to 41 hours. The difference of two hours out of forty would increase productivity by 5 per cent.
— A major reason for lower labour productivity is the excess capacity in Germany. An increase to a capacity utilisation to a level as in Spain or the UK would increase productivity by 10 per cent. The reason is that kilns are run with much the same number of employees whether capacity is at 70 per cent or 90 per cent.

Raw material costs

Raw material costs differed widely even within a country because they depend on various site specific factors (e.g. the location of the quarry, consistency of raw material, availability of additives). Table 6.3-7 shows the raw material cost variations found in the sample. These variations happen in each of the countries and often the average offsets these deviations. Costs for lime vary between 2.7 Euro per tonne and 3.8 Euro per tonne. These price differences include different qualities and different transport costs. Variations in prices for gypsum are much wider with some plants using recycled gypsum, if it is available and in the right quality. Prices for slag and fly ash as additives depend very much on availability and location. In our sample prices were relatively high in Germany except for fly ash.

Table 6.3-7 Average raw material costs in Euro per tonne of cement in the dry process

Average raw material costs	Germany	Spain	UK
Lime costs in Euro/t	3.8	3.3	2.7
Gypsum in Euro/t	12.6	8	23.5
Slag in Euro/t	14.1	7.2	n.a.
Fly ash in Euro/t	7.5	13.8	n.a.

Energy costs

Cement manufacture is a very energy-intensive process in which energy and electricity costs account for around 65 per cent of variable costs or 30-40 per cent of production costs excluding capital costs (EIPPC cement BREF, March 2000). Therefore, there are strong economic reasons to utilise energy and electricity efficiently.

The optimisation of energy use in existing plants has been largely achieved.[28] The choice between coal, fuel oil, gas and petcoke depends partly on the additional substances needed for the clinker process, but mainly on prices. Each energy type needs storage and feeding systems so that a fast response to changes in fuel prices is not possible in most plants. In the sample fuel oil was mainly used for the starting of the kiln only. Coal or fuel oil was usually bought on the world market. Nevertheless, fuel prices differed widely even within one country, de-

[28] The specific energy consumption in the production of clinker of the Cembureau members was reduced by approximately 30 per cent since the 1970s. It has been 6000 kJ/kg clinker in 1960 and slightly less than 5000kJ/kg in 1970. A modern dry kiln requires a specific energy consumption of about 3000 kJ/kg today (Buttermann 1997). The reduction in energy consumption was partly achieved with investment into modern technology which has led to an increase in capacity. Thus, the German average daily capacity increased from 348t/d in 1958 to 1900 t/d in West Germany and 2500 t/d in East Germany in 1999 (Buttermann 1997, Zahlen und Daten 1998/99).

pending on quantity and transport costs. Spanish plants used mainly petcoke, whereas some German plants relied on brown coal which is extracted in Germany. One of the reasons for very little petcoke use in Germany (only one plant used it in small quantities) was its higher sulphur content. Prices for petcoke with a high sulphur content have dropped significantly between January 1997 when it was 27 Euros per tonne FOB and June 1998 when it was just 5 Euros per tonne (International Cement Review, Sept. 1999). This explains the large price difference between Spain and the UK where plants referred to different years (see Table 6.3-8).

Table 6.3-8 Average energy and electricity costs in Euro

Average energy and electricity costs	Germany	Spain	UK
Petcoke in Euro/tonnes	55	50	35
Coal in Euro/tonnes	49.5	n.a.	36
Electricity in Euro/kWh	0.05	0.04	0.05

Most of the electricity prices refer to night tariffs because cement milling is frequently conducted during the night to use cheaper night tariffs. Prices were quite similar on average (see Table 6.3-8 above). Depending on their market power some plants within each country were able to negotiate more favourable terms.

Use of alternative fuels

Seven of the eight German plants burnt some alternative fuels, mainly tires but also plastics, paper waste, heavy viscosity fuel oils (HVFO) and other materials. Their use accounted for 10 per cent to 30 per cent of all fuel consumption at the time of the interviews, but some of the plants had applied for an expansion. In the UK 40 per cent of the plants used alternative fuels and applications for increased use have been made as well. Only one Spanish plant experimented with a small proportion of alternative fuels (tires) and aimed at its increased use. Spanish plants were still hesitant to use alternative fuels. This was mainly the case because the collection of waste like tires is not yet organised. Therefore a sufficient and economic supply cannot be assured at the moment.

Prices of alternative fuels vary according to type, quantity of supply and transport costs. The general experience in the sample was that costs were very low or that plants even made revenues when using alternative fuels. However, considerable investments have to be undertaken for preparing and stocking an additional type of energy and for feeding it into the burning system. To get a permit for alternative fuel took up to 24 months in Germany and the UK. The application procedures, test runs and monitoring devices were said to be costly. If a plant wants to burn alternative fuels, it can expect more stringent emission limit values in the permit and additional measurements of heavy metals.

Table 6.3-9 Percentage of plants in each country using alternative fuels, average share of alternative fuels of total energy consumption and average energy consumption in kcal/kg clinker*

Alternative fuel and energy indicator	**Germany**	**Spain**	**UK**
% of plants using alternative fuels or HVFO	87	20	40
Altern. fuels as av. % of all energy consumption*	26	4	20
Energy consumption in kcal/kg clinker**	790	850	810
* in plants using alternative fuels ** adjusted to the next 10			

Production cost structure

Most companies were willing to give a percentage breakdown of their production costs and some also added the absolute production costs. The latter were with 37 Euro per tonne of cement about equal in Germany and the UK and with 27 Euro substantially lower in Spain. Since raw materials from the own quarry were usually calculated very differently (total costs, processing costs or even not at all), they were not taken into account separately. Transport and distribution costs are not included either. Table 6.3-10 shows that only small differences concerning the percentage breakdown were found. Percentages for capital (depreciation) depend on investments and are treated differently according to tax law. Labour costs contribute a higher share to costs in Spain and in Germany than in the UK. Energy and electricity cost shares are lowest in German plants. Maintenance, purchased materials and services make up a very similar proportion in all countries.

Table 6.3-10 Average per cent of production costs

Cost type	**Germany**	**Spain**	**UK**
Labour	25	27	21
Energy	15	18	18
Electricity	13	14	16
Capital	15	10	10
Maintenance/purchased material and services	23	22	29
Other	9	9	6
Production costs per tonne in Euro	37	27*	37.5
*Data was not given at plant level, but at headquarter level.			

Customers, exports and imports

The customer structure indicates quality (and environmental) demands and possible impacts on value added. A comparison could only be made between German and UK plants because Spanish plant managers claimed not to know the destinations of sales. Their sales department was usually at the headquarters and marketing data was not forwarded to the plant.

Table 6.3-11 Average percentage distribution of sales according to type of customer in the dry process sample

Product destination	Germany, 8 plants	UK, 5 plants
Outlets	10	23
Construction sites	10	2
Ready mix	52	49
Pre-fabrication	22	26

German plants achieved more than 50 per cent of their sales with ready mix companies and 22 per cent with pre-fabrication producers (see Table 6.3-11). The quality demands by these large customers are high, but packing costs are low so that bulk cement is sold cheaper than packed cement. A similar high proportion is sold to the ready mix and pre-fabrication customers by the UK plants. Their percentage of sales made with outlets (which is usually packed cement) is more than double as high as in Germany, but only very little is sold directly to construction sites. Nowhere, environmental demands of customers were mentioned.

Exports did not play a dominant role. Only one Spanish plant reported exports amounting to 4 per cent, a relatively small amount of its sales. Other Spanish plants did not know their exports. Half of the German plants exported, but only small amounts to what was described as "old" markets, namely Belgium and the Netherlands. Most of the UK exports were delivered to Ireland and Denmark.

Imports are important as they might lead to price distortions if they are delivered at a much cheaper price. Imports to the Spanish market came from Bulgarian, Greek, Rumanian, Turkish and Tunisian competitors which reflects the Mediterranean location. Competition for the German plants came from Polish, Rumanian, Czech, Thai, Lithuanian and Ukrainian imports with cement often transported by river or sea. Polish imports to Germany have been very high in 1994 and were strongly felt in East Germany. Since then imports have been reduced partly because Polish demand has been buoyant. Another explanation for the reduction is joint ownership of Polish and East German plants. In the UK competition was mentioned from Lithuania, Tunisia, Letland as well as from Turkey and Greece.

Planned investment and its purpose

Plants were also asked whether they had investment plans for the next five years and if so, what the purpose of the investment was. Average planned investment expressed as a percentage of sales (see Table 6.3-12) did not differ markedly. Still, it is interesting to note that it was with 4.9 per cent highest in Spain, followed by Germany with 4.4 per cent and was lowest in the UK with 4.1 per cent. It seems that Spanish plants which invested less in the past, now try to catch up. In Spain both process improvements and environmental investments were named frequently as investment reasons. Several Spanish plants also planned an increase in production. The German plants did not stress environmental requirements very much in their investment plans, only twice investment into a SNCR installation for the reduction of NO_x was mentioned. They were more concerned with investment in alternative fuels, updating of old equipment and process improvements. Plants in the UK mainly planned to invest in their quarry and in energy saving; one plant mentioned an investment need related to the adjustment to the UK integrated pollution control (IPC) legislation which brings about stricter clean air standards for the plant.

Table 6.3-12 Average planned investment for each year during the next five years in per cent of sales and its purpose

Level and purpose of investment	Germany	Spain	UK
Planned investment for each year during next 5 years in % of 1998 turnover	4.4	4.9	4.1
Main purpose of investment	Alternative fuels, Process control, New mills, SNCR	Increase in production, decrease in emissions, clinker and raw material storage, process improvement, cement mills, bag filters, transport	Quarry, adjustment to IPC legislation, energy saving
Sample size	8	5	5

New Products and Prices

The average plant develops and improves cement types and mixtures. Although the cement type and quality is normalised by national and international standards, the differences within one norm are still so significant that it improves customer loyalty. New types of cement relate to the hardening behaviour, alkali contents, resistance to sulphur or new mixtures as for example with blast furnace ash. Except for Germany, the plants in each country produced on average four to five products of which one has been new or improved during the last five years. German plants produced on average more than twice as many products, of which four were new (see Table 6.3-13). All German plants regarded the development of new types as very important for their competitive situation. In one German plant new products accounted for 25 per cent of sales. The larger number of products requires a higher number of cement mills and silos.

Table 6.3-13 Average number of products, improved or new products in last 5 years, prices in 1998

Country	Germany	Spain	UK
Number of products	12	5	4
New or improved products in last 5 years	4	1	1
Development seen as important	High	Low	Medium
% of sales of 5 largest customers	22	26	46
Average price in Euro per tonne of cement	63	61	74

To see how dependent plants were on their customers the percentage of sales to the five largest customers was asked. German plants followed by Spanish ones have the lowest concentration of large customers. This means that sales and marketing require more intensive efforts. UK plants have the highest concentration. They could be much more affected by the power of the customers and pressures on prices. Prices ex works differed nationally and regionally. The highest prices were charged in the UK which can partly be explained by the high demand and partly by the exchange rate to the Euro. Prices in Spain and Germany were similar. In Germany there was a distinct price difference between East and West Germany because of the Polish competition mainly affecting the markets in East Germany (about 60 Euro per tonne of cement in East Germany and 65-70 Euro in West Germany).

6.3.5 Environmental characteristics of the cement sample

Energy consumption

All plants have three to five-stage cyclone preheaters. While all sample plants also had multi stage preheating they did not achieve the heat balance value of 3,000 MJ/t (about 720 kcal/kg) as mentioned in the newest technical reference document for best available technologies in the cement industry (EIPPC Cement BREF).

The average energy consumption in kcal per tonne of clinker was, at 790 kcal/kg, lowest in German plants but the gap relative to the consumption in the UK was just 20 kcal/kg. Highest specific energy consumption has been 850 kcal/kg in Spain (see Table 6.3-14). A low specific energy consumption does not only fulfil the aim to reduce fuel consumption but it has the additional and very important environmental effect to reduce $S0_2$ and CO_2 emissions.

Table 6.3-14 Energy consumption in kcal per kg of cement in sample plants

Energy consumption	Germany	Spain	UK
Kcal per kg cement	790	850	810

Electricity consumption

Electricity consumption is largely dependent on the wetness or mixture of raw material (higher need for drying and/or homogenising), the number and type of electrical devices for environmental equipment, the type and modernisation level of the cement mills and on the fineness grade of grinding which specifies the type of cement. A cement CEM 32.5 POZ (ordinary portland cement) requires about half of the specific electricity than a CEM 52.5 POZ. Therefore the plants were asked to refer for the electricity consumption to the grinding of CEM 32.5 POZ, the most common type of cement.

Table 6.3-15 Average electricity consumption in kWh/t for CEM 32.5 POZ in sample plants

Country	Germany	Spain	UK
Electricity consumption in kWh/t of CEM 32.5 POZ	101	113	127

The electricity usage of German plants was on average lowest and about 10 per cent lower than in Spain and almost 30 per cent lower than in the UK which has a relatively high specific consumption with 127 kWh/t (Table 6.3-15). Although the average figures are similar the values for single plants differ widely between 90 to more than 140 kWh/t. The highest individual consumption was found in the UK. Since equipment for environmental measures like electrostatic precipitators, bag filters etc. consumes electricity, an assessment of low specific electricity consumption is difficult. More modern equipment (mills, kilns etc.) saves on kWh but more environmental equipment leads to a higher consumption. In the last decade the specific consumption of electrical power has increased in many plants because more environmental measures have been installed.

Emission limit values and actual emissions

In this section first of all the influence of different emission measurement techniques is discussed. Then a comparison of emission limit values for the three major pollutants NO_x, SO_2 and dust as well as the data on the corresponding actual emissions collected in the sample will be presented. If a plant used alternative fuels, regulation concerning these pollutants became stricter and additional standards had to be achieved. Existing national environmental standards vary across the EU countries with respect to limits and also in averaging time, reference conditions, measurements techniques, compliance criteria, etc.

a) Impacts of emission measurement techniques

Differences in the emission measurement techniques can make a great difference to the emission results. There are various factors which can be of influence and which must be known in order to compare emission levels meaningfully. First of all it is important to know how often emissions are measured. If, for instance, emissions are measured only once or twice per year, many situations with higher or lower emissions may be missed. Secondly, the period of time over which is measured, is of crucial importance. If the measurement lasts only for a short time span (and is taken only a few times per year), the measurement cannot be representative of the full year. Therefore, in order to know the trend in emissions exactly, continuous measurement would be required. Continuous measurement also has the advantage that the burning process can be adjusted accordingly. Without continuous measurements a correct reaction by the kiln controller is difficult. Furthermore, it is important to know over what periods averages are calculated, e.g. half hour or daily averages. Emissions can vary to a great extent over a day. Daily averages allow an easier following of the limit values, as more time is given to compensate a short-time violation of the limits by lower values in the remaining period.[29] Finally, the reliability and tolerances of the measuring equipment are important to get comparable results.

---

[29] Thus, for example, German law has less stringent standards than in Sweden, but requires data collection averaged over every 30 min. Alternatively, Swedish inspection authorities

Table 6.3-16 presents the types of measurements found in the sample. For the interpretation of the actual emission levels of sample plants in the following paragraphs the impact of varying measurement techniques have to be kept in mind.

While in Germany emissions are continuously measured in all sample plants, only one Spanish plant has started to do so. The other measurements in the Spanish sample plants are undertaken in intervals between four times per year to once every two years. Half of the German sample plants reported directly on-line to the environmental agency. On-line monitoring sets a great incentive to avoid any unusual disturbances because authorities can immediately interfere in a limit value violation of a plant. The emission values are standardised in Germany according to an O_2 content of 10 per cent, whereas in the UK measurements are mostly based on an O_2 content of 11 per cent. However, UK emissions presented below have been adjusted to an O_2 content of 10 per cent.

Table 6.3-16 Frequency of various emission measurement techniques in percentage of all plants (NO_x, SO_2, dust) and O_2 content in the sample

Frequency of measurements	Germany	Spain	UK
Continuous	100 %	20 %	80 %
Other	–	4 times per year to once every 2 years	1- 4 times per year
O_2 content	10 %	Not measured	Mostly 11 %

b) Dust emissions

The main sources of dust are kilns, raw mills, clinker coolers and cement mills. We concentrated on measurements at kiln stacks. The lowest national dust limit was set in the UK with 40 mg/Nm^3. However, this was seen as a benchmark for future investments and official plant limits were with an average of 70 mg/Nm^3 much higher. In Germany the limits were set at 50 mg/Nm^3, whereas in Spain the limits were set at between 250-400 mg/Nm^3. One reason for the lax limits in Spain was that the regulations have not been changed since 1975. It is being discussed to reduce this value to 100 mg/Nm^3 which would still be much higher than in other EU countries. Regional authorities in Spain were in three of five cases more strict in setting dust emission limit values for the plants. Thus the actual plant emission limit values were in these cases below the values from the

---

require very stringent standards, but the calculation mode is a monthly average. This allows for peaks and troughs to be smoothed out over a longer period of time.

national regulation, but still three times as high as the average in the UK plants and about six times as high as the average in Germany where the plant limit was in all but one cases below the national limit of 50 mg/Nm^3 (see Table 6.3-17).

Although all plants in the sample had implemented commonly used dedusting equipment like electrostatic precipitators or bag filters, the actual emissions varied considerably. Average dust emissions were lowest in German plants with 22 mg/Nm^3. In some of these plants the lower emissions were required, when alternative fuels were introduced. In Spain and the UK actual dust emissions were three times respectively two and a half times as high as in Germany. In the UK the national benchmark limit was in most cases overrun by more generous official plant limits. However, for a new kiln installation, with 25 mg/Nm^3 an as strict emission limit value as in the strictest German case was set.

Table 6.3-17 Average dust emission limits and average performance at kiln stacks of the sample plants

Country	Germany	Spain	UK*
National limit	50	400/250 a) 100 b)	40 c)
Average official plant limit	37	220	70
Average actual emissions at plant	22	76	55
No. of plants in the sample	8	5	5

*This excludes one kiln with extremely high limits and high emissions
a) Current limits.
b) Limits under discussion.
c) Benchmark releases.
Averaging times, measurements techniques or frequencies etc. are not taken into account.

Source: national limits transferred from Annex 1 EIPPCB Draft BAT Document, 'Survey of national emission limits'; own data collection.

c) SO_2 emissions

Sulphur dioxide stems mainly from sulphur in the form of sulphides and organic sulphur contained in the raw meal. Plants which use raw materials with little or no volatile sulphur have low SO_2 emissions. It is beneficial both for environmental and operational reasons to control the sulphur content of the raw feed to the kiln, as major process disruption can occur due to high sulphur levels.

The national emission limit value is strictest in Germany as it is set at the 400 mg/Nm^3 level (see Table 6.3-18). In the UK a benchmark of 200mg/Nm^3 for the national regulation exists, but is exceeded by far by official plant limits. The only exception in the UK sample is a new kiln which is already subject to the benchmark value. In Spain the national emission limits allowed 2400 mg/Nm^3 which might be lowered to 600 mg/Nm^3 in the future. However, in Spain SO_2 emissions are very low due to the nature of the raw material.

The actual emissions were with an average of 350 mg/Nm^3 highest in the UK sample plants. In the German sample the average was 240 mg/Nm^3 and in comparison to the UK sample there was much less variation in the individual emission levels. Two German plants had implemented SO_2 reduction measures in order to reach the emission limit values. In Spain actual emissions were very low throughout the entire sample due to the nature of the raw material and not considered as a problem at all. Only two German and one British plant achieved as low SO_2 emission levels as all the Spanish plants.

Table 6.3-18 Average SO_2 emission limits and average performance in mg/Nm^3 at kiln stacks of the sample plants

Country	Germany	Spain*	UK*
National limit	400	2 400/6 000 (a) 600 (c)	200 (b)
Average official plant limit	340	2 400	700
Average actual emissions at plant	240	70	350
No. of plants in the sample	8	5	5

a) Current limits.
b) 'Benchmark releases' (Benchmark releases are, in particular, not applicable to existing plant but are a factor in considering appropriate limits).
c) Limits under discussion.
* higher limits proposed for SO_2 rich raw materials; Spain up to 1800 mg/Nm^3 and UK 600-2500 mg/Nm^3.
Averaging times, measurements techniques or frequencies etc. are not taken into account.

Source: Annex 1 EIPPCB Draft BAT Document and own data collection.

d) NO_x emissions

Control of oxygen content is critical to NO_x control. Generally, the lower the oxygen content at a cement kiln back end, the less NO_x is produced. This has to be balanced against increases in CO and SO_2 at lower oxygen levels (EIPPC Cement BREF, 2000).

Table 6.3-19 gives an overview about the national limits, the official plant limits and the actual average emission levels found in the sample. German plants have the strictest national limit at 800 mg/Nm^3, while Spain has no national limits for NO_x emissions. In the UK the benchmark is set at 900 mg/Nm^3 and might be set for new plants between 500-1200 mg/Nm^3. Some new installed plants in the UK already have emissions below 500 mg/Nm^3.

The official limits at the plant sites are sometimes lower, but close to the national limits. In Germany they are set with one stricter exception at the national limit of 800 mg/Nm³. The average in the UK sample amounted to 1070 mg/Nm³. Only one UK sample plant was subject to a NO_x limit of 800 mg/Nm³, all others had limits of close to 1000 up to 1300 mg/Nm³.

A comparison of the actual emissions shows that the German plants achieve with 650mg/Nm³ the lowest average value with a very even spread of individual levels. In Spain, where no limits exist, the emissions were highest. All of the Spanish sample plants emitted more than 800 mg/Nm³ with only two plants using continuous measurement. NO_x emissions in the UK sample plants were in the middle range, but individual values varied substantially. Four of the five UK plants had continuous measurements and took hourly averages.

Table 6.3-19 Average NO_x emission limits and average performance in mg/Nm³ at kiln stacks of the sample plants

Country	Germany	Spain	UK***
National limit	800*	n.a.**	900 (a) 500-1200 (b)
Average official plant limit	750	n.a.	1070
Average actual emissions at plant	650	1130	730
No. of sample plant	8	5	5
Benchmark releases. Benchmark releases are, in particular, not applicable to existing plant but are a factor in considering appropriate limits. * 500 mg/m³ for new kilns ** under discussion 1300-1800 mg/m³ *** adjusted to 10% O_2			

Source: Annex 1 EIPPCB Draft BAT Document and own data collection.

Number of environmental measures

During the interviews plant managers were interrogated on specific techniques used for reduction of emissions[30]. A list of currently commercially available techniques for the prevention or reduction of emissions and waste in the cement industry served as background to the interviews. This list was recently published in a background reference document as part of the work of the European IPPC Bureau in Seville (EIPPC Cement BREF and see also chapter four). This docu-

[30] The need for using these techniques depends on site specific conditions. In particular, SO_2 reduction measures may not be necessary, if there is availability of limestone with a low sulphur content.

ment refers to a series of general, primary (process-integrated) and secondary (end-of-pipe) best available techniques for emission reduction in the European cement industry (see Box 6.3-1 for the list of these measures). These best available techniques can trigger different environmental effects in the plants and different impacts on the business. All these effects have been recorded for each environmental measure during the interviews. A detailed analysis and discussion is given in chapter nine.

Plants were asked during the interviews which of the environmental measures shown in the box below were already in place and which they thought to implement in the future. The Table 6.3-20 below refers to the average number of each measure which has been installed by sample plants within the last ten years. The measures are shown as average numbers per plant. In Germany, on average, a total of 11 measures was counted against 5.4 and 5.8 measures in the sample plants in the two other countries. The composition of these environmental measures differed in each country. From the general primary measures used in German sample expert systems, automatic quality control and continuous measurement were almost uniformly installed. Also measures against fugitive dust like closing of raw material and clinker storage were very common. Only precalciners were not as widespread as in the British sample. Concerning primary NO_x reduction measures low NO_x burners were installed in all but one of the German sample plants since a long time. In the German plants also the use of secondary measures – and among them uncommon techniques like SNCR or absorbent addition – was higher than in the other two sample countries. Secondary measures are end-of-pipe technologies which increase production costs and therefore are generally only undertaken when a corresponding pressure exists. As shown above, legislative pressure in the sample under concern here, is strongest for the German plants.

Box 6.3-1 List of best available techniques for the reduction of NO_x, SO_2 and dust emissions used as background for the interviews in the cement industry

Process selection
Since the selected process has a major impact on the energy use and air emissions from the manufacture of cement clinker, the best available technique for the production of cement clinker for new plants and major upgrades is considered to be a dry process kiln with multi-stage preheating and precalcination.
Reduction of emissions, energy use and waste by general primary measures
A smooth and stable kiln process, operating close to the process parameter set points, is beneficial for all kiln emissions as well as the energy use. This can be obtained by applying:
Process control optimisation, including computer-based automatic control systems.
The use of modern, gravimetric solid fuel feed systems.
Minimising fuel energy use by means of: Preheating and precalcination to the extent possible, considering the existing kiln system configuration.
The use of modern clinker coolers enabling maximum heat recovery.
Heat recovery from waste gas.
Minimising electrical energy use by means of: Power management systems.
Grinding equipment and other electricity based equipment with high energy efficiency.
Careful selection and control of substances entering the kiln can reduce emissions.
When practicable selection of raw materials and fuels with low contents of sulphur, nitrogen, chlorine, metals and volatile organic compounds.
Reduction of oxides of nitrogen
The best available techniques for reducing NO_x emissions are the combination of the above described general primary measures and:
Primary measures to control NO_x emissions
Flame cooling
Low-NO_x burner
Staged combustion
Selective non-catalytic reduction (SNCR – easier for dry kilns than for semi-dry kilns)
Staged combustion and SNCR are not yet used simultaneously for NO_x reduction.
Reduction of oxides of sulphur
The best available techniques for reducing SO_2 emissions are the combination of the above described general primary measures and:
For initial emission levels not higher than about 1200 mg SO_2/m^3:
Absorbent addition.
For initial emission levels higher than about 1200 mg SO_2/m^3: Wet scrubber.
Dry scrubber.
Reduction of dust emissions
The best available techniques for reducing dust emissions are the combination of the above described general primary measures and:
Minimisation/prevention of dust emissions from fugitive sources
Efficient removal of particulate matter from point sources by application of:
Electrostatic precipitators with fast measuring and control equipment to minimise the number of CO trips
Fabric filters with multiple compartments and 'burst bag detectors'

Source: EIPPC Cement BREF, 2000.

Table 6.3-20 Average number of emission reduction measures implemented (within the last 10 years) per plant classified according to countries*

Type of BAT measure	Germany		Spain		UK	
General primary measures per plant	Average no.	Average % of all	Average no.	Average % of all	Average no.	Average % of all
Expert system	0.8	7.3	0.6	11.1	0.2	3.4
Automatic quality control	1.0	9.1	0.6	11.1	0.2	3.4
Precalciner	0.5	4.5	0.0	0	0.8	13.8
Modern clinker cooler	0.5	4.5	0.4	7.4	0.2	3.4
New or modernised mill	0.6	5.5	0.6	11.1	0.8	13.8
Raw material storage closed	0.8	7.3	0.2	3.7	0.2	3.4
Clinker closed	0.9	8.2	0.6	11.1	0.6	10.3
Paving, fugitive dust	1.0	9.1	0.4	7.4	0.2	3.4
Optimal fuel feeding	0.1	0.9	0	0.0	0.2	3.4
Optimal burning process	0.3	2.7	0	0.0	0.2	3.4
Continuous measurement	1.0	9.1	0.4	7.4	0.8	13.8
Average of general primary measures per plant	7.5	68.2	3.8	70.4	4.4	75.9
Primary measures to control NO_x emissions per plant						
MSC	0.3	2.7	0	0.0	0	0
low NO_x burner/ flame cooler	0.9	8.2	0.6	11.1	0.6	10.3
Sum of primary measures to control NO_x emissions per plant	1.2	10.9	0.6	11.1	0.6	10.3
Sum of total primary measures per plant	8.7	79.1	4.4	81.5	5.0	86.2
Average number of secondary measures per plant						
Bag filters installed after 1990	0.4	3.6	0.4	7.4	0.2	3.4
Electrostatic precipitators installed after 1990	1	9.1	0.6	11.1	0.4	6.9
SNCR	0.4	3.6	0	0.0	0	0.0
Absorbent addition	0.5	4.5	0	0.0	0	0.0
Wet scrubber	0	0.0	0	0.0	0.2	3.4
Sum of secondary measures per plant	2.3	20.9	1	18.5	0.8	13.8
Sum of all measures per plant	11	100	5.4	100	5.8	100
No. of plants	8		5		5	

*Figures in this Table are rounded to the first decimal number.

Only about 60 per cent of the Spanish sample plants had expert systems and/or automatic quality control. Spain shows the lowest number of general primary measures e.g. no precalciner kiln systems, low number of measures for fugitive dust avoidance and low importance of continuous measuring of emissions. Primary NO_x reduction measures are rarely in place and no secondary measures except dust reducing equipment are installed.

The main difference between the UK and the Spanish sample is that the UK plants used on average 0.6 general primary measures more than their Southern counterparts. In particular, the UK used a high number of precalciners and in many works new cement mills were installed. Furthermore, all but one UK plant measured emissions continuously. However, the UK plants were weaker than the German and the Spanish plants in having expert systems and automatic quality control. As primary NO_x measures low NO_x burners were installed. Secondary measures were almost as rare as in Spain: Only one plant had new bag filters installed and two had invested in electrostatic precipitators in the last decade. The only wet scrubber found in the sample was installed in the UK.

Representativeness of sample concerning environmental performance

A comparison of the NO_x and SO_2 abatement techniques in the sample data with the available data on a national level shows that the sample has been selected in quite a representative way (see Table 6.3-21).

Table 6.3-21 Percentage of full scale NO_x and SO_2 abatement installations in Germany, Spain and the UK on a national level and in the sample

Country	**Germany**		**Spain**		**UK**	
Distribution	% national	% sample	% national	% sample	% national	% sample
NO_x abatement						
Flame cooling	0	0	0	0	0	0
Mineralised clinker	6.1	0	0	0	0	0
Staged combustion	10.6	25	0	0	0	0
SNCR	22.7	37.5	0	0	0	0
SCR	1.5	0	0	0	0	0
SO_2 abatement						
Absorbent addition	16.7	50	0	0	0	0
Wet scrubber	0	0	0	0	5.5	20
Dry scrubber	0	0	0	0	0	0
Activated carbon	0	0	0	0	0	0

Source: EIPPC Cement BREF, Annex B, based on a Cembureau survey carried out in April 1999 and information provided by TWG experts.

Germany is leading concerning staged combustion, SNCR and absorbent addition not only concerning the sample countries, but also on a European level. In the German sample quite many plants with these techniques have been included. None of the Spanish plants has secondary NO_x or SO_2 abatement measures in place. Only one UK plant has a wet scrubber installed. This plant is included in the sample.

6.3.6 Matched plants and firms in the cement sample

Table 6.3-22 shows the number of plants matched with counterparts in the other country. As in the other case studies some firms could be matched twice. A more detailed matching according to size and environmental performance is undertaken in chapter nine (empirical results).

Table 6.3-22 Number of matched plants and firms in the cement industry

Country	Germany : Spain	Germany : UK
Total	8:5	7:4

6.4 Summary on sample description

The examination of the main economic features of the three samples has shown that sample plants in all sectors and countries have different strengths and weaknesses. Concerning the environmental characteristics, the scope of environmental measures is often similar in the sample countries, particularly in the food and the packaging case, but less so in the cement case. Even when there are similarities in the scope of environmental measures, there are differences in the depth of undertaking these initiatives and in the environmental impacts of these initiatives. In the following short summaries of the individual samples are given:

Food sample

The sample shows that plants are broadly comparable in terms of plant size distribution, legal form of organisation and main products. Skill levels are higher in German plants. Irish dairies clearly use more R&D resources. Age of machinery is about equal in the sample; only East German plants are very young due to the foundation of Greenfield sites and restructuring after reunification. Concerning exports Irish plants are leading, partly due to the small size of the home market.

German plants undertook the largest number of waste water initiatives with West German plants having a slightly higher number in place than their East German counterparts. Waste water standards in the Republic of Ireland/Northern Ireland were found to be both stricter and less strict than in Germany. On average, German standards tended to be more stringent.

Packaging sample

In the packaging case study strong multinationals are mainly active in the Republic of Ireland, although also in Germany quite a few plants belonged to larger groups. German plants clearly had a higher percentage of skilled personnel and on average a younger capital stock than plants in the other sample countries. Concerning R&D in dairying the Irish industry was the leader, for the drinks industry the German plants devoted most personnel and in packaging plants of the German metal sample spent most for R&D in terms of staff. On the export side the Irish dairy industry showed the highest export quota, in retailing the German industry was most active abroad. Employment growth rates in the Republic of Ireland were strongest for the dairy and retail industry, in Great Britain for the glass industry and in Germany for the plastic, paper and metal packaging plants. The glass sector in the German sample, however, suffered from the largest reduction in employment.

Table 6.4-1 Summary statistics in the packaging case study: distribution of initiatives (total, median, average) by industry and country

Initiatives		Dairy / drinks		Retail		Packaging	
		RoI/UK	Germany	RoI/UK	Germany	RoI/UK	Germany
Solid waste	Total	96	112	61	136	99	232
	Median	5	7	6	13.5	3	7
	Average	6.0	7.5	6.8	13.6	3.3	8.0
Env. Awareness	Total	16	35	11	39	22	44
	Median	1	2	1	4.5	1	1
	Average	1.0	2.3	1.2	3.9	0.7	1.5
All initiatives	Total	112	147	72	175	121	276
	Median	6	9	7	18	4	8
	Average	7.0	9.8	8.0	17.5	4.0	9.5

Table 6.4-1 summarises the discussion on the number of environmental initiatives in the packaging sample and shows that on average German firms undertook a larger number of all types of initiative than their Republic of Ireland/UK counterparts. There was, however, similarity in the range of initiatives undertaken in all three countries and some degree of consistency between sectors. There were suspected differences in the 'depth' and extent of individual initia-

tives undertaken. For example, the extent of recycling of bottles required by law in Germany and the breadth of waste segregation followed is more extensive than in the Republic of Ireland/UK. It is likely that initiatives undertaken, particularly with respect to packaging, are more comprehensive than is indicated by estimates of differences in the number of initiatives undertaken.

Cement sample

Whereas there were still some family owned plants in the German cement sample, multinational companies dominate the Spanish and UK sample. German cement plants clearly had a higher average capacity; however, the rate of capacity utilisation was lowest in Germany. This is the main reason for the relatively low level of labour productivity in comparison with the Spanish and UK plants. Kiln age was about equal in all countries. The degree of modernisation was higher both in Germany and the UK than in Spain. Production costs were higher in Germany and the UK. Simultaneously, from all sample plants, German cement plants showed the highest use of alternative fuels. Energy and electricity consumption as well as emission levels were lowest in Germany.

Concerning the number of environmental initiatives the clear picture emerged that German sample plants undertook more activities than the counterparts in the other sample countries. Especially, secondary abatement measures are more frequently in place in German sample factories and lead to the result that German sample plants reach the lowest emission levels of all sample countries.

Table 6.4-2 summarises the discussion on the average number of environmental measures undertaken in German cement sample plants against Spanish and British sample plants. In each German plant on average about double the number of environmental measures has been used than in the Spanish and British sample plants. The same sequence can be observed in comparing primary and secondary measures. The German plants used on average twice as many general primary measures as the Spanish counterparts and 1.7 as many as in the UK plants. Concerning primary measures for NO_x reduction German plants used with an average of 1.2 measures exactly double as many measures than their counterparts. Concerning secondary measures German sample plants even used 2.3 and 2.7 as many reduction techniques as the Spanish and British cement works examined.

Table 6.4-2 Ratio of average number of environmental measures undertaken in German sample plants against Spanish and British sample plants

Environmental measures	Germany/Spain	Germany/UK
General primary measures	2	1.7
NO_x primary measures	2	2
Sum of all primary measures	2	1.7
Secondary measures	2.3	2.7
Total measures	2	1.9

On average, German plants had the lowest dust and NO_x emissions. Lowest SO_2 emissions were found in the Spanish sample.

The empirical results on the impact of stringent environmental regulation on competitiveness in the German manufacturing industry will be analysed for the three case study industries in the chapters seven, eight and nine respectively.

7. Empirical results on the impact of stringent waste water regulation on competitiveness in the dairy and meat industry

This chapter presents the empirical results of the case study in the dairy and meat industry on the influence of waste water legislation on company competitiveness. The analysis is carried out along the set of hypotheses presented in chapter five and is – as a generic model for all three case studies of this thesis – undertaken in the following sequence: In a first step, it examines whether firm performance (competitiveness) is systematically related to the number of environmental initiatives in place. The approach chosen for the measurement of firm competitiveness examines among other things value added and physical productivity. In each sample country some best (national) practice plants were identified and sampled, but also some plants with less than best practice performance. These latter plants were selected in order to illustrate variations in the response to environmental challenges. Second, waste water costs (compliance costs) are put in relation to the economic performance indicators of sample plants. Third, waste water costs were connected to the number of waste water initiatives in order to check whether the initiatives are cost driven. As a fourth analytical step, further factors like plant size were introduced in order to trace possible relationships with waste water costs as well as with number of initiatives. Moreover, any employment and productivity effects of environmental initiatives were examined. Finally, the importance of environmental factors relative to general competitive advantages and disadvantages of sample plants is analysed.

7.1 Firm performance and number of waste water initiatives

In this section it is analysed for the dairy and meat sample whether high productivity plants are also in a position to use a large number of waste water initiatives and to what extent this may be related to the degree of an environmental management culture in place.

Firm performance: Measurement of international comparative productivity

Maybe the best summary indicator of competitiveness and firm performance is labour productivity and particularly the level relative to international best practice. It should be considered, however, that value added per head is a composite measure since international variations in value added per head levels can be attributed to differences in physical productivity (number of units per worker) and product quality (value added of each unit). For differences in physical productivity economies of scale would be explanatory and for product quality sources of differences are rather found in skills and R&D. Previous international matched plant studies have shown that both physical productivity and product quality play a significant role in explaining international productivity differences (see for instance Hitchens, Wagner, Birnie, 1990; Hitchens et al., 1995).

Measurement of international comparative productivity was based on sales and value added per employee data derived from the matched pairs of plants. Several times, more than one plant in one country could be matched with a plant in another country and vice versa, where each was particularly alike. Table 7.1-1 shows the numbers of plants matched eventually used for the analysis.

Table 7.1-1 Numbers of plants matched in the dairy and meat sample

Industry	Ireland : Germany
Dairy	13:10
Meat	8:9

Thirteen Irish dairy plants were matched with ten German plants, i.e. three Irish plants were matched twice. Eight Irish meat plants were matched with nine German meat plants, i.e. one German meat plant was matched twice. Sales and value added data for each pair of countries were deflated using appropriate purchasing power parities (PPPs) for that industry. The resulting values are obtained in a common currency and are divided by employment. From these data a comparative productivity index is computed for each pair of countries. Table 7.1-2 shows the average of these pairs of productivity indices. German sample plants achieved a considerable higher index in dairying than their Irish counterparts; differences in the meat industry were smaller.

Table 7.1-2 Comparative productivity performance of sample firms computed from sales data

Comparative productivity	Germany: Ireland (with Ireland = 100)
Dairy	139
Meat	113

Product quality was taken into account using a qualitative estimate as to whether a certain plant sold its products to certain markets or outlets only, e.g. high quality supermarkets. Table 7.1-3 shows that in dairying German plants produced more high value products than their Irish counterparts. In meat processing, comparative quality was about equal. In the vast majority of cases calculations of physical productivity indices per person employed were consistent in direction with the comparative productivity indices derived from sales data. That implies that low value added is almost always associated with lower physical throughput in these comparisons.

Table 7.1-3 Comparative product quality in the dairy and meat sample

% cases comparing quality in first area/second area	Germany : Ireland
Dairy:	
Greater than	35
Equal to	59
Less than	6
Meat:	
Greater than	17
Equal to	67
Less than	17

Although German plants had on average a higher productivity level than their Irish counterparts, there were seven Irish plants recording a high international productivity performance against German plants (see Table 7.1-4 for the number of international best practice producers).

Table 7.1-4 Numbers of dairy and meat plants identified as international best practice producers

Industry	Germany	Ireland
Dairy	4	3
Meat	4	4

Also, nationally plants were classified as best practice producers when they achieved sales levels which were 25 per cent and higher than the national sample average and when they achieved levels of physical productivity higher than 25 per cent than the national physical average per employee in the sample (see Table 7.1-5). Irish dairies showed the highest number of national best practice producers.

Table 7.1-5 Numbers of dairy and meat plants with relatively high productivity levels in national terms

Country	Dairy in terms of		Meat in terms of	
	Physical	Value	Physical	Value
Germany	2	5	5	6
Ireland	5	4	3	1

The question now arises whether the different definitions of competitiveness include the same firms. Therefore, for all indicators of competitiveness, the sample was divided in average, above average and below average performance in comparison to the mean value for each measure. As to what the relationship is between these individual measures of competitiveness, a correlation of 0.6 was found for the international competitive rankings and national physical productivity rankings, and correlations of 0.4 and 0.36 between the international rankings and sales ranking, and sales ranking and physical ranking respectively. As further indicators of competitiveness exports and employment growth were tested. However, only low and insignificant correlations were found for both indicators.

Number of waste water initiatives

Boxes 7.1-1 (a) and (b) and 7.1-2 (a) and (b) show the number of waste water initiatives undertaken in each individual plant in Germany and Ireland and also give examples of these activities. There is a tendency that the majority of German plants is ahead of Ireland concerning the implementation of initiatives. Several Irish plants still were at a consultancy stage as to whether or not to undertake a certain water reduction initiative. Also automatisation and monitoring is more widespread and for a longer time in place in Germany than in Ireland where monitoring devices were in several cases just about to be installed. However, especially among the Irish dairies there were four cases of extremely modern and active plants (two in Northern Ireland and two in the Republic of Ireland).

Box 7.1-1 Details on waste water initiatives in Germany

Dairy sample plants in Germany

Dairy – East Germany		
	No. of waste water initiatives	**Examples of types of waste water initiative**
Plant 1	3	Water reduction initiatives, e.g. the minimum amount of fresh water is used for cheese production.
Plant 2	3	Design of new waste water system. Recycling of wash water rinses. Recovery of whey concentrates from cheese production. Effluent plant.
Plant 3	4	Updating of plant. Reduction and recycling of wash water. Reduction of energy for hot cleaning water. Effluent plant.
Plant 4	4	Updating of plant through new filter systems. Reduction of waste water used during cleaning. Effluent plant.
Plant 5	5	Plant layout and drains updated. Wash water is recovered and then reused. Effluent plant.

to be continued

Dairy - West Germany		
	No. of waste water initiatives	**Examples of types of waste water initiative**
Plant 1	4	Waste water reduction through cleaning in place. Recycling of rinses.
Plant 2	6	No chlorine is used in detergents. Energy saving measures for heating of wash water.
Plant 3	4	Establishment of environmental management system planned. Water reduction/water monitoring. Installation of heat exchanging device to lower oil consumption for heating of water. Effluent plant of the size of a small town.
Plant 4	5	Automatic processing and control system for reduction of water consumption. Monitoring of fresh and waste water.
Plant 5	5	Separation of whey. Recycling of as much water as possible.

(b) Meat sample plants in Germany

Meat – East Germany		
	No. of waste water initiatives	**Examples of types of waste water initiative**
Plant 1	4	Mud from the sewage plant is delivered to agricultural sector. Fully automated washing installation for a plastic box system in order to reduce effluent and waste.
Plant 2	4	Waste water separation. Waste water reduction through high pressure. Energy saving measures. Flotation system for waste water pre-treatment.
Plant 3	3	Waste water separation. Energy saving.
Plant 4	4	Separation of mud for use as compost. Separation and collection of blood. Water recycling.
Plant 5	4	Coal heating system replaced with oil. Adoption of energy and water saving measures. Separation of blood waste.
Meat - West Germany		
Plant 1	5	Energy saving and energy optimisation measures. No chlorine in detergents. Separation of animal waste and blood.
Plant 2	4	Water reduction and recycling. Separation of animal waste and blood.
Plant 3	4	Automatic washing installation. Water recycling. Separation of animal waste and blood.
Plant 4	5	Heat from the cooling system is used to heat hot water. Controls on fresh water consumption. Separation of animal waste and blood.
Plant 5	4	Recycling. Automatic processing and control system of production in order to reduce fresh water demand. Automatic washing installation.

Box 7.1-2 Details on waste water initiatives in the Republic of Ireland and Northern Ireland

(a) Dairy sample plants in the Republic of Ireland and Northern Ireland

Dairy – Northern Ireland	**No. of waste water initiatives**	**Examples of types of waste water initiative**
Plant 1	1	Consultancy study of drainage system
Plant 2	3	Collection of engine oil.
Plant 3	4	Introduction of new speed motors thus reducing energy consumption and improving insulation.
Plant 4	4	Installation of pre-programmed water control systems replacing manual final rinse systems thus reducing water usage. Valve attachments introduced to reduce spillage.
Plant 5	4	Monitoring of COD levels. New ultra-filtration system reduces the amount of solids entering drainage. Installation of wet scrubbers to reduce air emissions. Collect pre-rinses and recover fats for further processing.
Plant 6	2	Various water recycling activities.
Dairy – Republic of Ireland		
Plant 1	3	Use of gas for hot water production; self imposed limits for air emissions. Separation of liquid waste.
Plant 2	5	Scrubbers used for odour removal. Consultancy report on disposal strategy.
Plant 3	3	Water recycling, maintenance of water usage systems, awareness for water consumption.
Plant 4	3	Upgrading existing effluent plant. Monitoring procedures.
Plant 5	2	Water recycling.
Plant 6	5	Segregation of effluent lines and stone waters. Installation and upgrading of tanks and drains. Monitoring of water pollution.
Plant 7	3	Waste water treatment. Drains survey.
Plant 8	0	–
Plant 9	3	Recycling of chemicals. Safe disposal of excess chemicals.

(b) Meat sample plants in the Republic of Ireland and Northern Ireland

Meat – Northern Ireland	**No. of waste water initiatives**	**Examples of types of waste water initiative**
Plant 1	2	Collection of blood using hollow knifes. Processing of animal wastes.
Plant 2	3	Separation of blood. Control electricity consumption. Controlled water consumption.
Plant 3	3	Separation of animal waste.
Plant 4	4	Separation of animal waste. Defrost chills and reuse water.
Plant 5	3	Boiler used to heat water as well as sewage sludge (used as fertiliser). Water recycling. Separation of blood.
Plant 6	4	Proposed treatment tank. Waste separated and offal processed as by-products.
Meat – Republic of Ireland		
Plant 1	5	Separation of animal waste. Collection of blood. Recovery of valuable wastes, e.g. sludge for fertiliser and hides sent for leather. Recycle rinses. Low volume water pump to reduce consumption.
Plant 2	2	Separation of blood and animal waste.
Plant 3	3	Change from water to steam cookers.
Plant 4	3	Tank purchased for blood collection. Separation of animal waste.
Plant 5	3	Separation of solid and liquid waste. Separation of blood.

Firm performance and number of waste water initiatives

In Table 7.1-6 below the relationship between output measures of firm performance (competitiveness) and the number of waste water initiatives in each country is presented. Best practice firms are defined by output performance as

a) internationally competitive plants as defined by the matched plant comparison, i.e. those plants performing equally or better than their comparators in the other countries;
b) firms with relatively high levels of value added per head, at least 25 per cent above the sample average;
c) firms with high levels of physical productivity, at least 25 per cent above the sample average;
d) high relative exports, minimally 25 per cent above the sample average;
e) growing firms in the period 1990 – 1995.

It is shown that there is in general a positive association between achieving average or above average use of waste water initiatives and above average competitiveness. When input measures of firm performance (plant age, skill level and R&D activities) are examined, a similar positive pattern emerges (Table 7.1-7).

Table 7.1-6 Number of waste water initiatives and firm performance: output measures in dairy and meat sample

For firms with above national average on this criteria	Germany		Ireland	
	Dairy	Meat	Dairy	Meat
International comparative productivity	4.8	3.8	3.3	3
Value added per head	4.6	3.3	3.3	3.3
Physical productivity	5	4.8	3	4
Exports	4.8	3.5	3	3.3
Plant growth	4.5	4.2	2.3	3.5
Sample Average	4.3	3.8	3	3.2

Best practice plants in terms of input measures of performance are those with a young plant age as well as a high level of both skills and R&D.

Table 7.1-7 Number of waste water initiatives and firm performance: input measures in dairy and meat sample

For firms with above national average on this criteria	Germany		Ireland	
	Dairy	Meat	Dairy	Meat
Plant age	4.3	4	2.5	3.3
Skills	4.5	4.3	3.3	2.6
R&D plant level	4.3	4.5	2.5	3.8
Sample Average	4.3	3.8	3	3.2

Therefore, the hypothesis that high productivity plants are in a position to use a high number of environmental initiatives is generally confirmed. The number of initiatives is higher in Germany than in Ireland.

Firm performance and environmental management culture

In addition to a record of the frequency of water initiatives a subjective variable concerning environmental management culture was constructed. This variable was based on the interviewer's assessment of the company's understanding of the impact of waste water on the environment, the knowledge about alternative treatment options available as well as their costs and benefits, the extent of environmental monitoring undertaken by the firm, knowledge of EU requirements, ethical attitudes and similar factors.

The variable "management culture" was scored on a five-point scale (5=high) and is strongly correlated with the number of waste water initiatives undertaken (0.69 across all meat firms and 0.65 across all dairy firms). However, it remains unclear whether the correlation can be viewed as cause or effect. Table 7.1-8 below shows that above average economic performance is associated with an above average score on the cultural variable.

Table 7.1-8 Degree of environmental awareness and firm performance (highest score = 5, indicating greatest commitment to environmental objectives)

For firms with above national average on this criteria	Germany		Ireland	
	Dairy	Meat	Dairy	Meat
International comparative productivity	4.8	4	3.5	3.3
Value added per head	4.2	3.5	4.5	3.3
Physical productivity	4.5	4.8	4.2	3.7
Exports	4.8	3.5	4	3
Plant growth	5	4.2	2.8	3.0
Sample Average	4.1	4	3.9	3.2

Again, on average, Germany showed a higher sample average concerning the environmental culture variable than Ireland.

To summarise the results of this section, the hypothesis that high productivity plants are in a position to use a high number of environmental initiatives is generally confirmed for all sample countries. The number of initiatives is higher in Germany than in Ireland.

7.2 Water consumption, water costs and firm performance

In this section it is examined whether best practice plants in terms of competitiveness can also minimise their effluent treatment costs. Although this study is primarily concerned with waste water costs, also fresh water costs are presented in this section because the actual consumption of fresh water determines the ultimate amount of waste water to be treated. Table 7.2-1 (a) shows water consumption per unit of output measured in litres of milk and kilogram of meat. Across the sample about half a litre of milk and a third of a kilogram of meat is processed per litre of water. But data in individual plants varied widely across as can be seen in Tables 7.2-1 (b) and (c). More German plants are slightly more efficient users of water, especially in dairying where fewer sample plants were drawing water from (low or no cost) wells.

Table 7.2-1 (a) Intensity of water use in dairy and meat sample

Kg meat or litre of milk per litre of water used	**East Germany**	**West Germany**	**Northern Ireland**	**Republic of Ireland**
Dairy	0.71	0.62	0.53	0.32
Meat	0.26	0.40	0.38	0.46

Table 7.2-1 (b) Distribution of dairy plants according to water use*

Litres of milk processed per litre of water used	**Germany**	**Ireland**
3+	2	
1-2.99	1	
0.5-0.99	4	7
< 0.5	2	8
*referring to sample plants where data was available		

Table 7.2-1 (c) Distribution of meat plants according to water use*

Kg of meat processed per litre of water used	**Germany**	**Ireland**
3+	1	
1-2.99	2	2
0.5-0.99	1	
< 0.5	7	9
*referring to sample plants where data was available		

Table 7.2-2 Source of water supply and water supply costs in the dairy and meat sample

Industry	**East Germany**	**West Germany**	**Northern Ireland**	**Republic of Ireland**
Dairy				
With wells	2	2	4	9
Mains water	3	4	4	0
Average cost per m^3 in Pound Sterling	0.77	0.48	0.27	0.02
Meat				
With wells	5	1	3	4
Mains water	1	4	6	1
Average cost per m^3 In Pound Sterling	0.23	0.71	0.32	Negligible

Table 7.2-3 Average cost of mains water in the dairy and meat sample

Country/Region	Average cost per m³ in Pound Sterling
East Germany	0.35-1.86
West Germany	0.86-1.37
Northern Ireland	0.50-0.65
Republic of Ireland	150 p.a. (flat rate) – 0.18

Table 7.2-4 shows data on waste water costs for those sample plants where the information was known. It shows that waste water costs were again higher in Germany than in Ireland, but it also shows that there is a wide range of costs in each country.

Table 7.2-4 Cost of waste water (number of sample plants)

Cost per m³ in Pound Sterling	Germany dairy	Germany Meat	Ireland dairy	Ireland meat
< 0.25			6	5
0.25 – 0.50			4	2
0.50 – 0.99	4	4	4	4
1-1.99	6	7	1	
1.99-3.00	1			
Sample average	1.43	0.93	0.36	0.39

Waste water costs depend on the way of discharging effluent. Table 7.2-5 shows the number of sample plants discharging to sewers or waters. It shows that German sample firms mainly discharged to sewers and that few German firms have their own effluent plant. This confirms the general preponderance of indirect dischargers in the German system. Only a quarter of Irish dairy firms sampled discharged to sewers compared with half those processing meat. Nevertheless all Irish meat firms and all but two dairy firms had effluent plants which were also used by indirect dischargers for treating the company's effluent to the required regulated level before the effluent reached the sewer. E.g. one plant in Northern Ireland reduced the pollution load by 50 – 60 per cent before the effluent went into the sewer system. The rigour with which an indirect discharger treated its effluent in an effluent plant was of course less than undertaken in an effluent plant of a direct discharger. Total waste water costs for an indirect dischargers are calculated as the sum of pre-treatment costs within the plant and sewage charges paid to the local authority.

Table 7.2-5 Numbers of direct and indirect dischargers in the dairy and meat sample

Industry	East Germany	West Germany	Northern Ireland	Republic of Ireland
Dairy				
Sewer	4	3	2	2
Waters	1	2	4	7
Effluent plant	1	2	5	8
Meat				
Sewer	3	5	4	1
Waters	2	1	2	3
Effluent plant	2	1	6	4

Costs for discharging to sewers were variable and were not necessarily greater than the costs of self treatment, although the usual reason given for installing an effluent plant instead of using the municipal sewer was to reduce costs. For example in the case of the German sample companies the installation of sewage plants was reported to be exclusively cost driven or simply required in order to meet national legislation. Similar reasons were given in the two Irish economies. Where possible costs per m^3 to firms treating their own effluent were compared with sewer charges. In Germany sewer charges amounted to a 49 per cent higher cost although there was a wide variation of between a 5 and a 128 per cent difference between sewer charges and average effluent plant costs. In one German plant with a very high pollution load, sewer charges were merely calculated on the basis of fresh water demand. In Ireland the differential varied a great deal too. E.g. in the Republic of Ireland costs per m^3 for two dairies discharging to sewers averaged at Pound 0.11 compared with Pound 0.36 for seven firms sampled discharging to rivers and so on. The sewer price differential arises from differing charging systems and differing pricing structures between countries. There are also differences between firms in the extent to which they pre-treat or clean their waste water and which influences the cost of water purification. However, the main conclusion is that individual firms undertaking the same activity in each country face a range of prices. Table 7.2-6 shows the ranges of sewer discharge costs faced by sample firms.

Table 7.2-6 Range of sewer discharge costs facing sample plants in Pounds Sterling

Industry	East Germany	West Germany	Northern Ireland	Republic of Ireland
Dairy	1.06-2.28	0.71-1.08	0.001-1.02	600- 8,000 p.a. (flat rate)
Meat	0.52-1.42	0.71-1.37	0.50-0.65	500-6,500 p.a. (flat rate)
Note: Using average market exchange rate of 1996 DM 2.26 = 1 Pound Sterling.				

Effluent plant costs

Table 7.2-7 shows the number of plants sampled having an effluent plant and the characteristics of these effluent plants. 75 per cent of treatment plants were fully depreciated at the time of the interviews. In Ireland several expansions had taken place or were expected. In East Germany several firms had built effluent plants post unification in order to comply with the national regulations.

Table 7.2-7 Characteristics of effluent plants in dairy and meat sample

Characteristics	Germany	Ireland
Number of plants	6	19
Built 1986 and later	5	5
Number fully depreciated	2	17
Number expanded	0	7
Expected expansions	0	7

Data could be collected on the costs of running effluent plants in all plants having such a treatment facility (see Table 7.2-8 and 7.2-9). Estimates concerning running costs were made of labour costs, chemical and detergent costs, electricity costs for pumping, maintenance and repairs; also depreciation costs were included where it was still necessary. Average effluent plant running costs – at 0.39 per cent of turnover – were highest in the German meat industry and at 0.14 per cent were lowest in the German dairy industry. In the Irish industries costs varied around 0.20 per cent of turnover in both industries, although the range in costs varied from 0.1 to 0.5 per cent of turnover (see Table 7.2-8).

Annualised capital costs of effluent plants were recorded only where available as percentage of turnover, they varied for Germany between 2.7 and 5.5 per cent of current turnover, in Ireland the variation was with 2.4 to 2.7 per cent smaller. Historical costs were measured in current values and only accounted for when incurred in the 1990ies, otherwise the data were too uncertain and distant to be meaningful (see Table 7.2-9 for the range of effluent plant capital costs).

Table 7.2-8 Average effluent plant running costs in dairy and meat sample

Effluent plant running costs (% turnover)	Germany	Ireland
Dairy	0.14	0.19
Meat	0.39	0.21

Table 7.2-9 Range of effluent plant capital costs in dairy and meat sample

Costs (% turnover)	Germany Dairy	Germany meat	Ireland Dairy	Ireland Meat
< 0.1	1		5	4
0.1-0.19		2	3	1
0.2-0.29			1	2
0.3-0.49			1	2
0.5 +	2		2	1

In comparison to the evidence on sewer discharge costs of indirect dischargers presented above it becomes clear that individual firms face a wide range of waste water costs irrespective of whether they are direct or indirect dischargers. To summarise, German firms faced the highest waste water costs per m^3.

Table 7.2-10 (a) and (b) show the relationship between waste water costs and best practice performance using output and input measures of efficiency. Table 7.2-10 (a) shows that the hypothesis that competitive plants incur low environmental costs is not generally confirmed. It can be supported for the Irish industries, but not for Germany. It can also be supported across the countries, when only internationally competitive or high export-oriented firms are examined. However, the lack of any clear evidence leads to the conclusion that a firm can hold a competitive position without being favoured by nationally or internationally low waste water costs.

As before, best practice plants in terms of input measures of performance are those with a young plant age as well as a high level of both skills and R&D. Again no clear relationship emerges between waste water costs and input performance.

Table 7.2-10 Waste water costs and firm performance

(a) Waste water costs as a percentage of turnover and firm performance: output measures in dairy and meat sample

For firms with above national average on this criteria	Germany		Ireland	
	Dairy	Meat	Dairy	Meat
International comparative productivity	0.69	1.28	0.07	0.4
Value added per head	0.87	1.08	0.1	0.26
Physical productivity	1.56	1.07	0.37	0.28
Exports	0.69	0.39	0.21	0.55
Plant growth	1.43	1.70	0.04	0.21
Sample Average	0.87	1.02	0.26	0.44

(b) Waste water costs as a percentage of turnover and firm performance: input measures in dairy and meat sample

For firms with above national average on this criteria	Germany		Ireland	
	Dairy	Meat	Dairy	Meat
Plant age	0.88	1.83	0.56	0.58
Skills	0.71	1.73	0.14	0.89
R&D plant level	0.72	1.6	0.16	0.29
Sample Average	0.87	1.02	0.26	0.44

To summarise, no clear evidence was found between waste water costs and firm performance. It can be concluded that a firm can hold a competitive position without being favoured by nationally or internationally low waste water costs.

7.3 Waste water costs, waste water initiatives and quality of effluent

The question now arises whether the actual number of waste water initiatives could be cost driven (see Table 7.3-1). Linking waste water costs to initiatives, a positive relationship emerged across all sample countries and a correlation of 0.3 was calculated. Similar small and positive correlations were found for the relationship between average water prices and average number of waste water initiatives. However, the statistically significant correlation disappeared, when the data were disaggregated to within country comparisons for any of these measures.

Table 7.3-1 Waste water costs and average number of waste water initiatives in the dairy and meat sample

Average cost of waste water in Pound Sterling per m³	Germany Dairy	Germany Meat	Ireland Dairy	Ireland Meat
< 0.25	–	–	2.6	2.8
0.25-0.49	–	–	3.3	5
0.50-0.99	5	4	3.5	3.2
1.00-1.99	4	3.7	4	–
1.99-3.00	4	–	–	–
Sample average	4.3	3.8	3	3.2

Quality of effluent discharge by individual sample plants

Waste water from the food industry is not intrinsically toxic. The polluting impact arises from its biodegradability which imposes a high biological oxygen demand (BOD) on the environment. The pollution intensity of effluent is mainly measured in units of biological oxygen demand (BOD), chemical oxygen demand (COD), suspended solids (SS), all expressed in milligrams per litre. Individual regulation often also includes pH value and limit values for heavy metals. The quality of the effluent at plant level depends on whether the plant is directly discharging into waters, in which case the effluent must be treated by the plant to a certain standard, or whether it is discharging into sewers, in which case the local authority undertakes the treatment to the required standard. Individual firms are then regulated by a licence in which the limit values are specified; in case of indirect discharging, the polluting strength of the effluent is specified.

In Germany two different national standards are set for direct and indirect dischargers, while the standards vary in Ireland on a case-by-case level according to the nature of the environment into which the discharge will occur. Irish standards can therefore be both stricter and weaker than the German limits. The strictness of the standards affects the costs of cleaning the effluent.

Sample plants provided evidence of their discharge limits through licences and laboratory analyses. Tables 7.3-2 (a) and (b) attempt to capture the variation in quality of the effluent standards for the individual plants. As a benchmark for the sample the fairly stringent German national limit for BOD of 110 mg/l and COD of 25 mg/l for direct dischargers was chosen. In the Tables below it is shown how many plants across the entire sample comply with this particular German regulation. It is also shown how many Irish plants fulfil the legislation set by their own authorities.

Table 7.3-2 Effluent performance of all sample plants

Number of all dairy sample plants complying with German standard; number of Irish dairy sample plants complying with Irish standard

Comply with	East Germany	West Germany	Northern Ireland	Republic of Ireland
German limit	2*	3*	2	4
NI and RoI limit	–	–	3	8
With effluent plant	1	2	5	8
Discharge to sewer	3	3	2	2
Discharge to river	2	2	4	7
Sample size	5	5	6	9
*One of these plants had due to its specific production process almost no waste water. Although the plant did not have a treatment plant, the waste water was very clean.				

Number of all meat sample plants complying with German standard; number of Irish dairy sample plants complying with Irish standard

Comply with	East Germany	West Germany	Northern Ireland	Republic of Ireland
German limit	4	2	1	3
NI and RoI limit	–	–	3	3
With effluent plant	2	1	6	4
Discharge to sewer	3	5	4	1
Discharge to river	2	1	2	3
Sample size	5	6	6	5

On average, the German standard is stricter than the Irish standard and the Table 7.3-2 (a) shows that three German dairies are direct dischargers and have an own treatment plant. Two of those plants show a better environmental performance than the limit values prescribed by law, while the third just meets the required standard. Two further German dairies can meet the standard for direct dischargers, although they do not have a treatment plant. Their specific production process does not trigger very polluted effluent. Other German dairies discharge into the sewage system and therefore are allowed to have higher values for BOD and COD. Overall, 50 per cent of German sample dairies conform with the German national standard compared with 40 per cent of Irish dairies. 60 per cent of Irish sample plants meet the standards which they have been set.

In the case of the meat sample three German meat companies have an own treatment plant and can therefore meet the national limit for direct dischargers. Other German meat companies discharge into the sewage system and have higher values for BOD and COD since the water will be treated in the municipal sewage plant. Overall, 54 per cent of German meat sample plants met the German national standard compared with 36 per cent of Irish companies. 45 per cent of Irish firms met their own regulatory standard.

Overall, no clear relationship between the number of waste water initiatives and waste water costs was found. Results on the quality of effluent are mixed. On average, more German than Irish sample plants complied with the German national standards.

7.4 Size of plant in relation to number of initiatives and waste water costs

Table 7.4-1 investigates the relationship between plant size and number of waste water initiatives. In dairying there is a clear relationship between size and number of initiatives, but less so in meat processing. While plant size itself may not be important in the meat industry, the overall scale of production is. There was one Northern Irish firm and one German firm which were able to produce a by-product from waste because of the concentration of other similar firms in the region.

Table 7.4-1 Plant size and number of waste water initiatives

Employment size	Germany Dairy	Germany Meat	Ireland Dairy	Ireland Meat
Under 100	4	3.9	2.1	2.8
101-300	4.3	4	3.8	3.3
300+	5	3.5	4	3.7

No clear relationship between waste water costs and size of plant could be found (see Table 7.4-2).

Table 7.4-2 Plant size and waste water costs (as average percentage of turnover)

Employment size	Germany Dairy	Germany Meat	Ireland Dairy	Ireland Meat
Under 100	1.24	0.39	0.16	0.50
101-300	0.60	1.85	0.11	0.27
300+	0.77	1.02	0.81	0.62

7.5 Waste water initiatives, their drivers and productivity and employment effects

The motivation for waste water investments was principally reported to be legislative requirements with German sample plants expressing more of this type of pressure than Irish plants. The second source of driver was found in cost reduction aspirations with an almost even distribution between industries and sample countries. However, often the cost considerations followed legislative requirements and consequent costs and charges. In several of these cases the adoption of waste water initiatives were subject to rigorous financial appraisals. Table 7.5-1 also shows that customer pressure, managerial attitude, safety reasons and public image were frequently quoted as reasons for undertaking environmental initiatives.

In a few cases also the availability of new technology and increased efficiency was mentioned.

Table 7.5-1 Drivers for undertaking waste water initiatives in the dairy and meat sample (number of plants)

Driver	Germany Dairy	Germany Meat	Ireland Dairy	Ireland Meat
Legislation	10	11	8	6
Cost	6	4	5	5
Customer pressure	5	2	3	1
Manager's attitudes	3	2	5	1
Safety of workers	1	3	2	3
Public image	3	2	1	1
New technology	1	0	1	2
Increased efficiency	1	2	1	–

Interviewees were also asked what the productivity and employment effects were of undertaking waste water initiatives. The answers are summarised in Ta bles 7.5-2 and 7.5-3.

In total, 25 per cent of plants indicated a positive labour productivity effect. This was normally based on the recognition of plants that more work was carried out by existing staff in order to undertake an environmental initiative with the extra work being absorbed within the company. Also, environmental initiatives gave rise to better understanding of how the company worked and productivity enhancing spin-offs were a by-product of these initiatives. However, there were little implications by way of productivity for the core business.

Table 7.5-2 Productivity effects arising from the reduction and treatment of waste water in the dairy and meat sample (number of plants)

a) Dairy

Productivity effect	East Germany	West Germany	Northern Ireland	Republic of Ireland
N/A	0	0	0	1
No change	4	4	6	7
Positive	1	1	1	1
Negative	0	0	0	0

b) Meat

Productivity effect	East Germany	West Germany	Northern Ireland	Republic of Ireland
N/A	0	0	0	0
No change	5	4	6	5
Positive	0	2	0	0
Negative	0	0	0	0

Plant managers were also asked to quantify the numbers of persons and hours worked to run the effluent plant or to undertake other minor waste water activities. Taking the two industries together there were eleven cases where additional work was insignificantly small and quantified as zero, undertaken as part of other duties. In twelve cases a part-time employment became necessary and in 15 cases one or more full-time persons were required. These jobs had to do with the management of large effluent plants.

Table 7.5-3 Effluent plant and other waste water employment in the dairy and meat sample (number of plants)

(a) Dairy

Employment effect	East Germany	West Germany	Northern Ire land	Republic of Ireland
N/A	4	2	1	1
0				
1 part-time		1	2	1
1	1	1	2	3
1-2			1	3
2		1		
3				
3+				

(b) Meat

Employment effect	East Germany	West Germany	Northern Ireland	Republic of Ireland
N/A	3	4		1
0	2			
1 part-time		1	1	3
1			2	1
1-2			3	
2				
3				
3+				

It was also specified that waste water initiatives required new skills. Examples of these skills were as follows (see Box 7.5-1).

Box 7.5-1 New skills required to carry out waste water initiatives

Skills	Details
Computer skills	The use of monitoring equipment for the control of water consumption required that one or more employees in many plants had to develop skills in order to effectively use such equipment.
Waste water reduction skills	Many plants had educated their workforce with regard to water reduction, e.g. tracing of leakages etc.
Cleaning skills	New cleaning methods such as cleaning in place (CIP) have resulted in employees recycling water from the last rinse. This requires more accuracy with regard to the quantities of detergents and water used. It also requires use of automated equipment.
Effluent plant monitoring	Companies with an effluent plant require that emissions are monitored and the plant is run smoothly. The job is highly skilled and requires a technically competent person who has to ensure that the pH balance is in balance at all times.

During the interviews and plant visits it was also stressed that a certain amount of management time is bound by environmental issues. Examples are given in Box 7.5-2.

Box 7.5-2 Examples of management time used for environment-related issues

Manager allocates one hour per week to various tasks.
Manager allocates five hours per week to various tasks.
Manager allocates eight hours per week to various tasks.
15 per cent of manager's weekly time are used for implementation of an environmental management system.
5 per cent of general manager's time are devoted to environmental issues.
7 per cent of general manager's time are devoted to environmental issues.
10 per cent of production manager's time are devoted to environmental issues.
20 per cent of production manger's time are devoted to environmental issues.
10 per cent of three managers' time are devoted to environmental issues.

It can be summarised that most additional work due to waste water measures is absorbed by the existing labour force and there is no pronounced productivity effect. Some training is required to undertake these activities and there is important use of skilled, technical and managerial time. These skills are also important when knowledge about new techniques has to be gained and in several cases it was stressed that it is more the bottleneck of time than capital or running costs of initiatives.

7.6 Main markets and customers, competitive advantages and disadvantages, constraints on growth

In this section the relative importance of environmental aspects in comparison to the general competitive position of a plant determined by main markets and customers as well as general competitive advantages and disadvantages and growth constraints are analysed. Results are based on the answers of plant managers.

Main markets and customer groups

Two-thirds of all sample firms were actively engaged in international competition with the principal export markets being the EU countries (see Tables 7.6-1 and 7.6-2). However, each sample country also successfully served export markets world wide. German firms are shown to have the narrowest export markets (especially in East Germany) and the Irish firms the widest. The small size of the Irish home market and the importance of the food industry there make exports particularly relevant. Overall, most sample firms were therefore open to international competition and to competitiveness effects of variations in environmental regulations and costs faced by firms.

Table 7.6-1 Numbers of dairy and meat sample firms exporting and exports as a percentage of all sales

Industry	East Germany	West Germany	Northern Ireland	Republic of Ireland
Dairy (no.)	2	3	5	8
Exports/sales (%)	19.6	20.6	45.2	57.8
Meat (no.)	1	3	4	4
Exports/sales (%)	1	10	26	68

Table 7.6-2 Main export markets in dairy and meat sample, number of firms

Industry	East Germany	West Germany	Northern Ireland	Republic of Ireland
EU	2	6	4	7
Eastern Europe and Russia	1		1	1
US and Canada				
Far East and Middle East		1		
"World", incl. Australia, Africa		1	2	

In addition to the examination of special markets it was also explored whether a certain group of customers was served in particular by the sample firms. This was done in order to identify particular pressures with respect to the environment. Table 7.6-3 shows the average importance of the main customers served by sample companies. It divides the data between wholesalers, retailers and other food processors. It turned out that firms were competing for similar customers and that major supermarket chains (retailers) represented with the exception of Northern Ireland the biggest customer group. Supermarkets were reported to exert pressure concerning food packaging, but not concerning waste water.

Table 7.6-3 Sales by type of customers in the dairy and meat sample

a) Dairy (percentage of sales)

Principal customer	East Germany	West Germany	Northern Ireland	Republic of Ireland
Wholesalers only			67	33
Retailers	80	25	33	33
Retailers and wholesalers	20	38		8
Other food processors		38		25

b) Meat (percentage of sales)

Principal customer	East Germany	West Germany	Northern Ireland	Republic of Ireland
Wholesalers only				
Retailers	20	43	50	
Retailers and wholesalers	40	28	50	
Other food processors	40	28		

Competitive advantages and disadvantages

Sample firms were facing competitive pressures on an international level. Therefore, firms were asked to specify the competitive advantages and disadvantages they were benefiting from or which they were facing in this struggle.

The main competitive advantages were related to product quality, plant location and labour related factors, in particular skills and costs. A variety of other advantages were associated with prices, costs, fiscal and credit advantages. Several firms also mentioned the technical aspects of their production processes to be favourable aspects for competing on the national and international markets (see Tables 7.6-4 (a) and (b)).

Table 7.6-4 Competitive advantages at sample firms

a) Dairy sample

Competitive advantages	East Germany	West Germany	Northern Ireland	Republic of Ireland
Product quality	2	4	4	1
Location	1	2	2	2
Labour related, e.g. skills, costs	1	1	3	5
Price			2	1
Cost advantages		1	1	2
Technical advantages	1	1		1
Flexibility	1	1	1	3
Fiscal and credit advantages				
Other	3	4		

b) Meat sample

Competitive advantages	East Germany	West Germany	Northern Ireland	Republic of Ireland
Quality	3	5	4	
Location	3	5		1
Labour related, e.g. skills	3	1	2	3
Price	1	1	1	
Cost advantages				1
Technical advantages		1	1	1
Flexibility	1	2		
Fiscal and credit advantages	4	1		1
Other	2	3		

Concerning competitive disadvantages pressures on price, location (for peripheral firms), labour availability, quality and labour relations were amongst the most important. Eight plants reported environmental costs to constitute a competitive disadvantage for them, seven of these plants were located in East Germany, the other in West Germany. The concern arose in new plants in East Germany where managers had made considerable capital investments to meet the new environmental regulation and the complaints arose about these investments and high treatment costs of waste water. However, in relation to the overall investment requirements at new Greenfield sites in East Germany, the environmental investments were not large. Managers did not state that the environmental investment represented a constraint on future growth. More centrally,

East German firms complained about peripheral location, market overcapacity and lack of high value added products. In indigenous East German factories no manager said that environmental regulation had proved a competitiveness or growth constraint. West German firms were more concerned about costs arising from compliance with hygiene regulation than from waste water regulation (see Tables 7.6-5 (a) and (b)).

Table 7.6-5 Competitive disadvantages in sample firms

(a) Dairy

Competitive advantages	East Germany	West Germany	Northern Ireland	Republic of Ireland
Pressure on price				
Location	1	1	4	2
Labour relations				
Lack of innovation				
Price of raw materials		1	2	
Supply raw materials				2
Unstable demand				1
Excess capacity		1	1	
Poor marketing resources	1	3		1
High environmental costs	3			
Inability to access markets	1		1	1

(b) Meat

Competitive advantages	East Germany	West Germany	Northern Ireland	Republic of Ireland
Pressure on price				
Location	2	1		
Labour relations	1		1	
Labour quality		2	1	
Lack of innovation				
Price of raw materials	1	1	2	1
No value added products	2	1	1	1
Overcapacity relative to market	3			
Unstable demand				
High environmental costs	4	1		
Hygiene costs	1	4		
Poor marketing resources		2		
Cost of credit, finance				
BSE	2	4	1	1
Age of plant		3	2	3

Growth constraints

Respondents were asked what factors constrained company growth. Both East and West German dairy firms referred to finance problems and the strength of the DM which limited export opportunities. Other factors included availability of raw materials, employee qualifications and co-operation. For meat processors, overcapacity, hygiene costs, BSE, raw material availability and price were the main limiting factors. Environmental factors were nowhere specified to be critical constraint for growth.

To summarise, this section has shown that dairy and meat companies are subject to strong national and international competitive pressures. Concerning environmental regulation only in companies located in East Germany environmental costs are a new and important costs of transition. However, environmental costs were not reported to influence competitiveness or to present a constraint on growth.

7.7 Conclusions on the impact of stringent waste water regulation on competitiveness in the German food industry

This case study examined the question whether German dairy and meat plants suffer from a negative impact on competitiveness caused by stringent waste water legislation in Germany. The results were contrasted with data collected in an Irish/UK counter sample of similar structure and size where the waste water legislation was on average less stringent at the time of the study.

It was shown that the stringent German waste water legislation implies relatively high compliance costs for German plants, but hardly any impact on their competitiveness. This can be explained as follows: There was support for a positive relationship between a high number of waste water initiatives and a high level of productivity both in the dairy and meat industry and both in Germany and the Republic of Ireland/Northern Ireland. Good economic performance was also associated with an above average score of environmental awareness. For the German dairy and meat sector a significant relationship between younger plant age and a higher number of waste water initiatives was found. A similar positive pattern emerges for skills and R&D and number of waste water initiatives. This was not the case in the Irish economies. More German plants are slightly more efficient users of water, especially in dairying where fewer sample plants were drawing water from (low or no cost) wells. Overall, both fresh water prices and waste water charges were higher in Germany than in Ireland, thus, providing a higher incentive to save water and to reduce compliance costs. More German firms were able to record positive productivity effects triggered by waste water activities, but these impacts were only marginal. Environmental aspects were only of minor importance when plants were asked for competitive advantages or disadvantages both in Germany and the Republic of Ireland/Northern Ireland. The detailed findings are summarised as answers to the hypotheses developed at the outset of this study.

Productivity and cost of compliance: The proportional cost of environmental compliance relative to turnover incurred by the firms was hypothesised to be a negative function of the productivity level because firms with the capabilities to achieve high productivity will also find it easiest to respond to environmental challenges without the penalty of reduced output and employment.

Findings: Waste water costs were found to be higher in Germany. Based on a statistical analysis of the relation between the cost of waste water and measures of firm competitive performance[31] the hypothesis that competitive plants incur low environmental costs is not generally confirmed. It can be supported for the Irish industries, but not for Germany. It can also be supported across the countries, when only internationally competitive or high export-oriented firms are examined. However, the lack of any clear evidence leads to the conclusion that a firm can hold a competitive position without being favoured by nationally or internationally low waste water costs.

Number of initiatives and firm performance, ability to absorb compliance costs, environmental awareness: The study also sought to link productivity and compliance costs to the number of environmental initiatives undertaken and the underlying environmental awareness.

Findings: Overall, there was support for a positive association between achieving average or above average use of waste water initiatives and above average competitiveness both in the dairy and meat industry and both in Germany and the Republic of Ireland/Northern Ireland. Above average economic performance is also associated with an above average score of environmental awareness. It was then tested whether the actual number of waste water initiatives could be cost driven. Linking waste water costs to initiatives, a positive relationship

[31] A number of indicators of competitiveness were used in the study. These included output side indicators (i.e. those that illustrate the consequences of the relative competitiveness of the firm): market share, exports, growth, productivity, influence on product price, managers' own assessment of competitiveness effects. There was also a set of indicators corresponding to the input side (i.e. representative of the likely explanations of competitiveness): these included physical and human capital (i.e. age of plant and skills) and R&D spending.

emerged across all sample countries and a correlation of 0.3 was calculated. Similar small and positive correlations were found for the relationship between average water prices and average number of waste water initiatives. However, the statistically significant correlation disappeared, when the data are disaggregated to within country comparisons for any of these measures.

Plant age: The age of a plant and its machinery is likely to have an impact on its environmental performance and on compliance costs. Modern plants can reach better environmental standards because the newest technology also embodies best environmental technology and allows to achieve environmental goals at a lower cost than with old equipment which would need expensive retrofitting.

Findings: Plant age was measured as one of the competitiveness indicators measured on the input side (others were skills and R&D). For the German dairy and meat sector a significant relationship between younger plant age and a higher number of waste water initiatives was found. This was not the case for the Irish economies. Concerning the hypothesis that low plant age would be connected to lower waste water costs, the opposite was found for Germany and results were mixed for the Republic of Ireland and Northern Ireland.

Plant size: Large plants are hypothesised to undertake more environmental initiatives because they often have more financial and human resources for the solution of environmental problems than small and medium sized plants and have also better opportunities to research cost efficient environmental measures. Thus the proportional cost of compliance relative to turnover is likely to be a negative function of the size of plants/firms.

Findings: In dairying there was a clear relationship between size and number of initiatives in all sample countries, but less so in meat processing. No relationship was found between plant size and waste water costs.

Green market pressure: Not only regulatory pressure forces firms to react, but also environmental pressure coming from the market place with respect to customers/consumers and suppliers. Relations within the supply/production chain are likely to be both an influence upon, as well as being influenced by, the level of environmental standards, e.g. a manufacturer may find it easier to increase the environmental standards of its products if it has a reliable and competent base of suppliers to draw on. A manufacturer may be forced to upgrade product and process environmental standards by pressure coming from the customers (incl. consumers) of plants in the three sectors under study.

Findings: The motivation for waste water investments was principally reported to be found on legislative requirements, with German sample plants expressing more of this type of pressure than Irish plants. The second source of driver was found in cost reduction aspirations with an almost even distribution between industries and sample countries. However, often the cost considerations followed legislative requirements and consequent costs and charges. Customer pressure and public image were far less frequently quoted as reasons for undertaking environmental initiatives. Supermarkets were not reported to exert pressure concerning waste water in the German food industry.

History of regulation: Plants in countries with a history of stringent environmental legislation will be in a better position to develop or gain access to environmentally and economically beneficial solutions than their counterparts in other countries with a history of less stringent regulation.

Findings: More German plants are slightly more efficient users of water, especially in dairying where fewer sample plants were drawing water from (low or no cost) wells. Overall, both fresh water prices and waste water charges were higher in Germany than in Ireland, thus, providing a higher incentive to save water. Also, the number of waste water initiatives was largest in Germany. Results on the quality of effluent are mixed. On average, more German than Irish sample plants complied with the German national standards.

Employment impact: The direction of the employment implication of adjustment to stringent environmental regulation is not clear a priori. The available macroeconometric evidence has shown that the overall employment effects of environmental policies are likely to be small.

Findings: In the two sample countries there were eleven cases where additional work was insignificantly small and quantified as zero, undertaken as part of other duties. In twelve cases a part-time employment became necessary and in 15 cases one or more full-time persons were required. These jobs had to do with the management of large effluent plants. It was not found that the employment impact in Germany was higher than in the Irish economies.

Influence of regulation and external pressures on the competitiveness of firms and plants: Compliance could affect productivity and competitiveness either positively or negatively. Regulation could place firms at a competitive disadvantage and lead to the loss of markets, particularly to countries with less stringent regulation. If the firm needs to redirect resources from other profitable opportunities, costs and prices rise and markets and customers are lost. On the other hand, high standards and strict enforcement, although they may represent a short-term cost and burden to the firm, could in the medium and longer term push firms on to a higher growth path by forcing them to make product and process changes which yield higher competitiveness. If this happened it would represent part of the so-called 'double dividend', i.e. gains in environmental performance would also be accompanied by increased economic performance.

Findings:
(a) Based on asking managers about the productivity effects of adopting waste water initiatives. In total, 25 per cent of plants indicated a positive labour productivity effect. This was normally based on the recognition of plants that more work was carried out by existing staff in order to carry out an environmental initiative with the extra work being absorbed within the company. Also, environmental initiatives gave rise to better understanding of how the company worked and productivity enhancing spin-offs were a by-product of these initiatives. However, there were little implications by way of productivity for the core business. More German firms were able to record positive productivity effects triggered by waste water activities, but these impacts were only marginal.

(b) Based on the importance of environmental factors influencing the firm's competitive performance relative to other factors: The main competitive advantages in the sample were related to product quality, plant location and labour related factors, in particular skills and costs. Concerning competitive disadvantages pressures on price, location (for peripheral firms), labour availability, quality and labour relations were amongst the most important. In particular new East German plants reported environmental costs to constitute a competitive disadvantage for them However, in relation to the overall investment requirements at new Greenfield sites in East Germany, the environmental investments were not large. Managers did not state that the environmental investment represented a constraint on future growth.

8. Empirical results on the impact of stringent packaging regulation on competitiveness in the supply chain of the packaging, food and retail industry

This chapter presents the competitiveness implications of investment in solid waste initiatives in the sample of plants visited in Germany, the Republic of Ireland and the UK. The focus of the analysis is set on the question as to whether the German plants suffer from an economic disadvantage because their waste legislation is the most stringent one in this sample.

First of all, the relationship between firm performance and number of solid waste initiatives is examined in Germany and then compared with the Republic of Ireland and the UK. In particular, the top quartile of each country sample in terms of economic performance (high productivity) is put in relation to the number of solid waste initiatives undertaken in this subsample and then compared with the results for the sample as a whole. In a second step, it is tested for each country sample whether high productivity firms are also able to minimise compliance costs better than the total sample in each country. Compliance costs are measured as waste costs which consist of recycling and landfill costs; in Germany also the DSD costs are included because they measure the additional cost of compliance with the prevailing stringent packaging legislation which the other sample countries did not yet incur at the time of the study. Thirdly, waste costs are put in relation to turnover and number of solid waste initiatives and are contrasted across all sample countries. Steps one, two and three are also performed on an international basis of matched pairs of firms.

Fourthly, the relationship between further firm characteristics like size, ownership, location and number of initiatives/solid waste costs is analysed. Fifthly, also the drivers and effects (including capital costs) of each initiative are examined and the motivation for undertaking the initiatives is analysed (mainly regulation or external pressure stemming from customers) and contrasted between the sample countries. Finally, an analysis of competitive advantages and disadvantages, constraints on growth as well as direct customer/supplier pressure is undertaken.

8.1 Firm performance and number of solid waste initiatives

In this section the relationships between the implementation of solid waste initiatives and firm performance are considered in Germany vs. the Republic of Ireland and Northern Ireland. Analyses examine the relationship between initiatives and selected output measures of performance (productivity, plant growth and exports), and selected input measures of performance (plant age, skills and R&D). The underlying hypothesis is that in each sample country more competitive firms are assumed to be better able to find ways to reduce waste by adopting a greater number of solid waste initiatives. At the end of the chapter also the influence of environmental awareness variables on the number of solid waste initiatives is examined.

Tables 8.1-1 and 8.1-2 firstly compare the average number of initiatives undertaken by those firms in the top quartile of the performance variable in each country, i.e. firms with the highest productivity performance, youngest plant age and so on with the average number of waste initiatives undertaken by sample firms as a whole. Secondly, an international comparison is made between matched pairs of firms. This asks the question whether there is a relationship between initiatives and measures of comparative success. Hence the relationship between ratio of firm performance (e.g. productivity, firm growth, R&D input) and the ratio of numbers of waste initiatives employed between matched pairs is examined across countries (Table 8.1-3). Results are presented below on a sector basis.

Dairy/Drinks

Table 8.1-1 (a) shows that in the case of Germany dairy/drinks firms in the top quartile of firm performance, on average, do undertake more solid waste initiatives than all German firms sampled (e.g. for value added 8.25 initiatives in the top quartile vs. 7.47 for all German plants), and there is a significant positive correlation shown between plant growth and numbers of initiatives undertaken.

Table 8.1-1 (a) Number of solid waste initiatives and firm performance: output measures in dairy/drinks and packaging

Firms in top quartile measured by	Dairy/drinks		Packaging	
	Solid waste initiatives	r	Solid waste initiatives	r
Germany				
Value added per head	8.25	0.25	7.57	0.08
Plant growth	8.75	0.65*	6.86	-0.37*
Exports	7.80	0.26	7.88	0.07
All	7.47		8.00	
Republic of Ireland/UK				
Value added per head	4.25	-0.56*	4.20	0.38
Plant growth	6.00	-0.089	3.80	-0.30
Exports	4.50	-0.56*	3.50	0.02
All	5.93		3.30	
Notes: **Significant at 1% level, * significant at 5% level (two-tailed p-value).				

Table 8.1-1 (b) Number of solid waste initiatives and firm performance: output measures in retailing

Firms in top quartile measured by	Solid waste initiatives	r
Germany		
Sales per head	17.50	0.43
All	13.60	
Republic of Ireland/UK		
Sales per head	8.50	0.20
All	8.71	
Notes: ** Significant at 1% level, * significant at 5% level (two-tailed p-value).		

Table 8.1-2 Number of solid waste initiatives and firm performance: input measures in dairy/drinks and packaging

Firms in top quartile measured by	Dairy/drinks		Packaging	
	Solid waste initiatives	r	Solid waste initiatives	r
Germany				
Plant age	6.50	-0.18	7.62	-0.16
Skills	8.17	0.49	8.11	0.15
R&D plant level	8.25	0.08	8.00	0.24
All	7.47		8.00	
Republic of Ireland/UK				
Plant age	7.00	0.68*	3.66	0.048
Skills	4.67	-0.071	4.00	0.40
R&D plant level	4.67	-0.49	3.75	0.061
All	5.93		3.30	
Notes: ** Significant at 1% level, * significant at 5% level (two-tailed p-value).				

Table 8.1-2 shows a broadly similar result for German firms based on all input measures of performance except for age where the number of solid waste initiatives in the sample as a whole is higher than in the top quartile. No consistent relationship is shown with output and input measures of performance for the Republic of Ireland/UK. Only concerning plant age, youngest Irish/UK firms undertake more solid waste initiatives than all firms on average (7 vs. 5.93) and this relationship is shown to be statistically significant.

The matched pairs comparison of differences in performance with differences in initiatives undertaken in the dairy/drinks sector between the Republic of Ireland/UK and Germany, shown in Table 8.1-3, shows that there is a negative correlation, but it is not statistically significant.

Packaging

The findings for the packaging sector indicate that only in the Republic of Ireland/UK, high performing firms in terms of output indicators undertake more initiatives. This is not the case in Germany where the top performers hardly reach the level of waste initiatives undertaken in the total sample, and this only concerning value added and exports. Correlations between comparative firm performance (between matched pairs of firms) and the comparative numbers of initiatives employed by firms are slightly positive, but not statistically significant (see Tables 8.1-1 (a), 8.1-2 and 8.1-3).

Retailing

The relationship between economic performance and numbers of initiatives employed by competitive retailing firms is especially in Germany positive, but not statistically significant (see Table 8.1-1 (b)). When firms are matched, there are generally positive correlations between differences in productivity performance and numbers of initiatives adopted (see Table 8.1-3).

Table 8.1-3 Correlations between comparative productivity and comparative numbers of initiatives for matched pairs, ROI/UK:Germany

Sector	Solid waste initiatives
Dairy/drinks	–0.35
Packaging	0.06
Retail	0.41

It was assumed that the highly developed environmental awareness culture in Germany measured by a high number of awareness initiatives would be related to a high number of solid waste initiatives as well. Table 8.1-4 shows that increased environmental awareness was associated with increased numbers of solid initiatives undertaken in all sectors of the German sample. The Table shows clearly that those firms with no environmental awareness characteristics employ fewer solid waste initiatives. There is also some evidence of a rising trend with the number of environmental awareness characteristics identified.

Table 8.1-4 Relationship between environmental awareness and solid waste initiatives by sector in Germany

No. of env. awareness init.	Dairy/Drinks		Packaging		Retail		All	
	Inits[a]	Firms[b]	Inits[a]	Firms[b]	Inits[a]	Firms[b]	Inits[a]	Firms[b]
0	6	2	8.3	7	–	0	7.1	9
1	6	3	7.8	10	9	2	6.9	15
2	7.7	3	7.6	5	10	1	7.9	9
3	8.9	4	8.8	4	15	2	8.4	10
> 3	8.0	3	10	3	13.4	5	10.5	11
All	7.5	15	9	29	12.7	10	8.9	54
Correlation coefficient*	0.44		0.38		0.56		0.50	

Notes:
[a] Average number of solid waste initiatives undertaken.
[b] Number of firms.
* Correlation is significant at 1% level.

Summary and discussion of results on firm performance and number of solid waste initiatives

This section has examined the relationship between the implementation of solid waste initiatives and firm performance. The tested hypothesis was that more competitive firms are better able to find ways to reduce waste by adopting a greater number of solid waste initiatives.

Overall, the support for a relationship between firm performance and the number of solid waste initiatives undertaken by the firm, are generally inconsistent and not statistically significant. The strongest linkage was found in the German dairy/drinks industry where top performers were with the exception of plant age as a performance indicator always able to implement a higher number of solid waste initiatives than the overall German sample. Moreover, also Germany's top retailers used a much larger number of initiatives than the average German retail firm. Results were, however, not statistically significant. In German packaging plants, only for high performing plants in terms of skills and R&D, a positive correlation with the number of solid waste initiatives was found. In the Republic of Ireland/UK, high-performing packaging firms in terms of all output and input indicators undertake more solid waste initiatives than the sample average. But correlations between numbers of initiatives employed and economic performance are not statistically significant in the packaging sector.

Even when correlations are in most cases statistically insignificant, the linkages shown for Germany could indicate that high productivity firms are at least in certain areas able to avoid/reduce/recycle waste to a higher degree through more initiatives. In the next section 8.2 the relationship between firm performance and waste costs is examined. A further statistical examination of the level of waste costs and number of solid waste initiatives is undertaken in section 8.3.

8.2 Firm performance and waste costs

This section examines the relationship between waste costs and firm performance between and across countries. It is hypothesised that economically well-performing firms are also enabled to minimise waste costs. Tables 8.2-1 (a) and (b) show the relationship between waste costs and plant and firm performance using output measures of efficiency. As before, performance is defined by productivity (value added per head), employment growth and export propensity. Solid waste costs are shown for firms and plants in the top quartile of each measure of efficiency and are discussed below on a sector basis for each country.

Dairy/Drinks

Table 8.2- 1 (a) shows that in the dairy/drinks sector there is a clear tendency in the Republic of Ireland and the UK for better-performing firms to be associated with lower waste costs in the majority of comparisons made. This is much less frequently the case for Germany. But correlations between solid waste costs and performance are generally not statistically significant, or are ambiguous or small. Table 8.2-2 examines the same relationships for inputs and similar conclusions apply. Overall, there is support for the hypothesis where well performing firms (in the top quartile of that performance measure) have in many cases lower than average waste costs, but the relationships lack statistical significance except for export propensity in the Republic of Ireland and the UK.

Although a number of correlations confirm the hypothesised direction of the relationship in these analyses, there is even among these findings an inconsistency in the direction of the relationship between solid waste costs and measures of performance. Table 8.2-3 shows correlations between relative waste costs (between matched pairs of firms) and relative performance. For the dairy/drinks industry the correlations are not statistically significant. Indeed, when individual cases are examined (for the set of matched pairs of firms) it can be shown that the contrary is true: relatively better productivity, growth and export performance can be achieved with relatively greater waste costs (and vice versa).

Table 8.2-1 (a) Solid waste costs and firm performance: output measures in dairy/drinks and packaging

Firms in top quartile measured by	Dairy/drinks		Packaging	
	Solid waste costs	R	Solid waste costs	r
Germany (include. DSD)				
Value added per head	0.68	0.08	0.46	0.09
Plant growth	0.16	-0.15	0.37	-0.12
Exports	0.97	0.03	0.25	-0.21
All	0.61		0.33	
Republic of Ireland/UK				
Value added per head	0.03	0.56*	0.05	0.38
Plant growth	0.005	-0.09	0.10	-0.3
Exports	0.02	-0.57*	0.05	0.02
All	0.05		0.12	
Notes: **Significant at 1% level, *significant at 5% level (two-tailed p-value).				

Table 8.2-1 (b) Retailing

Firms in top quartile measured by	Solid waste costs	Correlation
Germany		
Sales per head	0.05	-0.24
All	0.08	
Republic of Ireland/UK		
Sales per head	0.01	-0.05
All	0.01	
Notes: **Significant at 1% level, *significant at 5% level (two-tailed p-value).		

Table 8.2-2 Solid waste costs and firm performance: input measures in dairy/drinks and packaging

Firms in top quartile measured by	Dairy/drinks		Packaging	
	Solid waste costs	r	Solid waste costs	r
Germany				
Plant age	0.51	-0.31	0.36	-0.19
Skills	0.55	0.13	0.24	0.13
R&D plant level	0.69	0.087	0.08	-0.34
All	0.61		0.41	
Republic of Ireland/UK				
Plant age	0.03	0.68*	0.10	0.05
Skills	0.04	-0.07	0.17	0.40
R&D plant level	0.008	-0.49	0.08	0.06
All	0.05		0.11	
Notes: **Significant at 1% level, *significant at 5% level (two-tailed p-value).				

Packaging

The packaging industries are a heterogeneous group of industries. Table 8.2-1 (a) shows that average solid waste costs are higher for German firms in the top quartile compared with those of all firms in that sector, whereas they are lower in the Republic of Ireland/UK. Table 8.2-2, which examines the relationship between waste costs and input measures of performance, shows more instances of the hypothesised relationship in each country, but stronger in Germany. However, none of these relationships is supported by statistically significant correlations. Table 8.2-3, which correlates differences in economic performance with differences in waste costs between matched pairs, generates no statistically significant results.

Retailing

Table 8.2-1 (b) shows that German retailers in the top quartile of retail firm performance have lower waste costs than the average, but there is no statistically significant relationship. And Table 8.2-3, which correlates differences in productivity performance with differences in waste costs between matched pairs of firms, shows an insignificant and positive, rather than a negative relationship, as hypothesised.

Table 8.2-3 International comparisons: correlations between comparative waste costs and comparative firm performance, all matched pairs

Industry	Costs as % of turnover (incl. DSD)	Productivity	Exports	Growth
Dairy/ Drinks	Solid waste	0.15 n=17	–0.25 n=7	–0.37 n=11
Retail	Solid waste	0.04 n=6	–	–
Packaging	Solid waste	–0.31 n=17	0.07 n=12	–0.22 n=12

Summary and discussion of results on firm performance and waste costs

The tested hypothesis was that the factors which influence firm performance can also enable firms to minimise waste costs.

Overall, no statistically significant relationships between waste costs and firm performance have been found. While in many cases there is support for the hypothesis in terms of a relationship between low average waste costs and top performing firms, there are also many cases where low waste costs are associated with relatively poor economic performance. The hypothesis is therefore not confirmed.

In Germany, only top performers in the retail sector have lower waste costs than the average, but this is statistically not significant. By and large, in dairy/drinks and also packaging the German top performers have even higher costs than the average (exceptions are concerning plant growth in dairy/drinks and exports in packaging). Therefore the analysis implies so far that German top performing firms are able to undertake more solid waste initiatives in certain areas like dairy/drinks and retailing, but are hardly able to minimise waste costs by doing so. On the contrary, the impression arises that German firms may be disadvantaged through their waste costs stemming from the waste initiatives. This would be against the hypothesis that a higher number of waste initiatives reduces waste costs which will be tested in section 8.3 below.

8.3 Waste costs in relation to turnover and in relation to waste initiatives

This chapter analyses waste costs as a percentage of turnover in each industry and country and links number of waste initiatives with total waste costs.

Waste costs in relation to turnover

This paragraph examines the importance of solid waste costs as a percentage of turnover in each of the countries. Solid waste costs include net recycling costs and landfill costs.

Tables 8.3-1 (a) to (c) show the distribution of total solid waste costs for firms in each sector in the Republic of Ireland/UK and Germany. It shows that German firms more frequently incur greater average costs measured as a percentage of turnover. Figures in brackets show costs inclusive of DSD charges. DSD costs fall on the packer/filler – in our case this is the dairy/drinks industry – and to some small extent on the retailer in the form of secondary packaging passed on to the customer. Waste costs inclusive of DSD charges raise costs, particularly in the dairy/drinks sector. The cost differential between Germany and the Republic of Ireland/UK is small for the dairy/drinks sector (excluding DSD) and retailing, but greater in the packaging sector. Results based on matched pairs of firms, shown in the final row of each Table, again support the findings.

Table 8.3-1 Distribution of solid waste costs (landfill and recycling)

a) Dairy/Drinks[a]

% of turnover	No. of plants in Germany	No. of plants in the Republic of Ireland/UK
< 0.01	6	6
0.01–0.03	1	3
0.03–0.05	5	1
0.05+	3 (15)[b]	3
All	15	13
Mean	0.050 (0.61)[b]	0.040
Median	0.025 (0.30)[b]	0.012
Matched mean	0.08[b]	0.020

Notes:

[a] A distribution of solid waste costs including DSD is given for the German dairy/drinks industry in the Table below. The retail industry shows no change due to the negligible costs (as a % of turnover) faced by that sector.

[b] Including DSD.

German total waste costs incl. DSD as a % of turnover	No. of plants in Germany
< 0.3	7
0.3–0.5	5
0.5–1.0	–
1.0 +	3
Mean	0.61
Median	0.30

b) Retailing

% of turnover	No. of plants in Germany	No. of plants in the Republic of Ireland/UK
< 0.05	1	2
0.05–0.1	3	2
0.1–0.2	5	1
0.2+	1	1
All	10	6
Mean	0.11 (0.12)	0.10
Median	0.1035 (0.105)	0.0442
Matched mean	0.10	0.07

c) Packaging

% of turnover	No. of plants in Germany	No. of plants in the Republic of Ireland/UK
< 0.1	11	14
0.1–0.3	10	1
0.3–0.5	3	4
0.5+	5	1
All	29	20
Mean	0.36	0.12
Median	0.195	0.02
Matched mean	0.28	0.12

Table 8.3-2 shows that in Germany recycling costs are generally more important than landfill costs. This is due to the shift away from landfill to recycling. Landfill costs in absolute terms have risen greatly in Germany. Thus, recycling activities are still cheaper for an individual firm than sending the waste to landfill. The relatively high DSD costs for dairies and soft drinks producers reflect the fact that fillers pay the DSD fee for sales packaging under the German system. Dairies use a high percentage of throwaway plastic packaging for their products for which a high DSD fee must be paid. In the Republic of Ireland/UK samples recycling costs were negligible or yielded revenues.

Table 8.3-2 Percentage breakdown of total solid waste costs in German sample

Type of cost	Packaging	Dairy	Drinks	Retailing
Landfill costs	25	4	4	65
Recycling costs	63	1	25	32
DSD cost	12	95	71	3
Total solid waste costs	100	100	100	100

Waste costs in relation to waste initiatives

In the Tables 8.3-3 (a) and (b) the hypothesis is explored that waste costs are reduced by waste initiatives; in particular, that lower waste costs are associated with a greater number of initiatives. Tables 8.3-3 (a) and (b) show that in no country a statistically significant relationship was found between solid waste costs and the number of solid waste initiatives undertaken. In both dairy/drinks and retailing the hypothesised negative relationship is not indicated. In Germany a bit of the opposite was discovered in dairy/drinks and retailing: Higher waste costs go hand in hand with a high number of waste initiatives, although many plants have successfully reduced both the amount of waste as well as waste costs as far as possible (see Box 8.3-1 for relevant examples). Packaging firms have been excluded because of the wide mix of firms sampled in that sector.

Table 8.3-3 Solid waste costs and number of solid waste initiatives

a) Dairy/Drinks

% of turnover	**No. of initiatives in Germany[a]**	**No. of initiatives in the Republic of Ireland/UK**
< 0.01	7.67	5.33
0.01–0.03	6.00	6.00
0.03–0.05	7.00	9.00
0.5+	8.33	6.67
All	7.47	6.00
Correlation coefficient[b]	–0.034 (n=15) (0. 18, n=15 incl. DSD)	0.39 (n=13)

Notes:

[a] When DSD is included the average number of solid waste initiatives is as in the following Table.

[b] For all countries, the correlation coefficient is not statistically significant at the 5% level.

German total waste costs incl. DSD as % of turnover	**No. of initiatives in Germany[a]**
< 0.3	7.14
0.3–0.5	8.20
0.5–1.0	–
1.0 +	7.00
All	7.47
Correlation coefficient	0.18 (n=15)

b) Retailing

% of turnover	**No. of initiatives in Germany**	**No. of initiatives in the Republic of Ireland/UK**
< 0.05	10.00	8.00
0.05–0.1	14.67	4.50
0.1–0.2	13.80	9.00
0.2+	13.00	4.00
All	13.60	6.78
Correlation coefficient*	0.08 (n=10) (0.05, n=10 incl. DSD)	0.32 (n=5)

Note: *None of the correlation coefficients were statistically significant at the 5% level

Box 8.3-1 Number of solid waste initiatives, extent of waste reduction and corresponding level of waste costs in Germany

Drinks factories

In three plants with five, six and ten solid waste initiatives it was said that recycling has taken place for a long time and that there is no more remaining waste which would go to landfill, it was not considered a cost issue. Net recycling costs for cardboard, glass, paper labels and aluminium lids were 50,000 DM per year in one plant. Although some used resources could be sold, the revenues were used up for transport costs. In the plant with 10 solid waste initiatives there were revenues from recycling which amounted to several 10,000 DM.

A further large plant with nine solid waste initiatives had a recycling centre of its own which was organised as a profit centre. It cost 1 million DM per year, but it also earned 1 million DM per year. Finally another large plant achieved a reduction of 2,500 t of packaging material through the use of larger and reusable packs. Overall, this later plant used only six solid waste initiatives, but the mentioned one was especially successful in terms of waste reduction.

Dairy plants

Dairy plants were especially active in waste reduction and were able to compensate at least parts of higher DSD costs through these activities.

A large dairy in Bavaria with the highest number of ten solid waste initiatives was able to achieve cost savings of 100,000 DM through avoidance of transport packaging wherever possible. It was said that 800,000 boxes and 400,000 m^2 of foil were used less. Other waste initiatives were waste reducing, but cost neutral, like material switch or recycling of plastics.

A large dairy with nine waste initiatives avoided through recycling all waste which formerly went to landfill and made a cost saving. All waste was collected separately in a another dairy with nine solid waste initiatives. Total recycling costs were still about 30,000 DM and landfill costs amounted of 138,000 DM.

A yoghurt plant in East Germany with eight solid waste initiatives used for the filling of yoghurt plastic pots made from PS and PP foil in the dairy itself. 98 per cent of all yoghurt pots were extruded in the dairy. The plastic foil could not be light weighted anymore because otherwise the functions of the packaging (safe transport etc.) could not be fulfilled. However, it was said that the extruded pots are lighter than the pots bought in as ready-to-fill pots. A cost comparison between self-made and bought pots is not available. But it was also said that if pots were bought more space would be necessary and this would cause costs. The machinery for extruding pots was more expensive than just the filling machinery. Plastic production waste from pot making is collected, sorted and sold to a recycling company. There are small revenues, no costs. There was also incoming packaging e.g. cardboard etc. which became waste. This waste was again collected, sorted (containers for wood, metal scrap, cardboard) and sold to a recycling company. For some materials the company had to pay for recycling. There were about 365 m^3 of industrial waste which could not be recycled and went to landfill. Disposal per ton cost 280 DM. Total waste costs were estimated to be around 100,000 DM per year. This figure includes recycling and landfill costs, a breakdown could not be obtained.

In a medium sized dairy plant with only seven initiatives landfill costs were still 30,000 DM per year and recycling were 7,000 DM. In a very small cheese factory with only four initiatives where most of the cheese was sold directly in loaves without any packaging total recycling costs were just above 1,000 DM.

Retailers

One company with 16 solid waste initiatives in place stressed that since 1991 a revolution had taken place within the company in the area of waste. Until 1991 each shop belonging to the retail chain collected and pressed the waste itself, now it is much better organised and costs are saved. Thus, in 1991 total waste disposal costs were 11 million DM for the whole company. Until 1996 the number of outlets has doubled, there were new shops opened in particular in Southern and East Germany. However, the disposal costs were only 13 million DM. Disposal includes both recycling and final disposal.

Another large retail using 15 different solid waste initiatives chain specified its waste reduction strategy as follows:

Reduction: It is done in order to save costs. There is no investment cost. Current costs concern discussions with suppliers and shop workers. Productivity increases because handling is optimised. As a consequence employment is reduced. e.g. reusable transport packaging for yoghurt.

Recycling (of transport packaging): Again it is done to save costs. Recycling firms invest, so there are no costs for the retailer. Productivity effects depends on the fact whether own staff or personnel from the recycling firm sorts the material. But: Employment effect is very small.

Use of recyclable material (for products): it is discussed with suppliers. The retail company is not affected.

Use of reusable material: Reusable boxes for fruit and vegetables brought a cost saving of several million DM. Handling was improved (positive productivity effect). No employment effect. The use of reusable palettes for yoghurt saves a lot of cardboard packaging (several million pieces). It is not only ecologically superior, but also improves handling (productivity effect!) and food presentation. The use of reusable bottles for beverages is cost covering.

A small retail chain still sent some waste to landfill. The company had a special permit to bring everything which is not recyclable to landfill (incl. food waste). Total costs for landfill and recycling were 50,000 DM p.a., (of which landfill costs are 30,000 DM). This amounts to 0.12 per cent of turnover. In comparison, total disposal costs in small shops of a large retail chain were 0.24 per cent of turnover because the central take back system increased transport costs.

If DSD is the most visible cost in the German waste stream in the dairy/drinks sector, the question arises as to whether waste initiatives are related to the DSD costs alone. This is considered below in the Tables 8.3-4 and 8.3-5. The number of waste initiatives in relation to DSD costs alone is higher than if just waste costs without DSD charges are considered for Germany. However, the results are again statistically not significant.

Table 8.3-4 Distribution of the DSD cost for dairy/drinks firms in Germany

DSD costs as per cent of turnover	Number of plants
< 0.25	7
0.25–1.0	5
1.0 +	3

Table 8.3-5 Relationship between solid waste initiatives and DSD costs for dairy/drinks in Germany

DSD costs as a per cent of turnover	Number of solid waste initiatives
< 0.25	7.14
0.25–1.0	7.00
1.0 +	8.20
All	7.47
Correlation coefficient	0.06 (n=15)

The relationship is not statistically significant at the 5 % level.

Summary and discussion of results on waste costs in relation to turnover

This chapter has examined the importance of waste costs as a percentage of turnover in each of the countries. Waste costs are shown to be higher in Germany; inclusive of DSD they were much higher particularly in the dairy/drinks sector. A comparison of waste costs between countries was also made on a matched plant basis, and the findings supported these results.

The hypothesis that waste costs are reduced by waste initiatives was explored; in particular, that lower waste costs are associated with a greater number of initiatives. The relationship between numbers of solid waste initiatives undertaken and solid waste costs showed no statistically significant results and in Germany some positive linkage between high waste costs and high number of waste initiatives in dairy/drinks and retailing was observed.

In a next step it will be explored whether these results are generally true or whether there is some variation according to regional location, ownership and plant size in each country.

8.4 Size of plant, ownership and location in relation to number of initiatives and waste costs

This section examines first of all the relationship between size of plant, ownership and location and number of solid waste initiatives. In a few cases also a linkage with environmental awareness initiatives could be explored. Also, waste costs are related to plant size and location as far as possible.

Plant size and number of initiatives

Tables 8.4-1 (a) and (b) show the relationship between plant and firm size and numbers of solid waste initiatives for each industry.

Table 8.4-1 (a) Average number of solid waste initiatives by plant size; dairy/drinks and packaging

Plant size (employees)	Dairy/drinks		Packaging	
	No. of plants	Solid waste initiatives	No. of plants	Solid waste initiatives
Germany				
1-100	4	5.75	11	7.12
101-200	3	5.30	5	7.40
201+	8	7.88	13	7.89
Correlation coefficient		0.12		-0.11
Republic of Ireland/UK				
1-100	4	5.00	7	3.71
101-200	3	4.66	13	3.92
201+	5	7.20	4	2.75
Correlation coefficient		0.004		-0.20
Notes: **Significant at 1% level, *significant at 5% level (two-tailed p-value).				

Table 8.4-1 (b) Average number of solid waste initiatives by plant size; retailing

Company size (by employees)	No. of companies	Solid waste initiatives
Germany		
< 6000	3	3.30
6001-20000	1	13.00
20001+	6	16.30
Correlation coefficient		0.20
Republic of Ireland/UK		
< 6000	4	5.50
6001-20000	1	4.00
20001+	4	8.75
Correlation coefficient		0.60
Notes: **Significant at 1% level, *significant at 5% level (two-tailed p-value).		

With the exception of the packaging sector the relationship between plant size and number of solid waste initiatives is positive in the German sample. The relationship is weaker for the dairy/drinks industry, but is stronger for the retail sector across all countries (the correlation coefficients show the underlying plat data). In general, these relationships are not statistically significant.

Small firms in Germany and number of initiatives

For small German firms there is more evidence in most instances that they undertake fewer initiatives than their large firm counterparts. In Germany, concerning the dairy/drinks and retail industry a clear correspondence of small companies and low total number of average initiatives on the one hand, and large companies and high total number of activities on the other hand is recognisable (see Tables 8.4-2 (a) and (b)). For instance a small German cheese company with about 40 employees undertook a total of four solid waste initiatives, whereas a large dairy with 800 employees undertook nine initiatives. Within the sample large retailers undertook the highest number of solid waste initiatives. The link between size and activity is only dominant at the company level where environmental decisions are made. At the shop level it is much weaker (see Ta ble 8.4-2 (c)). In the packaging industry all plants were very active, but again the smallest ones undertook on average only 7.3 initiatives against 10 in the largest plants.

Table 8.4-2 (a) Average number of solid waste initiatives and plant size in the German dairy/drinks and packaging sector

Plant size	Dairy/drinks	Packaging
1-50	4	7.3
51-100	7.6	7.25
101-200	10	8.4
201-300	7.5	9
301-400	6.7	9
401-500	–	9.7
501-750	–	10
751+	8	10.7

Table 8.4-2 (b) Average number of solid waste initiatives and company size in the German retailing sector

Company size	Germany
1-3000	9.3
3001-6000	–
6001-20000	12
20001-50000	15.7
50000+	13.3

Table 8.4-2 (c) Average number of solid waste initiatives and outlet size in the German retailing sector

Outlet size	Germany
1-50	13.5
51-100	14.0
101-150	–
151-200	14.0

Waste costs and plant size

Are small firms disadvantaged by relatively high waste costs? Table 8.4-3 examines the relationship between waste costs and plant size. There is no evidence of any general relationship. Only in the case of German dairy/drinks firms sampled there is evidence of a negative relationship (when DSD costs are included) i. e. larger firms show lower waste costs. In fact, a number of relationships are positive especially in the case of the German retail firms sampled. One German retail shop manager calculated the relationship between waste costs and shop size. His analysis showed that costs rose with shop size (not firm size). This was because of the higher productivity and the greater range of products associated with larger shops and therefore more waste (including food waste) generated. A similar result was found in the Irish/UK retailing business.

Table 8.4-3 Waste costs and size

(a) Germany: Plant size, solid waste costs

Plant size (employees)	Dairy/Drinks	Packaging
< 100	0.02 (1.00)	0.70
101–200	0.02 (0.25)	0.80
201 +	0.01 (0.57)	0.19
All	0.02 (0.62)	0.49
Correlation coefficient	0.16 (-0.13)	–0.09
Note: In brackets DSD costs are included.		

(b) Retailing

Firm size (employees)	Germany	Republic of Ireland/UK
< 6 000	0.09 (3)	0.04
6 001–20 000	0.09 (1)	0.03
20 001 +	0.12 (6)	0.07
All	0.11	0.10
Correlation coefficient*	0.49	0.05
Note: *None of the correlation coefficients were statistically significant at the 5% level.		

Ownership and number of initiatives

Table 8.4-4 shows a breakdown of both the number of solid waste and environmental awareness initiatives adopted by German firms with different legal forms of ownership. While there are differences in the composition of firms included in each category by size and product detail, the Table shows that on average PLCs undertake more initiatives than limited companies. Since PLCs were larger firms than their limited liability counterparts (though not necessarily larger plants) the data lend support to a positive relationship between firm size and total numbers of environmental initiatives adopted. In addition PLCs tended to have environmental policies. About half those plants sampled within the PLC group were branch plants of multinational companies. Many of these plants reported that their environmental policy was based on strong company-wide policies to implement good environmental practices.

A positive relationship between firm size and total numbers of environmental initiatives adopted was also found in the Republic of Ireland. The average number of initiatives undertaken by multinational enterprises was 10.3 compared with 8.3 for all other PLCs sampled there.

Table 8.4-4 Average number of solid waste and awareness initiatives undertaken, broken down by legal form of firm ownership (number of firms shown in parentheses)

Germany	Dairy/Drinks		Retailing		Packaging	
	Solid waste	Awareness	Solid waste	Awareness	Solid waste	Awareness
Co-op	6.3 (3)	3.3 (12)	12 (1)	3 (1)	–	–
Ltd	2.6 (3)	7.8 (12)	10.8 (4)	2.5 (4)	8.3 (25)	1.5 (25)
PLC	–	–	14.4 (5)	5 (5)	9.3 (4)	2 (4)

Location and number of initiatives

Table 8.4-5 examines the number of initiatives adopted by sample firms in different locations. The Table shows a breakdown between the Republic of Ireland (ROI), Northern Ireland (NI) and Great Britain as well as East and West Germany. Data were available on this breakdown for dairy/drinks companies and packaging companies in all four countries, but only for the separate retail companies and chains in NI, ROI and Great Britain (there are no East German retail chains).

In retailing, the four companies located in Great Britain were the most advanced in terms of average number of solid waste initiatives. Concerning environmental awareness the Irish dairies are the most active within the entire sample in the Republic of Ireland/UK, but were still on a lower level than in West Germany.

Table 8.4-5 Plant location and average number of initiatives (number of firms shown in parentheses)

	Solid waste	Environmental awareness	Total no. of initiatives
Dairy/Drinks			
Germany			
West(10)	7.40	2.90	10.30
East(5)	7.60	1.20	8.80
Republic of Ireland/UK			
NI(6)	5.67	0.67	6.34
ROI(7)	5.00	1.43	6.43
GB(2)	6.00	0.50	6.50
Packaging			
Germany			
West(24)	19.50	4.10	23.6
East(5)	7.40	0.60	8.00
Republic of Ireland/UK			
NI(5)	4.40	0.60	5.00
ROI(9)	3.44	0.77	4.21
GB(11)	3.27	0.91	4.18
Retailing*			
Republic of Ireland/UK			
NI(2)	6.50	0.00	6.50
ROI(3)	4.33	0.67	5.00
GB(4)	8.75	2.25	11.00
Note: *No regional breakdown available for Germany.			

Overall, no consistent relationship is shown between numbers of different initiatives adopted by firms in these regions and locations. Only in the German packaging industries there is a clear difference in the number of initiatives adopted across all types of initiative considered. In this sector considerable lags were found, for example, between indigenous East German producers in the paper industry compared with their western counterparts.

Location and waste costs in Germany

In dairying West German plants incur lower total waste costs than plants in East Germany. Notably, the DSD costs are higher in East Germany where on average factories were larger and more specialised, e.g. on yoghurt packed exclusively in plastic pots for which DSD costs are high (see Table 8.4-6 (a)).

Table 8.4-6 (a) Waste costs and plant location in the German dairy/drinks sample, as percentage of turnover

Type of cost	Dairy EG	Dairy WG	Drinks EG	Drinks WG
Landfill costs	0.02	0.02	0	0.01
Recycling costs	0.02	0.01	0	rev.: 0.025
DSD cost	0.97	0.71	0	0.18
Total waste costs (net)	1.01	0.74	0	0.17

Within packaging comparisons between East and West Germany were only possible for glass, paper and metal. East German factories achieve lower waste costs for glass and metal packaging, but have higher waste costs concerning paper due to a high share of paper recycling costs (see Table 8.4-6 (b)).

Table 8.4-6 (b) Average waste costs and plant location in the German packaging sample, as percentage of turnover

Type of cost	Plastics WG	Glass EG	Glass WG	Paper EG	Paper WG	Metal EG	Metal WG	Mach. WG
Landfill costs	0.13	0.1	0.03	0.10	0.17	0	0.002	0.04
Recycling costs	0.07	0.34	0.3	0.88	0.55	rev.: 3.4	0.01	0.025
DSD cost	–	0	0.4	–	–	0	–	–
Total solid waste costs	0.20	0.44	0.73	0.98	0.72	revenues	0.012	0.065

Summary and discussion on results on the linkage between number of solid waste initiatives and size, ownership and location

This chapter has considered the distribution of the number of solid waste initiatives undertaken by firms in each of the sample countries in relation to size, ownership and location. It has also examined the relationship between waste costs and plant size as well as between waste costs and location.

Plant size, number of initiatives and waste costs

Although there is not always consistency between countries, industries and particular types of initiative, there is evidence of a relationship between numbers of initiatives performed and plant and firm size in all sample countries. It is particularly clear that small firms undertake fewer initiatives than their larger national counterparts. This was clearest in the German retailing sector when shop size was considered. The packaging sector includes firms producing different products in different size classes so comparisons must therefore be considered as inconclusive.

In general there was no evidence of a clear relationship between waste costs and plant size, although there was some evidence of higher costs associated with smaller size for firms in the German dairy/drinks sector.

Ownership and number of initiatives

A breakdown of the total number of initiatives adopted by firms with different legal forms of ownership showed that in all sample countries PLCs tended to undertake more initiatives than limited companies. PLCs tended to be larger firms with environmental policies and more environmental awareness initiatives, thus reinforcing the relationship between size and number of initiatives. There was also evidence that branch plants of MNEs undertook an above average number of initiatives. Many MNEs claimed to adopt strong company-wide environmental policies.

Location, number of initiatives and waste costs

In the sample, only for Germany some location observations can be made. In East German dairies DSD costs tended to be higher because more non-reusable packaging was used. A clear lag was also found concerning the number of solid waste initiatives, for example, between indigenous East German producers in the paper industry compared with their western counterparts.

For Germany all these results could imply that small plants potentially suffer from a competitive disadvantage through relatively high waste costs and that there could also be a negative impact on East German firms through stringent packaging legislation. This, however, can only be further proved through an analysis of the micro effects of each solid waste initiative on firm performance (see section 8.5). Individual effects are also linked to the motivation for undertaking each initiative. Therefore also the drivers of each initiative must be examined.

8.5 Solid waste initiatives, their drivers and effects on firm performance

This chapter first of all examines the drivers underlying the different solid waste initiatives undertaken by firms. Regulation was found to be a prominent driver in Germany, whereas in the Republic of Ireland/UK cost was much more important. Separate consideration is given to environmental pressures related to types of packaging arising from customers and suppliers. The chapter also considers in detail the economic effects of the adoption of these initiatives on resources and on the performance of the firm. Special emphasis is put on the additional requirements for capital and running costs in Germany separated by the different drivers. Details are also given on the employment effects in Germany.

In the case of the Republic of Ireland/UK the level of detail known about the initiatives varied. In some cases the initiative driver and economic effects were given, while in other cases only the initiative driver was known. The German managers were more aware of the detail and this may be a reflection of greater 'green accounting', stricter legislation or a more developed environmental management structure. The following analysis is based on initiatives where the associated drivers and effects were known.

8.5.1 Drivers for undertaking solid waste initiatives

Table 8.5-1 breaks down the different drivers for solid waste initiatives between legislation, cost, external pressure (including from customers and suppliers), health and safety, and public image. A further column contains the miscellaneous drivers. These represent a mixture of factors and unspecific drivers including, company policy, marketing and 'unknown' drivers. A number of the miscellaneous drivers in Germany referred to tradition and are evidence that many companies had well-established environmental management practices.

Table 8.5-1 shows that the most important driver in each of the sample countries was cost, followed by legislation and customer pressure. But stringent German legislation is responsible for significantly more initiatives than legislation in the Republic of Ireland/UK; customer or supplier pressure is also relatively more important in Germany and is itself often driven as a consequence of regulation.

Health and safety, public image and other drivers are relatively unimportant in any country.

Table 8.5-1 Percentage composition of drivers for undertaking solid waste initiatives, all sectors (in brackets number of initiatives)

Solid waste initiatives	Legislation	Cost	Customer/ Supplier	Health & Safety	Public Image	Misc.	No. of all initiatives
Germany	38.3 (184)	40.4 (194)	12.3 (59)	0.2 (1)	0.3 (1.5)	8.4 (40.5)	480
Republic of Ireland/UK	8.7 (16.5)	69.6 (131.5)	8.5 (16)	1.6 (3)	1.1 (2)	10.5 (20)	189
*Where figures are not whole numbers, they refer to cases where more than one driver was reported. e.g. between cost and customer/supplier.							

Table 8.5-2 includes an attribution of drivers to individual sectors and is to be read in conjunction with Table 8.5-3 which shows, for all industries sampled, the types of solid waste initiatives which were driven by regulation, cost and external (customer/supplier) pressures. These data are for Germany only where regulatory and external pressures on solid waste initiatives were most important. Overall, cost and legislation are the most important drivers across all industries.

Examination of the various sub-categories of solid waste initiatives shows a variation in the importance of different drivers in Germany. While reuse efforts have either been traditionally in place or are explicitly undertaken for cost reasons, recycling efforts are undertaken for cost reasons except by retailers who reported that their recycling activities were mainly driven by legislation. Special design and material switch, most frequently carried out in the packaging and dairy/drinks industry, were undertaken for cost reasons only in a minority of cases, the main reasons were legislative and subsequent customer/supplier pressure. Packaging recovery, packaging reduction and waste segregation were the activities with the highest frequency of legislative pressure. Light weighting was mainly undertaken for cost reasons (in order to save DSD fees as a filler in the dairy/drinks industry (and thereby indirectly driven by legislation) or to save material costs as a packaging producer); retailers felt that their light weighting of transit packaging was undertaken as a clear result of the new packaging legislation. Waste reduction (i.e. compacting of waste) and also waste targeting was clearly undertaken in order to save transport costs.

New designs to remove outer packaging, the introduction of larger packs and the elimination of excess packaging were undertaken for legislative reasons in nearly 50 per cent of cases and, to a lesser extent, for cost reasons.

Table 8.5-2 Percentage of solid waste initiatives attributed to different drivers, according to industries, in the German sample

% of initiatives driven by	Legislation	Cost	Customer/ Supplier	Miscellaneous*	TOTAL
Dairy/drinks	28.0	43.0	11.0	18.0	100
Packaging	39.2	42.7	12.1	6.0	100
Retail	45.6	34.6	14.0	5.8	100
*Includes health & safety and public image.					

Table 8.5-3 Percentages of solid waste initiatives attributed to different drivers, all industries, in the German sample

Initiative Description	Legislation	Cost	Customer/Supplier	Miscellaneous*
Reuse materials	8	29	28	35
Recycle materials	38	58	1	2
Special design	55	5	23	18
Material switching	40	17	43	0
Packaging recovery	65	33	2	0
Light weighting	20	49	31	0
Packaging reduction	50	32	15	3
Waste reduction	14	80	6	0
Waste segregation	61	34	5	0
Waste targeting	0	80	0	20
Total	34	42	13	10
*Includes health & safety and public image.				

8.5.2 Effects of solid waste initiatives

Managers were asked the following set of questions with respect to individual initiatives undertaken and their effects:

1. How much did it cost to implement this initiative?
   a) Capital costs?
   b) Running costs?

2. Did this initiative change machinery along your production line?

Quantify the cost savings in terms of potential unit cost or general savings from this initiative (in £ or DM per annum).

3. What are the effects of investigating this initiative in terms of:

   a) employment effects;
   b) productivity effects;
   c) product price;
   d) market share;
   e) competitive advantages.

Table 8.5-4 shows the economic effects of solid waste initiatives for each country. In total there were 990 effects from 480 initiatives in Germany and 346 effects arising from 190 initiatives in the Republic of Ireland/UK. These gave rise to 2.06 and 1.82 effects respectively.

Table 8.5-4 Solid waste initiatives, number and percentage of effects[a]

No. of effects	Germany	ROI/UK
Employment	46^{10}	28^{2}
Machinery	157	46
Additional capital/running cost	270	60
Percentage of total	47.8	39.3
Unit Cost Reductions	62^{3}	61^{2}
General Savings	229	84
Percentage of total	29.4	41.6
Competitiveness	118^{10}	11
Productivity	41^{32}	21^{9}
Price	39^{7}	24
Market Share	28^{2}	11
Percentage of total	22.8	19.7
Total no. of effects	990	346
Number of initiatives	480	189
No. of effects per initiative	2.06	1.83

Notes: [a] Assume all changes in productivity, price and unit cost are falls (negative). Assume all changes in competitiveness, employment, market share and machinery are rises (positive). A superscript shows the number of effects working in the opposite direction.

The vertical column divides these effects into the requirement for inputs: employment, machinery and capital cost. Financial savings arising from the initiatives are shown; these can give rise to unit cost savings, where savings are attributable directly to the product, and general savings, where departmental costs are effected. A company could respond positively to both questions for the same initiative. The competitiveness effects of the initiative is captured in four different measures (in answer to separate questions), namely the initiative gave rise to improvements in company competitiveness in the view of the respondent, or to changes in productivity, price or market share.

The Table above shows that a higher percentage of effects in Germany compared with the Republic of Ireland/UK are attributable to resource inputs (47.8 in Germany vs. 39.3 in the Republic of Ireland/UK). Fewer effects of initiatives are associated with cost savings (29.4 in Germany vs. 41.6 in the Republic of Ireland/UK), while a similar percentage give rise to competitiveness effects (22.8 in Germany vs. 19.7 in the Republic of Ireland/UK).

The Table also shows that in both countries, the employment implications of undertaking initiatives tend to be positive. Relatively few initiatives are associated with labour saving (as shown by a superscript) with more of these cases taking place in Germany than in the Republic of Ireland/UK (10 cases vs. 2). There is a general implication of a requirement for additional or modified machinery and of additional costs with the need for more machinery in Germany being about three times as strong as in the Republic of Ireland/UK, and the requirements for capital costs being even 4.5 times as strong.

Unit cost savings are less frequent than general savings (though the two can overlap). Interestingly, almost half of all German solid waste initiatives were accompanied by general savings, in general of landfill and/or recycling costs through reduced amounts of waste. Management's view of the competitiveness effects of initiatives are in many cases positive and supported by price reductions and improved market shares. Where there is an effect on productivity this is generally negative in the Republic of Ireland/UK, while in Germany it is on average beneficial, reinforcing the positive competitiveness effects reported.

Effects of solid waste initiatives in relation to drivers

Table 8.5-5 shows a breakdown of these effects by driver for each industry in Germany. It shows:

(a) that in the case of the packaging and dairy/drinks industries, initiatives driven by regulation were more likely to require additional resources (except labour), were less likely to create both unit cost and general financial savings and especially in the dairy/drinks sector were less likely to improve the competitiveness of the firm compared with non-regulatory drivers (costs and

Table 8.5-5 The Number of Effects arising from Solid Waste Initiatives[a] (Germany)

No. of effects	Cost			Legislation			External Pressures			All		
	P	D/D	R	P	D/D	R	P	D/D	R	P	D/D	R
Employment	$4^{0.5}$	6	7.5^{2}	4^{3}	1	11	$5^{3.5}$	1	4.5	13^{7}	10	23^{2}
Machinery	32.5	24	18	19	11.5	24.5	10.5	7	6.5	62	45	50
Capital/ running cost	42.5	33	33	52	18.5	34	10	11	8	114	78	78
Percent of total	38.5	55.6	50.9	68.0	68.1	48.9	28.8	31	51.4	43.4	52.0	50.5
Unit Cost	24	3	1.5	5	2	3.5	15.5^{3}	6	–	45^{3}	12	5
Savings	63.5	33	27	19.5	9	36	11.5	5	13	99	53	77
Percent of total	47.6	31.7	24.8	19.6	24.2	27.8	30.5	18	35.1	33.1	25.4	27.4
Competitiveness	$26.5^{0.5}$	5	23	13^{4}	0.5	24	$8.5^{5.5}$	9^{1}	3	50^{10}	18^{1}	50
Productivity	3^{3}	5^{4}	3^{3}	4^{1}	2.5^{2}	$4.5^{4.5}$	9^{7}	7^{7}	$0.5^{0.5}$	16^{11}	17^{13}	8^{8}
Price	6.5	1	3.5	5^{1}	–	3.5	14.5^{4}	6^{2}	–	26^{5}	8^{2}	7
Market Share	2.5	$3.5^{0.5}$	0.5	$3.5^{0.5}$	$0.5^{0.5}$	1	$4^{0.5}$	9	1.5	10^{1}	15^{1}	3
Percent of total	18.8	12.8	26.1	20.4	7.7	23.2	39.5	50.8	13.5	23.4	22.6	22.1
Total effects	205	113.5	117	125	45.5	142	89	61	37	435	256	301
No. of Initiatives	99.5	48	47	80	31	62	28	12	22	234	112	136
Avg. effects	2.06	2.36	2.45	1.56	1.47	2.29	3.16	5.08	1.68	1.86	2.29	2.21

P= Packaging industries D/D Dairy and Drinks industries R= Retailing

Notes: [a] As for Table 8.5-4

external pressures). There was little difference between the effects of initiatives driven by regulatory and non regulatory drivers for retail firms.

(b) that initiatives driven by external pressure (customer/supplier pressure) were for the packaging and dairy/drinks industries less likely to require additional resources than legislation driven initiatives, more likely to generate unit cost savings, especially in the packaging sector, and more likely to create positive competitiveness effects compared with cost driven initiatives. In contrast retailing firms were more likely to achieve general cost savings through external pressures, but these were not high enough to achieve overall competitive benefits.

In the following first the additional cost effects incurred in Germany are explored in greater depth. Furthermore, the employment and competitiveness impacts are examined.

Machinery effects and related capital costs

Tables 8.5-6 (a) and (b) provide a quantification of the importance of machinery and other investments required to undertake various initiatives in Germany and the Republic of Ireland/UK where these were identified. Investments are classified into major and minor. Major includes large-scale projects such as depots, new lines and new machines. Minor includes compactors, new tools, pallets, filters and so on.

Table 8.5-6 (a) Breakdown of Machinery Effects by Sector, Germany* (percentage in brackets)

No. of Machinery effects	Dairy/Drinks	Retail	Packaging
Solid waste			
Major	17 (36)	6 (38)	8 (22)
Minor	30 (64)	16 (62)	29 (78)
Total	47	22	37

Note: *The number of machinery effects shown in this Table differs from those in shown Table 8.5-4. The data here refer to where specific investments were known. Occasionally more than one initiative was associated with a single investment.

Table 8.5-6 (b) Breakdown of Machinery Effects by Sector, Republic of Ireland/UK*(percentage in brackets)

No. of Machinery effects	Dairy/Drinks	Retail	Packaging
Solid waste			
Major	4 (27)	5 (56)	4 (22)
Minor	11 (73)	4 (44)	14 (78)
Total	15	9	18

Note: *The number of machinery effects shown in this Table differs from those in shown Table 8.5-4. The data here refer to where specific investments were known. Occasionally more than one initiative was associated with a single investment.

Tables 8.5-6 (a) and (b) show that solid waste projects required only minor investments in the majority of cases. More German firms invested in plant and machinery and they invested more heavily in major projects in the dairy/drinks industry. Irish and UK plants invested on average less frequently across all sectors, but more heavily in the retail industry than the German plants. Major investments in the packaging industry accounted for about the same percentage share of all investments both in Germany and the Republic of Ireland/UK.

Extent of additional expenditure in Germany

In this section the size of the required additional investments for implementing solid waste initiatives including their enhanced running costs is analysed for German firms for which such costs were more significant and where data were more complete than for the Irish/UK sample firms.

Table 8.5-7 shows the capital investments that were associated with solid waste initiatives. The Table shows that the median investment was greatest for legislation driven investments in Germany. In Table 8.5-7 it is illustrated that much more initiatives required running costs and this was about evenly distributed between legislation and cost driven initiatives.

Table 8.5-7 Drivers and associated capital costs, solid waste initiatives, Germany

Driver:	Median cost (DM mn)	Number	Percentage
Legislation	300	5	17
Cost	264	15	50
External pressure	150	6	20
All non-legislative	240	25	83
All pressures	150	30	100

Table 8.5-8 Drivers and associated running costs, solid waste initiatives, Germany

Driver:	Number	Percentage
Legislation	46	43
Cost	52	48
External pressure	4	0.04
All non-legislative	62	57
All pressures	108	100

Table 8.5-9 shows a detailed distribution of additional capital and running costs for German firms arising from all solid waste initiatives, broken down by industry and by size.

Table 8.5-9 Costs arising from solid waste initiatives in Germany

Costs:	Dairy/drinks		Packaging		Retailing		All industries	
	No.	%	No.	%	No.	%	No.	%
Capital costs (DM)								
< 1 000								
1 000-10 000	1	9			1	17	2	7
10 001-50 000	1	9	3	23			4	13
50 001-100 000	2	18	4	31	2	33	8	27
100 001-500 000	5	45	2	15	2	33	9	30
> 500 000	2	18	4	31	1	17	7	23
All solid waste	11	100	13	100	6	100	30	100
Running costs (DM)								
< 1 000	2	7	2	3			4	4
1 000-10 000	14	50	27	46	3	14	44	41
10 001-50 000	9	32	19	32	9	43	36	33
50 001-100 000	1	4	3	5	2	10	6	6
100 001-500 000			6	10	3	14	9	8
> 500 000	2	7	2	3	4	19	9	8
All solid waste	28	100	59	100	21	100	108	100

The top half of Table 8.5-9 shows that about half of the German capital projects involved expenditure of above DM 100,000. Most were undertaken in the dairy/drinks and packaging industries. The bottom half of Table 8.5-9 shows annual running costs associated with solid waste initiatives. Nearly one-half involved expenditures of less than DM 10,000 and one-fifth above DM 50,000. Most of these German initiatives were concerned with recycling. In the Republic of Ireland/UK waste collection and disposal were the main class of solid waste initiatives causing running costs. Material recycling in the Republic of Ireland/UK was undertaken where this was collected free or produced revenue.

Additional expenditure in Germany due to legislation

In the following these raised costs are analysed by driver for Germany. Table 8.5-10 focuses on raised costs arising from legislation in Germany. Costs are divided between capital and running costs. It shows that only a small number of regulatory driven initiatives (similar to the Republic of Ireland/UK case) involved capital expenditure. Most initiatives involved raised running costs, of which just over half were eventually cost neutral to the firm. These additional costs are generally recycling costs. These explicit costs do not occur in the Republic of Ireland/UK samples. Some, but not all, of these recycling initiatives were cost neutral in Germany depending on the market for secondary materials.

Table 8.5-10 Legislation driven solid waste initiatives in Germany and their associated costs

Cost effects	Dairy/drinks		Packaging		Retailing		All industries	
	No.	%	No.	%	No.	%	No.	%
Capital costs	1.0	3.4	3.5	5.0	1.0	2.9	5.5	4.0
Running costs	10.5	35.6	36.0	51.1	11.5	33.9	58.0	43.3
No cost effects	18.0	61.0	31.0	43.9	21.5	63.2	70.5	52.7
Total	29.5	100.0	70.5	100.0	34.0	100.0	134.0	100.0

Additional expenditure in Germany due to external pressures

Table 8.5-11 shows that of 10 initiatives driven by external pressure in Germany one-quarter involved capital expenditure. The remainder involved raised running costs of which virtually all were cost neutral. This arose through light weighting, material reuse, material switching and so on. Details on the additional expenditure in Germany are shown in the appendix to this chapter.

Table 8.5-11 Externally driven solid waste initiatives in Germany and their associated costs

Cost effects	Dairy/drinks		Packaging		Retailing		All industries	
	No.	%	No.	%	No.	%	No.	%
Capital costs	6	85.7	4	13.8	0.5	5.9	10.5	23.6
Running costs			4	13.8			4.0	9.0
No cost effects	1	14.3	21	72.4	8	94.1	30.0	67.4
Total	7	100	29	100	8.5	100.0	44.5	100.0

Employment and competitiveness effects in the German sample

In the next two sections details on both employment and competitiveness effects arising from solid waste initiatives in the German sample are presented.

Employment effects arising from solid waste initiatives in the German sample

Table 8.5-12 compares the additional employment effects arising from initiatives between Germany and the Republic of Ireland/UK (for which complete and comparable data were available).

Table 8.5-12 Employment implications of solid waste initiatives

Employment impact	Germany			
	Dairy/drinks	Packaging	Retail	Total
Negative	2	7	2	11
Positive	3	1	12	16
Part-time				
1	1	3		4
2	1		1	2
3	1			1
4				
5				
5-10				
10-100	2	2	3	7
100+		1	4	5
1000+			1	1
Total	10	14	23	47
	Republic of Ireland/UK			
	Dairy/drinks	Packaging	Retail	Total
Negative			1	1
Positive		1	4	5
Part-time		1		1
1	5	3	1	9
2		1		1
3				
4				
5		1		1
5-10				
10-100				
100+		1		1
1000+				
Total	5	8	6	19

The Table does not show where initiatives involved inputs from the established workforce or the R&D department. The larger number of projects engaged in by German companies gave rise to about two and a half times the number of employment effects of their Republic of Ireland/UK counterparts. German initiatives also gave rise to relatively large employment effects. The net employment effect in all sectors was clearly positive. In cases where a fall is shown, these were always small (if not precisely quantified).

Details of the sources of these German employment effects are given in the Appendix A.8 to this chapter.

Competitiveness effects arising from solid waste initiatives in the German sample

There are a total of 118 competitiveness effects arising from solid waste initiatives in the German sample with only 11 of these effects being negative. The majority of effects was found in the packaging and retail industry. There were only a few impacts on market share and productivity. Price effects were most frequent in the packaging sector. Details of the sources of the nature of any competitiveness effects are given in the Appendix A.8 to this chapter.

8.5.3 Summary and discussion on results on drivers and effects of solid waste initiatives

Chapter 8.5 has shown that legislation was a much more important driver for solid waste initiatives in Germany than in the other countries considered. Comparisons between countries showed that initiatives undertaken by German firms were more likely to require additional resources and more major investments (because the majority was legislation driven), were less likely to bring about unit cost savings, but more likely to bring about general savings and overall more likely to give rise to positive competitiveness effects compared to the other two countries. The larger number of solid waste projects and initiatives engaged in by German companies gave rise to two and a half times the number of employment effects of their Republic of Ireland/UK counterparts.

In particular, the detailed micro level analysis of cost effects of solid waste initiatives broken down by the different drivers has shown that hardly any negative competitiveness impacts were triggered in Germany. Frequently higher capital costs could be compensated by general savings from the solid waste projects.

8.6 Competitive advantages and disadvantages, constraints on growth and direct customer/supplier pressure

Respondents were asked to specify the competitive advantages and disadvantages they faced; this was done in order to measure the importance of environmental factors influencing the firm's competitive performance relative to other factors. Results are shown in Table 8.6-1 (a) and (b). A wide range of different advantages and disadvantages was given. Companies were also asked concerning their constraints on growth and direct customer/supplier pressure.

8.6.1 Competitive advantages and disadvantages

Competitive advantages

In the sample, most advantages related to non-price factors especially product quality and variety, design and presence in niche markets. Environmental factors occurred infrequently.

In Germany e.g. dairy/drinks companies consider their product quality as their main competitive advantage. Only two companies in the German dairy/drinks sample stated that their environmental initiatives conferred important competitive advantages, in particular their eco-audits were seen as helpful for marketing as well. In the packaging sample manufacturers of machinery and also metal can manufacturers saw their technology as their most important competitive asset. In East Germany metal and glass manufacturers had occupied niche markets and considered this as the source of their high performance; in one case special metal cans for Marks & Spencer were produced. Only one German paper and board and three plastics firms stated that the environmentally friendly quality of their products conferred an advantage. In the paper plant case the product was made from 100 per cent recycled paper, in the plastics producing plants the products were reusable. A glass bottle factory thought of its own recycling plant as a competitive advantage. None of the plants mentioned light weighting as a special competitive advantage, it was reckoned to be a necessary adjustment measure in a changing regulatory and market climate. From all Irish/UK sample plants only three packaging firms claimed an advantage for environmentally driven waste pressures. Retailers in all sample countries considered their size, variety, service and price levels as most important competitive advantages. No retailer mentioned environmental benefits (see Table 8.6-1 (a)).

Competitive disadvantages

Environmental factors were mentioned on even fewer occasions as a competitive disadvantage. Two German companies reported losses in market share because

Table 8.6-1 (a) Competitive Advantages: main factors and numbers of responses

	Dairy/Drink		Retail		Packaging	
	Irl/UK	Germany	Irl/UK	Germany	Irl/UK	Germany
Service	7	3	4	2	4	5
Price	5	1	4	1	3	1
Wages						
Hygiene					1	
Variety	6	1	3	4		1
Image						
Convenience			1			
Monopoly					3	3
Size			1	4	3	
Suppliers						
Location	2	4	2		2	4
Marketing	4	2	1	1	3	2
Versatility	2	4	1	1	4	9
Environmental factors		2			3	5
Quality	7	9	6	3	8	
Labour quality	11	1	1	2	2	
Design		1			4	11
Part of Group		1				1
Exports		1				
Own Brand		1				
No. of Depots				2		
Niche Products						6
Established	1				3	
Networks					2	

of strict environmental regulation. Both had produced PVC which was banned for health reasons. One paper company had lost market share on folding cartons because the Packaging Ordinance gave incentives to reduce transport and outer packaging; it was also said that light weighting created a competitive disadvantage because it raised quality problems for the packer/filler which would eventually impact on the carton producer. No firm in the Republic of Ireland/UK reported that environmental matters constituted a competitive disadvantage. Price was stated to be an important competitive disadvantage, especially in the dairy/drinks industries. One German drinks company thought that the Packaging Ordinance with its high quotas for reusable packaging was a disadvantage. However, another drinks company strongly favoured the German system of reusable packaging and feared that the softer EU legislation could undermine the German system. German packaging companies complained about low prices and high wages. Retailers in all sample countries faced marketing problems and difficulties in competing with discount stores (see Table 8.6-1 (b)).

Table 8.6-1 (b) Competitive Disadvantages: main factors and numbers of responses

	Dairy/Drink		Retail		Packaging	
	Irl/UK	Germany	Irl/UK	Germany	Irl/UK	Germany
Service		1		1		
Price	4	5	1	2	3	6
Wages						8
Hygiene	1					
Variety	1	1	2			
Image		1	1			
Convenience		1			3	
Size	1	3	3	1	3	4
Suppliers					2	
Location	3				4	4
Marketing	2	3		3	1	
Versatility	1				1	
Environmental factors		2				3
Quality				1		
Labour quality	2	1	2	1	2	1
Design	1				2	3
Shrinking Mkt.		1			4	
Transport costs	3					

8.6.2 Constraints on growth

Firms were also asked to list factors that constrained growth. Not a single firm in the entire sample cited environmental regulation or costs as a constraint on growth. In the German dairy/drinks sector overcapacity, transport costs and raw material costs were said to be significant constraints on growth. In the packaging industry raw material costs, fierce competition, low margins as well as overcapacity were stated to be constraints on growth.

industry raw material costs, fierce competition, low margins as well as overcapacity were stated to be constraints on growth.

8.6.3 Direct customer/supplier pressure

Interviewees were also asked to state the importance of direct environmental pressures from customers or suppliers, and pressures placed on suppliers, relative to other pressures put on or exerted by the interviewed plant.

It became clear that although pressure concerning packaging was present, it was just one of many other pressures put on the sample plants. These other plants included price, hygiene, quality and so on, see Table 8.6-2). Both in Germany and in the Republic of Ireland/UK pressures by customers on packaging companies included light weighting and downgauging, reuse of trays and reduction of transit packaging.

Retailers in the UK were under little environmental customer pressure, while German retailers were subject to a variety of environmental pressures including the wish for reusable bottles, labelling, animal rights etc.

Detailed examples of the extent of environmental pressures in the various German industries are given below at the end of this chapter in Box 8.6-1.

Finally, the question must be asked whether retailers themselves exerted pressures on their suppliers. An analysis of the number of solid waste initiatives undertaken by dairy/drinks firms serving major retailers compared with all other dairy/drinks firms is shown in Table 8.6-3 below. No evidence was found that such suppliers undertook more solid waste initiatives.

Table 8.6-2 Pressures on firms, number of responses

	Packaging		Dairy/drinks		Retail	
From customers/consumers	Irl/UK	Germany	Irl/UK	Germany	Irl/UK	Germany
Service	3		2			
Price	8	9	7	9	1	
Delivery	5		3	3		
Hygiene	2		1			
Convenience			5		2	
Quality	6		1			
Environmental of which						
Lightweight	10	11	6	7	2	3
Information	6					
Packaging Size	3	7	3	1	2	
Total	34	20	25	19	5	11
On Suppliers						
Service	1					
Flexibility		1				
Variety		1				
Quality	2	7	3			
Price	1	13	2	1		10
Delivery (JIT)	1					
Hygiene					4	10
Convenience			2		1	
Environmental of which						
Lightweight	8	6	9	9	4	7
Information	1					
Packaging size	1	6	6	4		
Total	13	28	16	10	9	27

Table 8.6-3 Numbers of initiatives undertaken by dairy/drinks firms serving major retailers

Average number of initiatives	UK	Germany
Solid waste initiatives:		
Supplying major retailers	5.2	6.8
Others	5.6	8.3
Environmental awareness initiatives:		
Supplying major retailers	2.6	2.2
Others	2.0	2.5

Box 8.6-1 Case study examples of environmental pressures on German firms

In general, price was felt to be the source of strongest customer/consumer pressure in all industries; environmental issues were of less importance.

In the German drinks industry sample, three of five firms reported that retailers were especially interested in efficient handling of materials flow. Two mineral water and soft drinks companies felt pressure towards installing a new filling line for reusable PET bottles. They planned to fill soft drinks in 1-litre bottles made of PET and abandon 0.7-litre bottles made of glass. The major reason for this is that PET bottles are much lighter than glass bottles. Subsequently the products can then be distributed over a larger area. Retailers push the new bottling systems because they save shelf space. The drinks industry transforms the pressure that it receives from retailers in terms of new packaging into a co-operative relationship with packaging retailers. Thus the drinks industry develops, together with the packaging industry and other consulting bodies such as the Association of the German Mineral Water Industry, new forms of packaging. One company regularly checks the company that takes away its waste for recycling. Furthermore, large drinks companies do audits in the factories of their packaging suppliers in order to ensure quality.

Customer pressures in the dairy industry were related very strongly to price. In terms of packaging one company had made an effort to introduce reusable glass bottles for UHT milk. However, the product was too highly priced, and due to a lack of demand retailers withdrew the product. Other pressures reported were a growing demand for family size packs and also for organic products.

Dairies urge the packaging producers to lightweight and also ask for price reductions for the lighter packaging. There was one exceptional instance of backward pressure where a supplier of cleaning substances urged a dairy to rent reusable containers for the cleaning substances from the supplier. The dairy needs to pay a deposit for the containers, but ultimately it will profit from this solution since landfill costs for the disposal of throwaway containers no longer have to be paid.

In particular, in the packaging industry the call for price reductions was closely linked to customers' wishes for lighter materials. According to the German packaging regulation the

customers of the packaging industry (i.e. fillers like the dairy and drinks industry) have to pay the DSD licence fee. Since the fee depends partly on the weight of the packaging material, the fillers have a high interest in lighter materials. However, in the plastics industry, for example, the room for light weighting seems to be used up by the current technology and materials available. Further light weighting would negatively effect the handling of the containers and would raise a quality issue.
Within the packaging industry paper and board converters try to pass light weighting pressures upstream to the paper mills. One paper mill showed concern about the quality of its inputs which consisted of 99 per cent of used paper. The paper mill pushed the suppliers of the used paper to deliver material which was better sorted, for instance without having any metal clips in it. Also glass companies were very interested in using secondary material of high sorted quality. Plastics companies were extremely concerned about the price of their inputs; metal companies needed high-quality steel and aluminium to make cans. Machine manufacturers showed a high degree of self-sufficiency because they produced many of the machine parts themselves.
When the German Packaging Ordinance was implemented, the large retailers developed packaging checklists in order to express clearly to their suppliers what sort of packaging they were looking for. For instance, exact sizes for transit packaging would be given. Composite packaging is banned as much as possible in the checklists. Because retailers were more concerned with hygiene, when checking production sites there was a positive response if their suppliers were environmentally certificated (e.g. ISO 14000 or EMAS). Smaller retailers said they had not enough power to have an environmental influence on their suppliers. Many of the retailers who were interviewed reported that they were currently more concerned with issues such as larger stores, provision of family packs and labelling of novel food than with packaging.

8.6.4 Summary and discussion

As was shown in the sample description in chapter six sample firms in each country export widely and are therefore facing international competition and are potentially vulnerable to international variations in environmental costs. German firms were shown to have the highest waste costs, simultaneously they export more in the retailing and packaging sector than their Irish/UK counterparts (though less in the dairy sector). In relation to their national and international ability to compete sample firms were asked for an estimation of their competitive advantages and disadvantages, possible constraints on growth and direct pressures from customers/suppliers.

Most competitive advantages related to non-price factors, especially product quality and variety, design and presence in niche markets. Environmental factors occurred infrequently. Environmental factors were mentioned on even fewer occasions as a competitive disadvantage.

Concerning constraints on growth no firm in any country cited environmental regulation or costs as a constraint on growth.

Firms were also asked to list the direct pressures they faced from customers and demanded from suppliers. Price, quality and service were not only frequently cited, they were also considered to be stronger and more important pressures than environmental factors.

8.7 Conclusions on the impact of stringent regulation on competitiveness in the German supply chain of packaging, dairy/drinks and retail industry

This study considered the question whether for individual firms at various points along the supply chain environmental compliance was a cost and if so to what extent or whether in fact it represented a competitive disadvantage. The analysis of the interview data collected in German plants in various industries was contrasted with results found in an Irish/UK sample of similar structure and size where the packaging legislation was less stringent at the time of the study. It was shown that the stringent packaging legislation implies relatively higher compliance costs for German plants in comparison with the other sample countries, but hardly any impact on their competitiveness. The most obvious reason is that the absolute cost of complying with packaging regulation measured as percentage of turnover is very small. Further findings are mixed.

The average number of solid waste initiatives was highest in Germany, reflecting that German plants had undertaken a lot of efforts to minimise waste and also to reduce waste costs. However, no consistent relationship between number of environmental initiatives and compliance costs was found. Neither could the hypothesis be confirmed that high productivity plants always incur low waste costs. Only the top performers in the German dairy/drinks and retail industry were able to implement a higher number of solid waste initiatives than the overall German sample. But this was not statistically significant. It could be shown for Germany that increased environmental awareness was associated with increased numbers of solid initiatives undertaken in all sectors of the German sample. Younger plant age was linked to lower waste costs in Germany. A positive correlation between skills and R&D and the number of solid waste initiatives could only be found in German packaging plants. This is not astonishing because design changes etc. do require special skills and R&D, whereas waste efforts like segregation of waste at the level of the filler or the retail industry are not very skill intensive. The hypothesis that small plant size is linked with higher compliance costs could not be confirmed, partly because small plants undertake significantly less waste initiatives. The most important driver for solid waste initiatives in each of the countries was cost, followed by legislation and customer pressure. In Germany legislation was responsible for significantly more initiatives than in either of the other countries. Initiatives undertaken by German firms were in total more likely to require additional resources (although in most cases only small amounts), were less likely to bring about unit cost savings, more likely to trigger general financial savings and more likely to give rise to positive competitiveness effects compared with those undertaken in the other sample countries. The majority of solid waste initiatives, however, was cost driven in every country and these initiatives triggered the largest number of general savings effects for the plants under consideration. Most expenditures were cost neutral and few required capital expenditure. In Germany several companies stated that their environmental initiatives conferred important competitive advantages. No Irish or UK firm quoted environment as a competitive advantage. Environmental factors were mentioned on very few occasions as a competitive disadvantage in Germany. Overall, there was little evidence of packaging regulation as leading to a competitive advantage or disadvantage.

The detailed findings are summarised as answers to the hypotheses developed at the outset of this study.

Productivity and cost of compliance: The proportional cost of environmental compliance relative to turnover incurred by the firms was hypothesised to be a negative function of the productivity level because firms with the capabilities to achieve high productivity will also find it easiest to respond to environmental challenges without the penalty of reduced output and employment.

Findings: Waste costs were found to be higher in Germany. In the dairy/drinks sector, an important element of the raised costs was DSD fees. However, based on a statistical analysis of the relation between the cost of waste and measures of firm competitive performance[32] no statistically significant relationships have been found. While in many cases there is support for the hypothesis in terms of a relationship between low average waste costs and top performing firms, there are also many cases where low waste costs are associated with relatively poor economic performance. The hypothesis is therefore not confirmed. In Germany,

[32] A number of indicators of competitiveness were used in the study. These included output side indicators (i.e. those that illustrate the consequences of the relative competitiveness of the firm): market share, exports, growth, productivity, influence on product price, managers' own assessment of competitiveness effects. There was also a set of indicators corresponding to the input side (i.e. representative of the likely explanations of competitiveness): these included physical and human capital (i.e. age of plant and skills) and R&D spending.

only top performers in the retail sector have lower waste costs than the average, but this is statistically not significant. By and large, in dairy/drinks and also packaging the German top performers have even higher costs than the average.

Number of initiatives and firm performance, ability to absorb compliance costs, environmental awareness: The study also sought to link productivity and compliance costs to the number of environmental initiatives undertaken and the underlying environmental awareness.

Findings: Overall, the support for a relationship between firm performance and the number of solid waste initiatives undertaken by the firm, are generally inconsistent and not statistically significant. The strongest linkage was found in the German dairy/drinks industry where top performers were able to implement a higher number of solid waste initiatives than the overall German sample. Moreover, also Germany's top retailers used a much larger number of initiatives than the average German retail firm. It could also be shown that increased environmental awareness was associated with increased numbers of solid initiatives undertaken in all sectors of the German sample. It was also tested whether lower waste costs are associated with a greater number of initiatives. The relationship between numbers of solid waste initiatives undertaken and solid waste costs showed no statistically significant results and in Germany some positive linkage between high waste costs and high number of waste initiatives in dairy/drinks and retailing was observed.

Plant age: The age of a plant and its machinery is likely to have an impact on its environmental performance and on compliance costs. Modern plants can reach better environmental standards because the newest technology also embodies best environmental technology and allows to achieve environmental goals at a lower cost than with old equipment which would need expensive retrofitting.

Findings: Plant age was measured as one of the competitiveness indicators measured on the input side (others were skills and R&D). While the hypothesised direction of the relationship between younger plant age and lower waste costs is confirmed for the German dairy/drinks and packaging industry, the relationship is not found to be statistically significant.

Plant size: Large plants are hypothesised to undertake more environmental initiatives because they often have more financial and human resources for the solution of environmental problems than small and medium sized plants and have also better opportunities to research cost efficient environmental measures. Thus the proportional cost of compliance relative to turnover is likely to be a negative function of the size of plants/firms.

Findings: Although there is not always consistency between countries, industries and particular types of initiative, there is evidence of a relationship between numbers of solid waste initiatives and plant and firm size in all sample countries. It is particularly clear that small firms undertake fewer initiatives than their larger counterparts, especially in the German retailing sector. In general there was no evidence of a relationship between waste costs and plant size, although there was some evidence of higher costs associated with smaller size for firms in the German dairy/drinks sector.

History of regulation: Plants in countries with a history of stringent environmental legislation will be in a better position to develop or gain access to environmentally and economically beneficial solutions than their counterparts in other countries with a history of less stringent regulation.

Findings: In each country the most important solid waste initiatives were packaging reuse and recycling. In Germany a wider range of solid waste initiatives was important, especially packaging recovery, light weighting, packaging reduction and waste segregation. Overall in Germany 480 solid waste initiatives or 8.9 per firm were identified, while they were only 256 or 4.7 per firm in the Republic of Ireland/UK sample. There was also some indication that initiatives undertaken by firms in Germany were broader and deeper. The larger average number of initiatives undertaken by German firms arose mainly from the greater percentage of firms engaged in each type of waste initiative examined. It was particularly notable that a higher percentage of German firms were engaged in packaging design, material switching, packaging recovery and waste segregation, and that these were activities strongly driven by packaging regulation pressures. In total, German sample plants achieved more positive competitiveness effects per packaging initiative than plants sampled in the other countries.

There was also greater environmental awareness among German companies. Environmental policies, audits, managers and meetings were taken as indicators of environmental awareness/efficiency. German firms undertook more environmental awareness initiatives than their counterparts in Republic of Ireland/UK firms. As already shown above, increased environmental awareness was associated with increased numbers of solid waste initiatives undertaken by the firm.

Green market pressure in the supply chain relations: Not only regulatory pressure forces firms to react, but also environmental pressure coming from the market place with respect to customers/consumers and suppliers. Relations within the supply/production chain are likely to be both an influence upon, as well as being influenced by, the level of environmental standards, e.g. a manufacturer may find it easier to increase the environmental standards of its products if it has a reliable and competent base of suppliers to draw on. A manufacturer may be forced to upgrade product and process environmental standards by pressure coming from the customers (incl. consumers) of plants in the three sectors under study.

Findings: (a) The most important driver in each of the countries was cost, followed by legislation and customer pressure. In Germany legislation was responsible for significantly more initiatives than in either of the other countries. Health and safety, technology and other drivers were relatively unimportant. In Germany recycling, special design, material switching, packaging recovery, and packaging reduction (especially removal and reduction of labels and removal of outer packaging, use of larger packs and the elimination of excess packaging) were important initiatives driven by legislation. External pressures gave rise to other initiatives, particularly increased material reuse, but also special design, material switching, light weighting and the removal of outer packaging.

(b) Firms were also asked about the importance of direct environmental pressures from customers or suppliers in comparison to other pressures. While customer/supplier environmental pressures on packaging were present they were only one of many pressures. The main pressures firms faced were with respect to the more usual competitive aspects of price, quality and service.

(c) The hypothesis that those firms supplying major retailers would face a greater environmental pressure and would therefore undertake more initiatives was tested. However, there was no evidence of a greater number of initiatives being undertaken by those firms supplying major retailers in the sample.

Ownership and location: Multinational companies are likely to show better environmental performance than other enterprises because they export both the environmental and the efficiency standard of their home country. Concerning location, plants in peripheral sites are likely to be less advanced on environmental matters, to undertake less environmental initiatives and to be potentially subject to higher environmental costs because of lack of access to suppliers, language barriers, higher transport costs etc. than plants in centrally located regions.

Findings: A breakdown of the total number of initiatives adopted by firms with different legal forms of ownership showed that on average PLCs in all countries undertook more initiatives than limited companies. PLCs tended to be larger firms with environmental policies and more environmental awareness initiatives, thus reinforcing the relationship between size and number of initiatives. There was also evidence that branch plants of MNEs undertook an above average number of initiatives. Many MNEs claimed to adopt strong company-wide environmental policies. In the sample, only for Germany some location observations can be made. In East German dairies DSD costs tended to be higher because more non-reusable packaging was used. A clear lag was also found concerning the number of solid waste initiatives, for example, between indigenous East German producers in the paper industry compared with their western counterparts.

Employment impact: The direction of the employment implication of adjustment to stringent environmental regulation is not clear a priori. The available macroeconometric evidence has shown that the overall employment effects of environmental policies are likely to be small.

Findings: The larger number of projects engaged in by German companies gave rise to about two and a half times the number of employment effects of their Republic of Ireland/UK counterparts. German initiatives also gave rise to relatively large employment effects. The net employment effect in all sectors was clearly positive.

Influence of regulation and external pressures on the competitiveness of firms and plants: Compliance could affect productivity and competitiveness either positively or negatively. Regulation could place firms at a competitive disadvantage and lead to the loss of markets, particularly to countries with less stringent regulation. If the firm needs to redirect resources from other profitable opportunities, costs and prices rise and markets and customers are lost. On the other hand, high standards and strict enforcement, although they may represent a short-term cost and burden to the firm, could in the medium and longer term push firms on to a higher growth path by forcing them to make product and process changes which yield higher competitiveness. If this happened it would represent part of the so-called 'double dividend', i.e. gains in environmental performance would also be accompanied by increased economic performance.

Findings: (a) Based on asking managers about the economic effects of adopting waste reduction initiatives. The economic effects of undertaking the different solid waste were analysed. Comparisons between countries showed that initiatives undertaken by German firms were in total more likely to require additional resources (because more initiatives were driven by legislation which frequently caused recycling costs), were less likely to bring about unit cost savings, more likely to trigger general financial savings and more likely to give rise to positive competitiveness effects compared with those undertaken in the other sample countries. The majority of solid waste initiatives, however, was cost driven in every country and these initiatives triggered the largest number of general savings effects for the plants under consideration. Most expenditures were cost neutral and few required capital expenditure.

(b) Based on the importance of environmental factors influencing the firm's competitive performance relative to other factors: Most competitive advantages related to non-price factors, especially product quality and variety, design and presence in niche markets. Environmental factors occurred infrequently. In Germany several companies stated that their environmental initiatives conferred important competitive advantages. No Irish or UK firm quoted environment as a competitive advantage. Environmental factors were mentioned on very few occasions as a competitive disadvantage in Germany. Again no Republic of Ireland/UK firm reported that environmental matters constituted a competitive disadvantage.

No firm cited environmental regulation or costs as a constraint on growth. In Germany overcapacity, raw material costs, fierce competition, low margins and transport costs were more frequently cited. In addition, German sample firms were shown to export more in both the retailing and packaging sectors, despite having the highest compliance costs (but less in the dairy/drinks industries than their Irish counterparts).

Overall, there was little evidence of packaging regulation as leading to a competitive advantage or disadvantage in this case study of a supply chain. At the micro level there was some evidence of competitiveness benefits arising from regulation and external pressures in Germany but these advantages were not evident at the firm level. Probably the most important reason is that the impact of regulation and the cost of complying with regulation is relatively insignificant in relation to other factors which influence the competitive performance of firms.

9. Empirical results on the impact of stringent clean air regulation on competitiveness in the cement industry

This chapter presents the competitiveness implications of investment in pollution abatement technologies in the sample of cement plants visited in Germany, Spain and the UK. The focus was on clean air pollution which had been found to be the most essential environmental problem in the cement industry.

Particular emphasis is put on the question whether the German plants suffer from an economic disadvantage because they are most stringently regulated in this sample. In contrast to the food and packaging case study, it was possible to gain access to detailed data on the emission situation of sample plants. This had an impact on the analytical approach insofar as a concise classification of the cement sample according to environmental parameters was possible. Within this framework it was possible to ask the question whether the top environmental performers were economically any different from their counterparts with a lower environmental performance. Analytically, the following steps are undertaken in order to examine the relationship between economic and environmental performance:

First of all, the cement sample is divided in plants of different environmental qualities within each country and between the sample countries (see section 9.1). As a basis for this classification the actual plant emission levels as well as the number of environmental measures are used. Then, in section 9.2, the method for analysis and a matching framework both within Germany and between Germany and the other sample countries are defined. The statistical analysis in section 9.3 is centred around the hypothesis that cement plants with many environmental initiatives and low emissions have stronger economic performance than their counterparts in the categories with lower environmental performance. This would be because high productivity plants would find it easiest to respond to environmental challenges and to deliver high environmental performance. In order to test this hypothesis, plants with a favourable environmental performance are compared with less environmentally friendly classified plants of the same size concerning key economic indicators like productivity, age of kiln, investment strategies, sales per head, production costs and prices. This matching is first of all undertaken within Germany in order to check whether most stringently regulated plants are economically any different from on average regulated plants in Germany. Second, the matching is also undertaken between Germany and Spain as well as Germany and the UK in order to examine any differences on an international level. In a next step, in section 9.4, the individual effects of the implementation of environmental measures on a plant (micro) level are analysed. Finally, general competitive advantages and disadvantages are examined across the plants in order to check how environmental expenditure affects the plants' competitiveness relative to other factors (section 9.5).

9.1 Classification of cement sample according to emissions and number of environmental measures

In order to examine the impact of stringent regulation on competitiveness, strongly regulated plants already achieving low emissions and using a high number of abatement techniques are compared with other plants in the sample having higher emissions and using less pollution reduction measures. Therefore each of the 18 plants in the cement sample was classified according to its environmental performance measured as a combination of the actual emissions and the number of environmental measures implemented.

First of all the classification of the plants according to their emissions of NO_x, SO_2 and dust was undertaken. Thereby the categories low, medium and high emissions were defined according to the actual data found in the sample plants. Plants were regarded to have low emissions, if they stayed below the relatively strict limit values found in new German and UK sample plants which were more stringently regulated than the corresponding national average. These strict limit values found in the sample were for NO_x 500 mg/Nm^3, for SO_2 200 mg/Nm^3 and for dust 30 mg/Nm^3. Thus, the classification "low emissions" refers to NO_x emissions of lower than 500 mg/Nm^3, to SO_2 emissions of lower than 200 mg/Nm^3 and to dust emissions of lower than 30 mg/Nm^3 (see Table 9.1-1 below). The category "medium emissions" was defined according to the second most stringent range of regulation found in the sample which was the current German clean air regulation for existing plants. A level of NO_x emissions classified as "medium" refers to NO_x emissions between 500 and 800 mg/Nm^3, to SO_2 emissions between 200 and 400 mg/Nm^3 and to dust emissions between 30 and 50 mg/Nm^3. Finally the class of "high emissions" is defined as more than 800 mg/Nm^3 for NO_x emissions, for SO_2 emissions more than 400 mg/Nm^3 and more than 50 mg/Nm^3 for dust emissions.

In a second step the emission data of all sample plants were compared with the ranges shown in Table 9.1-1. Plants did normally not achieve all the emission ranges for all pollutants within one group. The most common ranking exercise was similar to the following one: For example, if a plant stayed well within the ranges for low emissions concerning NO_x and dust (or NO_x and SO_2), but was slightly beyond the SO_2 (or dust) range, it was still qualified as having "low emissions" overall. Thus, if a plant met a specific class for two of the three pollutants, the entire plant was classified according to this dominant effect. No weighting of emissions was carried out except for two cases where SO_2 and dust values were extremely low, but NO_x values with between 1000 and 2000 mg/Nm^3 extremely high. In this case a trade-off between the two extremes was made and the plants were ranked as having "medium emissions".

Table 9.1-1 Classification of emission data in low, medium and high and number of plants in each country falling into these categories

Emission values	LOW*	MEDIUM*	HIGH*
NO_x in mg/Nm^3	< 500	500-800	> 800
SO_2 in mg/Nm^3	< 200	200-400	> 400
Dust in mg/Nm^3	< 30	30-50	> 50
Germany	2	6	0
Spain	0	1	4
UK	1	2	2
Total	3	9	6
*if majority of emissions fell into that group.			

Thirdly, the number of environmental measures in place was examined and separated according to their nature, being primary (process-integrated) or secondary (end-of-pipe) measures. The sample data was again divided in three groups for the two types of measures, indicating that plants had a "low", "medium" or "high" number of primary and/or secondary abatement measures installed (see Table 9.1-2 and 9.1-3).

The distribution of primary measures varied widely between three and ten initiatives, but most plants used between five and nine measures. Only one British and two Spanish plants had fewer than five primary measures in place and a total of two German plants had more than nine primary measures installed. Thus, less than five primary measures was ranked as "low number of primary measures", five to nine measures was considered as a "medium number of primary measures" and more than nine initiatives was seen as a "high number of primary measures" (see Table 9.1-2).

Concerning secondary measures most cement factories in the sample had one or two measures installed. There was one British plant using none and one German plant using three secondary measures. Therefore less than one secondary measure was ranked as "low number of secondary measures", one measure was considered as a "medium number of secondary measures" and two or more initiatives were seen as a "high number of secondary measures" (see Table 9.1-3).

Table 9.1-2 Classification of number of primary measures in the cement sample and number of plants in each country falling into these categories

No. of primary measures	LOW < 5	MEDIUM 5-9	HIGH > 9
Germany	0	6	2
Spain	2	3	0
UK	1	4	0
Total	3	13	2

For the ranking of the total number of environmental measures, the numbers of primary and secondary measures were added up, with secondary measures being the constraint for the classification (see Table 9.1-4). Nine of the total of 18 plants were classified as having a medium number of environmental initiatives in place, three plants as having a low number and six plants as having a high number.

Table 9.1-3 Classification of number of secondary measures in the cement sample and number of plants in each country falling into these categories

No. of secondary measures	LOW < 1	MEDIUM 1	HIGH 2 and more
Germany	0	2	6
Spain	0	5	0
UK	1	4	0
Total number of plants	1	11	6

Table 9.1-4 Classification of total number of environmental initiatives and number of plants in each country falling into these categories

No. of total environmental measures	LOW < 6	MEDIUM 6-10	HIGH > 10 or more
... of which secondary measures	< 1	1	2 and more
Germany	0	2	6
Spain	2*	3	0
UK	1	4	0
Total no. of plants	3	9	6
*Note: There were two cases in Spain where one secondary measure was installed. However, the total number of primary measures was so much lower than 6 so that these two plants were classified as having a low total number of environmental measures.			

Finally, the ranking of emissions and total number of abatement measures was combined. Since emission levels were reached with varying number of environmental measures, the decisive criteria for the classification of environmental performance was to have low, medium or high emissions. Low emissions could be reached either with, as expected, a high or a medium number of environmental measures. Plants with this combination were called "group 1". Medium emissions were even possible with a low, medium or high number of abatement techniques ("group 2"). High emissions logically went hand in hand with a low number of pollution reduction measures. However, sometimes also a medium level of initiatives still resulted in high emissions. The latter combination of high emissions and low/medium number of environmental measures was called "group 3". Table 9.1-5 presents an overview of this final classification for all plants in the different sample countries and their sizes.

The German sample shows, as expected due to the stringent regulation, a relatively more favourable environmental performance than sample plants in the UK and Spain. From the eight German sample plants two plants have low emissions and are ranked as group 1. One of these two plants is a small cement factory having a medium number of environmental initiatives in place. The other plant is a large one with a high number of environmental initiatives. The remaining six German plants all have medium emissions with one of them using a medium number of abatement measures and five plants having a high number of these techniques in place. Five of these six plants are of medium size, the sixth one is a large one. Two of the total of the five medium sized German plants ranked to have medium emissions had remarkably lower emissions than their counterparts within the same emission range. These two plants were analysed separately within the German group 2 and called group 2a. The other three medium sized plants in group 2 were called group 2b. For ease of illustration the remaining large plant in group 2 with average emissions was also included in group 2b (see also section 9.2).

In the Spanish sample only one plant is classified as a group 2 performer with medium emissions and a medium number of environmental initiatives. The other four plants all have high emissions and are ranked as group 3. Two large plants have installed a low number of pollution reduction measures, two medium sized Spanish plants used a medium number.

One new large plant of the UK was ranked as having low emissions and a medium number of environmental initiatives. Furthermore, two plants achieved medium emissions. These were one large and one medium sized plant, with a low and medium number of pollution reduction measures. The remaining two UK plants were ranked to have high emissions, although they had a medium number of environmental initiatives in place. One of these plants was large, the other of medium size.

Table 9.1-5 Number of cement plants in different environmental categories and size classes*

Environmental category	Germany		Spain		UK	
	No. of plants	Size class	No. of plants	Size class	No. of plants	Size class
Group 1:						
Low emissions and medium number of env. Measures	1	Small	0	–	1	large
Low emissions and high number of env. measures	1	Large	0	–	0	–
Group 2:						
Medium emissions and low number of BATs	0	–	0	–	1	large
Medium emissions and medium number of BATs	1	Medium**	1	small	1	medium
Medium emissions and high number of BATs	5	4 medium***, 1 large	0	–	0	–
Group 3:						
High emissions and low number of BATs	0	–	2	2 large	0	–
High emissions and medium number of BATs	0	–	2	2 med.	2	1 large, 1 med.
Total number of plants	8		5		5	

* Size classes are defined as follows: small size: < 600,000 tonnes of cement per year; medium size: 600,000 – 1,000,000 tonnes of cement per year; large size: > 1,000,000 tonnes of cement per year.

** This plant has remarkably lower emissions than the average of all medium sized plants.

*** One of these plants has remarkably lower emissions than the average of all medium sized plants. These two marked medium sized plants required a further differentiation of group 2 later on and were analyzed in a group called group 2a, whereas the other three medium sized plants are called group 2b. For ease of illustration group 2b also includes the remaining large plant of group 2 with average emissions.

In summary, the total of 18 cement sample plants was divided in three groups of different environmental quality according to their emissions and numbers of environmental measures. As expected, German plants fell into the groups with higher environmental quality (two plants in group 1 and six in group 2). Four of the five Spanish plants were classified as group 3 performers. In the UK sample plants were to almost even parts both group 2 and 3 performers.

9.2 Method of analysis

In this section it is examined whether the German cement plants suffered from any systematic economic disadvantages because they were stringently regulated. To this end, matching of all cement plants of the same size, but of varying environmental performance was undertaken both within Germany and between Germany and Spain as well as Germany and the UK. First of all, the group 1 plants located in Germany were compared with the group 2 plants in Germany for each size group as far as the data allowed it; then the total German sample was compared with the Spanish and UK plants of similar size, but with a less favourable environmental performance.

Analyses examined the relationship between environmental performance (measured by emissions and numbers of environmental initiatives) and competitiveness measured by output and input measures. Output measures taken into consideration are productivity, sales per head, price (together with production costs) and plant growth; input measures are age of equipment, skills and also environmental strategies like the use of environmental management systems. Furthermore the role of ownership, market structure and location is analysed in the context of environmental performance and competitiveness. It is hypothesised that plants with favourable environmental performance (group 1) are likely to be the more competitive plants both in terms of the output and input indicators. This would be the case because competitive factors also influence the ability of a plant to absorb or minimise any additional costs arising from investment in and operation of environmental measures.

Table 9.2-1 shows the number of plants which could be matched with counterparts within Germany, between Germany and Spain and between Germany and the UK. First of all, within Germany two medium sized plants of relatively favourable environmental performance were compared with three other medium sized plants with higher emissions. The matching of the medium sized German plants within Germany required a further differentiation of the German group 2 in group 2a and group 2b as already described in section 9.1 (see also notes to Tables 9.1-5 and 9.2-1). Furthermore, one large German plant in group 1 was compared with another large German plant in group 2b. For the small German plant in group 1 there was no direct comparison possible within the same size group in Germany. However, a direct matching between this small German and the small Spanish plant in group 2 was possible. Moreover, all the five German medium sized plants in group 2 were compared with their two Spanish counterparts of the same size ranked as group 3. All large German plants in group 1 and 2 were matched with the large Spanish plants of group 3.

Table 9.2-1 Distribution of sample plants according to environmental performance and size

Country/Size	Germany			Spain			UK		
Env. Performance	Group 1	Group 2	Group 3	Group 1	Group 2	Group 3	Group 1	Group 2	Group 3
Small	1	0	0	0	1	0	0	0	0
Medium	0	2 + 3*	0	0	0	2	0	1	1
Large	1	1	0	0	0	2	1	1	1
Total	2	6	0	0	1	4	1	2	2

*Two of the total of the five medium sized German plants ranked to have medium emissions had remarkably lower emissions than their counterparts within the same emission range. These two plants were analysed separately within the German group 2 and called group 2a. The other three medium sized plants in group 2 were called group 2b. For illustration purposes group 2b also includes one large plant with average emissions. For ease of illustration group 2b also includes the remaining large plant of group 2 with average emissions.

Finally, all the five medium sized German plants of group 2 were matched with the British plants of group 2 having slightly higher emissions than their German counterparts in the same environmental category. All large German plants of group 1 and 2 were matched with the large British plants of group 2 and 3.[33]

The number of individually matched pairs is shown in Table 9.2-2. A full set of comparisons was possible between Germany and Spain; due to the lack of small sample plants in the UK no comparison of small British and German plants was possible.

Table 9.2-2 Number of individually matched pairs in the cement sample*

Env. group/ size	Germany : Germany	Germany : Spain	Germany : UK
Small	–	1:1	–
Medium	2:3	5:2	5:2
Large	1:1	2:2	2:2
Total	3:4	8:5	7:4

*First number always refers to plants in higher environmental category than second number.

[33] Also, the large German plant in group 1 was matched with the large British plant in group 1 in order to find out how the equally stringently regulated plant in the UK dealt with the situation. This was undertaken separately since it was the only case where matching of an identical level of environmental performance was possible across countries. Both plants were new and highly productive. Results for the UK plant confirm the German strategy (shown later on in this document) of coping with high compliance costs through increased use of process improving measures and use of alternative fuels.

9.3 The relationship between economic and environmental performance

In this section the relationship between economic and environmental performance is explored. To this end data on output and input measures of competitive performance are linked with the environmental performance on a matched pairs basis. The environmental performance is already captured in the classification of plants as group 1, 2 or 3 performers (emission levels and numbers of environmental initiatives are the decisive criteria for the classification). As output indicators of competitive performance data on productivity, capacity utilisation, sales and prices were used. Input measures of competitive performance mainly consisted of age of kiln, skills and investment levels. Furthermore, environmental and economic performance were put in relation to the level of compliance costs measured as environmental investment. Finally, the influence of other company characteristics like location and ownership as well as the use of alternative fuels and of environmental management systems on the relationship between competitiveness and environmental performance was examined.

The section starts with the relationship between output measures of competitive performance and environmental performance.

Productivity and environmental performance

The hypothesis that productivity and environmental performance are positively related could not be confirmed across all size bands of environmental categories. Table 9.3-1 indicates that the small German plant in group 1 has a higher productivity than the counterpart in group 2 in Spain. Within Germany the large plant in group 1 also shows a higher productivity level than an equally large plant in group 2b. The same is true for large plants in the German/British comparison. However, in the comparison of the medium sized plants the German plants in group 1 and group 2a never reach the productivity level of their counterparts in group 2b within Germany. Moreover, all medium sized German plants are with 8,700 tonnes of output per plant and year less productive than their Spanish and UK counterparts of the same size with higher emissions (9,100 and 11,300, respectively).

Table 9.3-1 Average productivity in tonnes of cement per plant and year

Country	Germany	Germany	Germany	Spain	Germany	UK
Env. Category	Average of group 1 and 2a	Average of group 2b	Average of group 1 and 2 (total)	Average of group 2 and 3	Average of group 1 and 2 (total)	Average of group 2 and 3
Small plants	–	–	6500	5000	–	–
Medium plants	6500*	10200**	8700	9100	8700	11300
Large plants	9600	9000	9300	12600	9300	8700
Total n° of matches	3	4	8	5	7	4

Figures are calculated as the output in the kiln and cement area incl. maintenance; figures are rounded to the next hundred.

* Average of group 1 and 2a. The latter are medium sized plants with remarkably lower emissions than the average of the total group 2.

** Average of remaining three medium sized plants of group 2 called group 2b. For ease of illustration group 2b also includes the remaining large plant of group 2 with average emissions.

Capacity utilisation and environmental performance

Productivity is closely related – among other factors like labour intensity - to capacity utilisation. The rate of capacity utilisation in the German plants has on average been lower than in the Spanish or British plants due to market reasons and not because of environmental legislation. Large Spanish plants show the highest degree of capacity utilisation; this would also explain their high productivity. Within Germany the large East German plants show the lowest degree of capacity utilisation (see Table 9.3-2). This would be because the construction boom after the German reunification has slowed down.

Table 9.3-2 Average percentage of capacity utilisation over the last five years

Country	Germany	Germany	Germany	Spain	Germany	UK
Env. category	Average of group 1 and 2a	Average of group 2b	Average of group 1 and 2 (total)	Average of group 2 and 3	Average of group 1 and 2 (total)	Average of group 2 and 3
Small plants	–	–	80	79	–	–
Med. plants	83*	82**	83	90	83	95
Large plants	65	69	67	93	67	100
Total n° of matches	3	4	8	5	7	4

* Average of group 1 and 2a. The latter are medium sized plants with remarkably lower emissions than the average of the total group 2.

** Average of remaining three medium sized plants of group 2 called group 2b. For ease of illustration group 2b also includes the remaining large plant of group 2 with average emissions.

Sales per head and environmental performance

Contrary to the basic hypothesis that plants with high environmental performance are also strong economic performers, plants with low emissions and many pollution abatement measures do not have higher sales per head than their counterparts with less favourable environmental performance. Medium and large German plants in group 1 and 2 tend to have even lower sales per head than their counterparts in group 2 and 3 in Spain and the UK (see Table 9.3-3).

Table 9.3-3 Sales per head in M Euro

Country	Germany	Germany	Germany	Spain	Germany	UK
Env. category	Average of group 1 and 2a	Average of group 2b	Average of group 1 and 2 (total)	Average of group 2 and 3	Average of group 1 and 2 (total)	Average of group 2 and 3
Small plants	–	–	0.28	0.23	–	–
Medium plants	0.25*	0.27**	0.26	0.34	0.26	0.37
Large plants	0.25	0.33	0.29	0.39	0.29	0.27
Total no. of matches	3	4	8	5	7	4

* Average of group 1 and 2a. The latter are medium sized plants with remarkably lower emissions than the average of the total group 2.

** Average of remaining three medium sized plants of group 2 called group 2b. For ease of illustration group 2b also includes the remaining large plant of group 2 with average emissions.

Prices and environmental performance

As for productivity and sales, no systematic relationship between environmental performance and price as output measure of competitive performance could be discovered. Plants with low emissions and many abatement techniques do not achieve higher average prices for their products. This is clearly the case within Germany. Between Germany and Spain there is a small difference in favour of Germany which, however, is in most cases small. UK plants achieve higher prices than in Germany (see Table 9.3-4). But this is to a large extent due to the strong Pound Sterling.

Table 9.3-4 Average price in Euro per tonne of cement

Country	Germany	Germany	Germany	Spain	Germany	UK
Env. category	Average of group 1 and 2a	Average of group 2b	Average of group 1 and 2 (total)	Average of group 2 and 3	Average of group 1 and 2 (total)	Average of group 2 and 3
Small plants	–	–	75	60	–	–
Medium plants	60*	70**	65	60	65	72
Large plants	55	65	65	60	65	75
Total no. of matches	3	4	8	5	7	4

* Average of group 1 and 2a. The latter are medium sized plants with remarkably lower emissions than the average of the total group 2.

** Average of remaining three medium sized plants of group 2 called group 2b. For ease of illustration group 2b also includes the remaining large plant of group 2 with average emissions.

In the following the relationship between input measures of competitive performance and environmental performance is examined.

Age of equipment and environmental performance

It was hypothesised that modern plants can reach better environmental performance because the newest technology also embodies best environmental technology. To summarise, in all countries new kilns have a high environmental performance and produce high productivity levels. In Germany also relatively old kilns can reach a favourable environmental performance. This would be due to the national approach to environmental standards. Therefore also old kilns are maintained to a high standard (see also section below on investment). The detail is as follows:

Within Germany younger kilns always show better environmental performance. All plants in the German group 1 and 2a have newer kilns than the plants in group 2b having higher emissions. Concerning the comparisons between Germany and Spain/the UK, only the large new plants in East Germany are younger than their Spanish and British counterparts and therefore reach a higher environmental performance. The medium sized plants in Spain are as old as in Germany, but the small German plant in group 1 is even older than the small Spanish plant in group 2. All of the German medium sized plants in group 1 and 2 are older than their British counterparts with higher emissions and less pollution abatement measures (see Table 9.3-5).

Table 9.3-5 Age of kiln in years and environmental performance

Country	Germany	Germany	Germany	Spain	Germany	UK
Env. category	Average of group 1 and 2a	Average of group 2b	Average of group 1 and 2 (total)	Average of group 2 and 3	Average of group 1 and 2 (total)	Average of group 2 and 3
Small plants	–	–	44	30	–	–
Medium plants	24*	30**	28	28	28	13
Large plants	5	8	7	25	7	20
Total no. of matches	3	4	8	5	7	4

* Average of group 1 and 2a. The latter are medium sized plants with remarkably lower emissions than the average of the total group 2.

** Average of remaining three medium sized plants of group 2 called group 2b. For ease of illustration group 2b also includes the remaining large plant of group 2 with average emissions.

Production costs and environmental performance

Competitive plants are also assumed to reach high efficiency standards on the input side, this would include efforts to minimise the costs of the factors of production. However, no influence of environmental performance on production costs could be discovered (see Table 9.3-6).

Within Germany better environmental performance is not always related to lower production costs: German plants with a higher environmental performance have equally high and only sometimes lower production costs than their counterparts with lower environmental performance. In comparison to the UK and Spain, the large plants of the German group 1 and 2 have indeed lower production costs than their counterparts in the UK with lower environmental performance, but hardly higher costs than their counterparts in Spain (because these German plants are new and use the most cost efficient production technology). There is hardly any cost difference between medium sized German and UK plants. Cost differences between small and medium sized German and Spanish plants are greatest; this is also reflected in the comparison of the averages on a national level.

Table 9.3-6 Production costs in Euros per tonne of cement

Country	Germany	Germany	Germany	Spain	Germany	UK
Env. category	Average of group 1 and 2a	Average of group 2b	Average of group 1 and 2 (total)	Average of group 2 and 3	Average of group 1 and 2 (total)	Average of group 2 and 3
Small plants	–	–	39	30	–	–
Medium plants	38*	38**	38	27	38	37.5
Large plants	30	34	32	30	32	35
Total no. of matches	3	4	8	5	7	4

* Average of group 1 and 2a. The latter are medium sized plants with remarkably lower emissions than the average of the total group 2.

** Average of remaining three medium sized plants of group 2 called group 2b. For ease of illustration group 2b also includes the remaining large plant of group 2 with average emissions.

Expert systems and related skills and environmental performance

In chapter 6.3 it was shown that 80 per cent of German plants had expert systems in place. In Spain and the UK it was only 60 per cent of sample plants. The assumed positive impact of skills (here measured indirectly in the use of expert systems) on environmental performance can be demonstrated only for small and large German plants and in the international comparison between Germany and UK for medium sized plants (see Table 9.3-7). The skills related to the use of expert systems were stressed by the German plants. The operators in the control room had a special training where steadiness both of process efficiency and emissions was taught.

Table 9.3-7 Skills Percentage distribution of expert systems

Country	Germany	Germany	Germany	Spain	Germany	UK
Env. category	Average of group 1 and 2a	Average of group 2b	Average of group 1 and 2	Average of group 2 and 3	Average of group 1 and 2 (total)	Average of group 2 and 3
Small plants	–	–	100	0	–	–
Med. plants	50*	100**	60	100	60	50
Large plants	100	100	100	50	100	0
Total no. of matches	3	4	8	5	7	4

* Average of group 1 and 2a. The latter are medium sized plants with remarkably lower emissions than the average of the total group 2.

** Average of remaining three medium sized plants of group 2 called group 2b. For ease of illustration group 2b also includes the remaining large plant of group 2 with average emissions.

Total annual investment and environmental performance

As an indicator of the general degree of modernisation total annual investment over the last five years as percentage of sales was taken and compared across the sample (see Table 9.3-8).

There is a small tendency that the German plants with low emissions and many pollution reduction initiatives invest on average a little bit more than their counterparts with higher emissions and less pollution reduction measures. This becomes clearest in the international comparison of medium sized and large plants between Germany and Spain and also for the medium sized plants between Germany and the UK. Within Germany and also in the comparison of large plants between Germany and the UK annual investment as percentage of sales during the last five years is about equal for all types of environmental performers. One small Spanish plant has invested more than its German counterpart with better environmental performance. This was done because the plant was very much run down and needed substantial modernisation.

Table 9.3-8 Total annual investment as % of sales over the last five years

Country	Germany	Germany	Germany	Spain	Germany	UK
Env. Category	Average of group 1 and 2a	Average of group 2b	Average of group 1 and 2 (total)	Average of group 2 and 3	Average of group 1 and 2 (total)	Average of group 2 and 3
Small plants	–	–	1.9	3.7	–	–
Med. plants	5*	5.7**	5.4	1	5.4	4
Large plants	4.9	4.8	5	3.8	5	5
Total no. of matches	3	4	8	5	7	4

* Average of group 1 and 2a. The latter are medium sized plants with remarkably lower emissions than the average of the total group 2.

** Average of remaining three medium sized plants of group 2 called group 2b. For ease of illustration group 2b also includes the remaining large plant of group 2 with average emissions.

Table 9.3-9 Future investment in % of sales

Country	Germany	Germany	Germany	Spain	Germany	UK
Env. Category	Average of group 1 and 2a	Average of group 2b	Average of group 1 and 2	Average of group 2 and 3	Average of group 1 and 2	Average of group 2 and 3
Small plants	–	–	5.4	0	–	–
Med. plants	3.5*	3.7**	3.6	4.4	3.6	3.3
Large plants	6	0.75	3.4	2.9	3.4	2.9
Total no. of matches	3	4	8	5	7	4

* Average of group 1 and 2a. The latter are medium sized plants with remarkably lower emissions than the average of the total group 2.

** Average of remaining three medium sized plants of group 2 called group 2b. For ease of illustration group 2b also includes the remaining large plant of group 2 with average emissions.

Future investment and environmental performance

In all but one case German plants plan higher future investment than in the other countries. These German investment plans refer to maintenance and updating and also investment in alternative fuels and end-of-pipe techniques. Only one medium sized Spanish plant expressed its plans for growth and simultaneous decrease in emissions. Investment in the UK mainly referred to environmental issues (see Table 9.3-9 above).

Environmental investment as a proxy of compliance costs

In addition to total annual investment over the last five years also detailed data for all environmental investments undertaken during the last 10 years were asked. Data were scarce for general primary measures. However, investments for NO_x primary (process-integrated) measures and secondary (end-of-pipe) measures for the reduction of NO_x, SO_x and dust emissions were well recorded. Since secondary measures and also NO_x primary measures are much more expensive than general primary measures, they reasonably reflect the additional burden created by environmental regulation and can be regarded as a proxy of total compliance costs. Most of the environmental measures under concern were undertaken during the last five to 10 years, were converted into prices of 1998 and put into relation of sales in 1998.

Since data on running costs could not be obtained on a large scale, they are not considered in the following analysis. The available database is also used in section 9.4 when individual effects of the abatement measures are presented.

As far as the data are available there is a trend that German plants incur the highest compliance costs measured as capital costs of environmental measures. All German plants in group 1 and 2 have on average invested more in environmental initiatives than their Spanish and British counterparts with higher emissions. Medium sized German plants in group 1 and 2a have invested less than their counterparts in the German group with a little bit higher emissions. This is the case because several plants of the group with higher emissions had high emissions before heavily investing in abatement measures. It also explains why these plants still have relatively high emissions despite of the relatively high environmental investment (see Table 9.3-9). Environmental investments in the large new German plants cannot be separated out of total investment. Large British plants had a need for environmental upgrading and show quite high investment levels.

From this analysis it is also clear that Spanish plants have invested least and have the weakest environmental performance with four of the five sample plants being classified as group 3. The only Spanish plant in group 2 has with 5.4 per cent of compliance costs invested more than three times as much as a Spanish average medium sized plant of group 3 and about seven times more than an average large plant of group 3. Thus, in the Spanish sample compliance costs fall with plant size and rise with environmental performance. In the UK compliance costs rise with plant size.

The economic implication of the higher environmental investment in Germany despite lower productivity can only be understood in the context of the analysis of the use of alternative fuels (see below).

Table 9.3-10 Environmental investment in primary NO_x measures and secondary measures for NO_x, SO_x and dust reduction as % of sales*

Country	Germany	Germany	Germany	Spain	Germany	UK
Env. Category	Average of group 1 and 2a	Average of group 2b	Average of group 1 and 2 (total)	Average of group 2 and 3	Average of group 1 and 2 (total)	Average of group 2 and 3
Small plants	–	–	n.a.	5.4	–	–
Med. plants	4**	8***	6.4	1.6	6.4	6
Large plants	new plant	New plant	New plants	0.8	new plants	14.1
Total no. of matches	3	4	8	5	7	4

* The investments under concern were undertaken between 1988 and 1998, were converted into prices of 1998 and put into relation of sales in 1998.

** Average of group 1 and 2a. The latter are medium sized plants with remarkably lower emissions than the average of the total group 2.

*** Average of remaining three medium sized plants of group 2 called group 2b. For ease of illustration group 2b also includes the remaining large plant of group 2 with average emissions.

Alternative fuels and economic/environmental performance

Because of its high-energy costs, the cement industry has been searching for alternative fuels. Among the types of alternative fuels most frequently used are used tyres, rubber, paper waste and paper sludge, waste woods, waste oils, sewage sludge, plastics and spent solvents.

Alternative fuels are a cost saving device. Their use is more frequent in the German plants than either in the Spanish or UK plants. All but one German plant of group 1 and 2 use more alternative fuels than their national or international counterparts in group 2 and 3. One large German plant in group 1 covers at the moment 25 per cent of its energy consumption by means of alternative fuels, whereas its counterpart in group 2 already uses 50 per cent of alternative fuels. However, the plant in group 1 has applied for a new permit with a higher share of alternative fuels. Simultaneously, plants reported that permits to use alternative fuels were dependent on the implementation of stricter environmental standards.

Table 9.3-11 Use of alternative fuels in % of total energy consumption

Country	Germany	Germany	Germany	Spain	Germany	UK
Env. Category	Average of group 1 and 2a	Average of group 2b	Average of group 1 and 2 (total)	Average of group 2 and 3	Average of group 1 and 2 (total)	Average of group 2 and 3
Small plants	–	–	0	0	–	–
Med.plants	17*	11.7**	13.8	1.8	13.8	10
Large plants	25	50	32.5	0	32.5	15
Total no. of matches	3	4	8	5	7	4

* Average of group 1 and 2a. The latter are medium sized plants with remarkably lower emissions than the average of the total group 2.

** Average of remaining three medium sized plants of group 2 called group 2b. For ease of illustration group 2b also includes the remaining large plant of group 2 with average emissions.

There is an overall trend that absolute compliance costs and also the use of alternative fuels are highest in Germany (see Tables 9.3-9 and 9.3-11 above). Thus, there is a high potential to offset the additional costs of environmental compliance. In Box 9.3-1 examples of savings through the use of alternative fuels are presented for selected German and British plants.

Box 9.3-1 Examples of costs savings through the use of alternative fuels in Germany and the UK

Because of its high energy costs, the cement industry has been searching for alternative fuels. Among the types of alternative fuels most frequently used are used tyres, rubber, paper waste and paper sludge, waste woods, waste oils, sewage sludge, plastics and spent solvents. The change from primary to secondary (alternative) fugitive materials is technically relatively easy, although the use of alternative fuels triggers more stringent environmental standards. This has been found in all the German cement sample plants which were already fairly stringently regulated before they introduced alternative fuels. Still, the cost reduction through alternative fuels is sufficiently large to be offsetting the compliance costs for clean air standards in the cement industry. The German waste market is such that if a company uses tyres, it receives about 30 Euros per tonne from the supplier. Moreover it saves 75 Euros per tonne of coal which would have been needed instead of the tyres. This leads to an enormous annual saving of fuel costs depending on the amount of fuel substitution. Some German companies planned to substitute up to 75 % of their fuel by secondary materials until the year 2001. Individual examples of cost savings in the German sample are as follows:

A large German plant obtains on average 37.5 Euros per tonne of alternative fuel. About 2 tonnes of alternative fuels are needed to reach the same heat value as produced by a tonne of coal which costs about 50 Euros per tonne. Consequently per tonne of replaced coal the plant saves

125 Euros (2x37.5 + 50). Altogether the plant can save energy costs amounting to 7.5 per cent of its annual turnover. The plant was modernised and incurred the highest compliance costs in the sample. It was planned to reduce energy costs down to zero through the increased use of alternative fuels.

Another large German plant uses 17 per cent of alternative fuels and obtains 10 Euros per tonne. Price for brown coal costs usually lies around 25 Euros per tonne. The plant can save energy costs amounting to 1.7 per cent of its annual sales.

A large British plant used 20,000 tonnes of tyres and obtained revenues of 18 Euros per tonne. Coal would costs the plant 57 Euros per tonne. It was reported that savings through alternative fuels sum up to 93 Euros per tonne of replaced coal and to 2.5 per cent of turnover.

Timing of investment in relation to timing of permits for alternative fuels

The paragraph above showed that high compliance costs are softened through alternative fuels. In addition, a comparison of the time sequence of important investments in primary NO_x reduction, secondary measures and upgrading efforts on the kiln showed that permits for the (increased) use of alternative fuels were nearly always obtained in a phase with increasing environmental compliance costs. This was the case both in Germany and the UK and also confirmed in the single Spanish case where alternative fuels were used (see Table 9.3-12 for an overview). This environmental strategy was also confirmed during interviews with environmental directors at headquarter levels. But it was stressed that permits for the use of alternative fuels are frequently accompanied by stricter limit values which triggers investment and emission measurement efforts. In the sample there were cases where NO_x and dust levels became stricter, and in particular emissions of heavy metals were monitored more intensely. In Spain, the use of alternative fuels triggered the installation of continuous measurement devices.

Table 9.3-12 Timing of the use of alternative fuels and major items of environmental investment

Country	Germany	Germany	Spain	UK
Env. Category	Group 1 and 2	Group 2	Group 2 and 3	Group 2 and 3
Small plants:				
First year of use of alt. fuels	n.a.	–	1999	–
Date and type of important env. Investment	n.a.	–	Continuous measurement, 1999	–
Medium plants:				
First year of use of alt. fuels	1997/98* 1980*	1985** 1985**	No alt. fuels	1995
Date and type of important env. Investment	EP*** 97/98 Low NO_x, 1985	Low NO_x 1991 and 1985		EP*** 1992
Large plants:				
First year of use of alt. fuels	1991	1994, increased 1999	No alt. fuels	1995, 2000
Date and type of important env. investment	All measures 1991	All measures between 1994 and 1999		Kiln upgrade 1996 Expert system 1999
Total number of matches	3	4	5	4

* Data refer to medium sized plants with remarkably lower emissions than the average of the total group 2.

** Data refer to medium sized plants with higher emissions than the average of the total group 2.

*** EP = electrostatic precipitator.

Environmental management systems and environmental performance

No direct relationship between environmental awareness measured through the use of environmental management systems and environmental performance can be seen. Only one of the medium sized group 2b performers in Germany has such a system in place (see Table 9.3-13).

Table 9.3-13 Existence of environmental management systems

Country	Germany	Germany	Germany	Spain	Germany	UK
Env. category	Group 1 and 2	Group 2	Group 1 and 2 (total)	Group 2 and 3	Group 1 and 2 (total)	Group 2 and 3
Small plants	–	–	1 none	1 none	–	–
Medium plants	1 planned* 1 none*	1 planned** 1 none** 1 in place**	2 planned 2 none 1 in place	2 planned	2 planned 2 none 1 in place	1 none 1 in place
Large plants	1 in place	1 none	1 in place 1 none	2 planned	1 in place 1 none	2 in place
Total no. of matches	3	4	8	5	7	4

* Data refer to medium sized plants with remarkably lower emissions than the average of the total group 2.

** Data refer to medium sized plants with higher emissions than the average of the total group 2.

Ownership and location

Table 9.3-14 shows that two of the German plants in group 1 and 2 are independent plants. That means that also independent plants, which are often assumed to have less financial power, can reach a high level of environmental performance. It of course also reflects the German legislation which is uniform across different types of ownerships or economic capacities. Multinationals in Spain and the UK showed a less favourable environmental performance than multinationals located in Germany. The Spanish plant with the lowest emissions was owned by a German multinational. Although this plant was the "best" in terms of emissions in Spain, it did not emit as little as its German sister plant. Many primary measures were in place in that Spanish plant, but less secondary pollution reduction techniques than in a comparable German plant. That implies that multinational companies can somehow benefit from softer legislation than in their home country.

In no region location mattered in terms of access to suppliers of environmental technologies. None of the sample plants mentioned difficulties in getting technical help for environmental problems.

Table 9.3-14 Ownership distribution

Country	Germany	Germany	Germany	Spain	Germany	UK
Env. category	Average of group 1 and 2a	Average of group 2b	Average of group 1 and 2	Average of group 2 and 3	Average of group 1 and 2	Average of group 2 and 3
Small plants	–	–	Ind.	MNE	–	–
Medium plants	2 ind. *	3 MNE**	2 ind 3 MNE	2 MNE	2 ind 3 MNE	2 MNE
Large plants	MNE	MNE	2 MNE	2 MNE	2 MNE	2 MNE
Total number of matches	3	4	8	5	7	4

* Average of group 1 and 2a. The latter are medium sized plants with remarkably lower emissions than the average of the total group 2.

** Average of remaining three medium sized plants of group 2 called group 2b. For ease of illustration group 2b also includes the remaining large plant of group 2 with average emissions.

Summary: The relationship between economic and environmental performance

The German plants with low emissions and many pollution abatement measures classified as group 1 and 2 perfomers show in some respects a better economic performance than their national and/or foreign counterparts with less favourable environmental performance, but not in all aspects.

In a national comparison the German small and large plants in group 1 and 2a reach a higher productivity level and their kilns tend to be more modern than in German plants of group 2b. All German plants tend to higher investments in the past and plan higher future investment than their foreign counterparts. Moreover, environmental investment levels are highest in the German sample. Simultaneously, the German plants in group 1 and 2 clearly use more alternative fuels than the Spanish and UK plants which gives rise to an economic advantage due to lower energy costs. That is why all sample plants using alternative fuels have obtained the necessary permits in times of increasing environmental investment. In Germany independent plants are also among the best environmental performers. This implies that location matters in terms of stringent regulation. Expert systems (and the related skills of the persons in the control room using the system) show a positive impact on environmental performance in small and large plants in Germany. Also medium sized German plants in group 1 and 2 use expert systems more frequently than their British counterparts in group 2 and 3.

However, production costs, price per tonne and sales per head do not give the German plants an economic advantage over the plants in group 2 and 3 of other countries. Concerning capacity utilisation German plants even have the worst performance of the sample.

9.4 Drivers and effects of air pollution abatement efforts in the cement industry

In this section the air pollution abatement measures used in this cement sample are examined in detail and it is analysed whether plants in different environmental categories implement the individual measures differently. The motivation for each initiative is examined and as far as the data allowed it, their timing and the investment costs was recorded and potential changes arising for the business. These changes included any impacts on operating costs, environmental performance, employment and training needs, payback times, maintenance, process efficiency, and impacts on capacity, output and profitability.

All these effects are analysed below and particular emphasis is given on the question how German firms have implemented the abatement measures. At the end of this section the effects of each measure are summarised in a tabulated format.

First of all some measures which were exclusively used in German plants are examined. These are end-of pipe solutions for both NO_x and SO_x abatement. Then low NO_x burners are analysed as a commonly used measure of a process integrated solution for NO_x abatement. In the area of dust abatement electrostatic precipitators and bag filters are examined. Also the impacts of energy saving and process optimisation measures as well as the use of expert systems are explored. Moreover, the use of alternative fuels is considered since it was a very widespread and economically important measure. Finally, noise abatement is added to the BAT list of abatement measures because plants frequently mentioned them.

Abatement measures used in Germany only:

(Selective non-catalytic reduction) (SNCR), multi stage combustion (MSC) and absorbent addition

There were some NO_x and SO_x reduction measures which were only recently implemented in German plants because of legislative pressure. Concerning NO_x reduction these measures were selective non-catalytic reduction (SNCR) and multi stage combustion, with respect to SO_x reduction it was absorbent addition.[34] Two German plants in the group 2 with higher emissions had installed SNCR. Both belonged to multinational companies and undertook the measure for legislative reasons in 1995 and 1999. Capital costs were 1.25 M Euros in a plant producing 0.75 M tonnes of cement per year and 1.5 M Euros in a plant producing 1.5 M tonnes. The larger plant only had a production cost increase of 0.15 Euro per tonne, whereas it was 0.25-0.5 Euros per tonne of cement in the smaller plant. Both plants achieved an emission reduction, it was even 400 mg in the larger plant. This large plant also achieved permanently lower NO_x emissions than the other small plant with SNCR. With the exception of training effects no other impacts on business were stressed.

Another German plant in the group 2 with higher emissions and with an annual output of 0.7 M tonnes had invested in multi staged combustion (MSC) in 1995. Again the project was motivated by legislative pressure. The plant managed to bring NO_x levels down by 300 mg to a level of 700-800 mg and got a grant of 5 per cent of the total investment sum. Alternatively the plant could have chosen SNCR, but it was argued that an increase in operating costs (through the ammonium costs) should be avoided. The capital costs were higher than if MSC had been built in from the beginning. Running costs stayed unchanged, there was only a small increase in monitoring cost. Maintenance costs amount to 5 per cent of the investment cost. Initial problems with MSC have been overcome, in part by training for the kiln supervisors and precise monitoring.

Two further, large plants in group 1 and in the part of group 2 with lower emissions did not meet the SO_x standards and were forced by legislation to implement absorbent addition. The group 1 plant with lower emissions invested 0.45 M Eeuros in 1995 and has now an increase in operating costs by 0.3 Euros per tonne of clinker. SO_x emissions went down from 1000 to 350 mg. The other large plant invested only in 1997/98 and needs 30-40 kg of lime per hour to operate the absorbent addition installation. Finally, in 1998 a plant with an output of 0.9 M tonnes in group 2 has invested 0.35 M Euro in absorbent addition, but manages to keep the increase in production cost at 0.08 Euro per tonne of clinker (see Table 9.4-1).

Concerning SNCR, MSC and absorbent addition no loss in competitiveness was reported in the German sample plants. With regard to SNCR and absorbent addition those plants with the highest investment costs achieved the largest reduction in emissions.

---

[34] See section 4.3-2 for a technical explanation of these measures.

Low No_x burner

Six German firms had invested in low NO_x burners and had done this mostly in the mid eighties, earlier than their counterparts in Spain and the UK had got active in this respect.

In all but one cases the initiative was motivated by strict NO_x standards. One plant mentioned cost savings as a reason for investment. The plant which had invested most with 0.75 M Euros also achieved the largest decrease in emissions from 1600 to 750 mg of NO_x. Other plants only achieved a reduction of 10-15 per cent with smaller investments, but started also at lower NO_x levels. None of the Spanish plants had to invest in low NO_x burners because of regulation; they did it all voluntarily for process efficiency reasons, upgrading or in a plant owned by a German company because of company policy. Spanish plants invested later than the German plants and have on average spent less for this investment. The majority of three British plants with a low NO_x burner had also invested for process efficiency reasons. In one British example it was stressed that the decrease in energy has led to a reduction in energy costs and that the payback period was only 3-4 years. Employment was reduced by one person and there was a positive impact on profitability (see Table 9.4-2).

Overall, although German plants were forced to invest in this measure early on, no negative impact on competitiveness was reported.

Electrostatic precipitators

The German plants in group 1 and 2a have voluntarily replaced their electrostatic precipitators, whereas the plants in group 2b with higher emissions have waited for the regulatory pressure to invest. Capital costs varied between 0.5 and 7.5 M Euro per filter. With the highest investment the largest reduction in dust emissions could be achieved. Plants did not mention an increase in operating costs or an impact on profitability. It was explained that the exchange of an old electrostatic precipitator for a new one helps to reduce electricity so that plants with a more modern filter have in the long run not only lower dust emissions, but also better economic results.

Spanish and British plants have invested in or modernised their electrostatic precipitators exclusively for regulatory reasons. One plant in the Spanish group 2 reported that it needed more electricity for the new filter, but dust emissions were lower (see Table 9.4-3).

To summarise, the environmentally most favourable performing German plants have voluntarily invested in electro filters and could achieve more profitable solutions for their business than the plants in Spain and the UK which undertook the investment because of legislative pressure.

Bag filters

88 per cent of the German plants and 80 per cent of both the Spanish and British plants in the sample had invested in new bag filters. The driver was environmental regulation exclusively. The German plants undertook the investments in the early and mid nineties, whereas plants in Spain and the UK only started in the late 1990ies. Investment costs did not vary very much between countries and environmental performers. Operating costs in the German group 1 and 2a were reduced after the updating. The plants in the German group 1 and 2a achieved the largest reduction in emissions. Finally plants mentioned some maintenance costs, but no other significant impact (see Table 9.4-4).

German plants undertook this initiative more frequently than plants in other sample countries, but did not show a major difference in adopting bag filters than elsewhere.

Energy saving

All plants undertook permanent measures to reduce energy costs. However, larger changes to the machinery like a preheater were only undertaken in two German and one British plant. The Spanish plants had invested in smaller projects like energy monitoring and have spent only a small percentage of the capital costs which their German and British counterparts have incurred. All plants achieved a reduction in energy costs, but the reduction of energy consumption was strongest in Germany. One of the German plants achieved 820 kcal per tonne of cement, whereas a British plant still needed 850 kcal after installing a preheater. The respective share of energy costs in total production costs was 13.5 per cent in the German plant against 16 per cent in the UK plant. One of the German plants was subsidised (see Table 9.4-5).

Overall, German plants with particular energy initiatives have reduced their energy consumption more than plants in the other countries. Also their share of energy costs in total production costs is lowest.

Process optimisation

Several plants have introduced particular process optimisation measures to increase the process efficiency and to achieve cost reductions and quality improvements. One German plant in group 2b has reduced NO_x levels from 1000 to 800 mg by means of trials, constant audits and measurements. In another

German and also in a British plant new scales and a new feed system have improved process efficiency. Also in two Spanish plants the introduction of a computer system and new mills has improved the production flow and decreased energy consumption (see Table 9.4-6).

All plants have optimised the process, no particular economic or environmental advantage is visible for the German plants.

Expert systems

Three German plants have introduced experts systems in the year 2000. Two plants have invested in expert systems for process efficiency reasons and in one case it was also required by the authorities before getting a permit for burning alternative fuels. The installation of expert systems costs in large plants five times as much as in small plants. Process efficiency was always improved, in one case operating costs were reduced. NO_x emissions could be reduced by as much as 20 per cent. People in the control room were especially trained in order to use the expert system. The expert system decreased the need for employment and in these cases there was a positive profitability impact.

Spanish and UK plants have introduced the expert systems earlier, but with less capital cost. No significant impact on emissions was recorded. In one case operating costs were higher than without the expert system (see Table 9.4-7).

Overall, German plants spent more on expert systems and achieved more beneficial results both with respect to economic and environmental consequences of this measure.

Alternative fuels

88 per cent of the German plants used alternative fuels, whereas none of the Spanish plants and only 25 per cent of the British plants used them. The use of alternative fuels is motivated by cost reducing reasons, and has started in Germany already in the mid 1980ies. Investment can be several million Euros, but operating costs are reduced because of lower energy costs. Plants achieve a reasonable payback and report a positive impact on profitability. In one German case also NO_x levels were reduced. A British plant achieved a payback of three years and could also increase its process efficiency by 2 per cent. Profitability even increased by 3 per cent (see Table 9.4-8). It was stressed by the managers that permits for alternative fuels are often linked with stricter limit values and are not a pure cost saving alone.

Noise

German plants have implemented noise abatement measures most frequently in the sample. They did it because of regulation and often over a long time period, in one case during 8 years. They have invested most in these measures and have also reduced the noise levels most effectively. No negative impact on the business was reported (see Table 9.4-9).

Future environmental investment and its impacts

When asked for possible environmental investment in the future, two medium sized German factories mentioned SNCR and said that it would be expensive. Several upgrading measures with mixed effects were said to be necessary in a Spanish plant. Another Spanish plant wanted to invest in a precalciner, but only if a capacity enlargement could be realised at the same time. One UK plant thought about new bag filters and said it would be neutral for the business. In another one a low NO_x burner could be installed; this was not considered to be economically problematic. However, SNCR was thought of being costly for the plant (see Table 9.4-10 at the end of this section).

Summary on drivers and effects of pollution abatement measures in German plants relative to Spanish and British plants

German firms have on average implemented more abatement measures than their counterparts in Spain and the UK and have done this in the majority of cases with either economically or environmentally more beneficial effects. There were also some cases where both the environment improved and some efficiency increase took place as well.

Measures with economic and environmental benefits:

The environmentally most favourable performing German plants have voluntarily invested in electro filters and could achieve more profitable solutions for their business than the plants in Spain and the UK which undertook the investment because of legislative pressure. Moreover, German plants with particular energy initiatives have reduced their energy consumption more than plants in the other countries doing it also for cost reducing reasons. Also their share of energy costs in total production costs is lowest in the entire sample. German plants spent more on expert systems and achieved more beneficial results both with respect to economic and environmental consequences of this measure than plants in other countries investing also for process efficiency reasons.

Measures with economic benefits, but environmentally neutral:

Almost all German plants used alternative fuels since the mid eighties and thus could reduce their energy costs substantially. Only in exceptional cases emissions went down. This initiative was not used frequently in the other sample countries at the time of the study.

Measures with environmental benefits and economically neutral:

Concerning the purely legislation driven measures SNCR, MSC and absorbent addition which are undertaken only in the German sample and nowhere else, it does not seem that the German plants suffer from a loss in competitiveness. With regard to SNCR and absorbent addition those plants with the highest investment costs achieved the largest reduction in emissions. All plants using these secondary abatement techniques, also used alternative fuels and could at least partly compensate the increase in production costs due to secondary measures with a decrease in energy costs.

Although German plants were forced to invest in low NO_x burners early on, no negative impact on competitiveness was detected. German firms have invested earlier in bag filter replacements and have reduced emissions more effectively. This is also the case for noise measures.

Neither environmental nor economic advantages:

Concerning process optimisation measures there is no particular advantage for German firms.

Table 9.4-1 NO_x and SO_x reduction measures used only in Germany and their impacts on business*

Country	Germany*		Germany		Germany	
NO_x and SO_x reduction measures	SNCR		MSC		Absorbent addition	
Env. Category	Group 1and 2a	Group 2b	Group 1 and 2a	Group 2b	Group 1 and 2a	Group 2b
Total number of plants using the initiatives	0	2	0	1	1	1
Driver		Legislation		Legislation	Legislation	Legislation
Year		1995/1997		1995	1995/1997	1998
Capital cost in M EUR		1.25/1.5		3.5	0.45	0.35
Operating cost per t of clinker in EUR		0.15/0.5		0	0.3	0.06
Reduction of NO_x or SO_x in mg		By 400 mg		From 1000 to 800	From 1000 to 350	Reduced
Subsidies		0		30%	0	50%
Maintenance		0		5% of investment sum	0	Little
Process efficiency		0		0	0	0
Employment		0		0	0	0
Training		Little		For 3 supervisors	0	0
Capacity		0		0	0	0
Output		0		0	0	0
Profitability		0		0	0	0
Total number of plants in each group	4	4	4	4	4	4

*Two of the total of the five medium sized German plants ranked to have medium emissions had remarkably lower emissions than their counterparts within the same emission range. These two plants were analysed separately within the German group 2 and called group 2a. The other three medium sized plants in group 2 were called group 2b. For illustration purposes group 2b also includes one large plant with average emissions.

Table 9.4-2 Low NO_x burners and their impacts on business

Country	Germany*		Spain		UK	
Env. Category	Group 1 and 2a	Group 2	Group 2	Group 3	Group 2	Group 3
Total number of plants using the initiatives	3	3	1	2	2	1
Driver	Cost, 2 x legislation	Legislation	Company policy for NO_x reduction	Process efficiency, upgrading	Process efficiency, legislation	Process efficiency
Year	2 x in 80ies	1985/1990	1994	1985/1993	1995	
Capital cost in M EUR	0.75	0.1, 0.175, 0.5	0.24	0.07	0.32	0.32
Operating cost per t of clinker in EUR	Not measured	0.01, 0.5	n.a.	Reduced (Less energy)	Reduced	Reduced (Less energy)
Reduction of NO_x in mg	Small, from 1600 to 750	200 mg less, 10-15 % less	Reduced	Decreased	Reduced	Reduced
Subsidies	0	0	0	0	0	0
Payback	0	0	0	0	0	3-4 years
Other capital items	0	0	0	0	0	0
Maintenance	0	0	0	0	0	0
Process efficiency	0	0	0	0	0	Improved
Employment	0	0	0	0	0	1 person less
Training	0	Little	0	0	0	For all operators
Capacity	0	0	0	0	0	0
Output	0	0	0	0	0	0
Profitability	0	0	0	0	1 x positive	Positive
Total number of plants in each group	4	4	1	4	2	2

*Two of the total of the five medium sized German plants ranked to have medium emissions had remarkably lower emissions than their counterparts within the same emission range. These two plants were analysed separately within the German group 2 and called group 2a. The other three medium sized plants in group 2 were called group 2b. For illustration purposes group 2b also includes one large plant with average emissions.

Table 9.4-3 Electrostatic precipitators and their impacts on business

Country	Germany*		Spain		UK	
Env. Category	Group 1 and 2a	Group 2b	Group 2	Group 3	Group 2	Group 3
Total number of plants using the initiatives	5	3	1	4	2	1
Driver	Replacement	Regulation	Regulation	Regulation	Regulation	Regulation
Year	1991, 1998	1990, 1994	1996	1990	1990	n.a.
Capital cost in M EUR		0.5-7.5	1.5	1.2, 1.6		0.15
Operating cost per t of clinker in EUR	Less energy	125,000; 50% less energy	More electricity	Electricity	Electricity	Electricity
Reduction of dust in mg	From 18 to 10, from 40 to 24	10 less, now 15, reduction by 20%	Less	n.a.	n.a.	n.a.
Subsidies	0	0	0	0	0	0
Payback	0	0	0	0	0	0
Other capital items	0	0	0	0	0	0
Maintenance	0	0	0	0	0	0
Process efficiency	0	0	0	0	0	0
Employment	0	0	0	0	0	0
Training	0	0	0	0	0	0
Capacity	0	0	0	0	0	0
Output	0	0	0	0	0	0
Profitability	Negligible	0	0	0	0	0
Total number of plants in each group	4	4	1	4	2	2

*Two of the total of the five medium sized German plants ranked to have medium emissions had remarkably lower emissions than their counterparts within the same emission range. These two plants were analysed separately within the German group 2 and called group 2a. The other three medium sized plants in group 2 were called group 2b. For illustration purposes group 2b also includes one large plant with average emissions.

Table 9.4-4 Bag filters and their impacts on business

Country	Germany*		Spain		UK	
Env. Category	Group 1 and 2a	Group 2b	Group 2	Group 3	Group 2	Group 3
Total number of plants using the initiatives	4	3	0	4	2	1
Driver	Regulation	Regulation	–	Regulation	Regulation	Regulation
Year	1993, 1994-96	1993	–	1994, 1997, 1998	n.a.	1999
Capital cost in M EUR	0.25, 0.45, 2.25, 30	n.a.	–	0.2-2.7	n.a.	0.16
Operating cost per t of clinker in EUR	Reduced	n.a.	–	0.048	n.a.	n.a.
Reduction of dust in mg	From 150-200 to 20-50	Not quantified	–	Reduced	Reduced	Reduced
Subsidies		0	–	0	0	0
Payback	0	0	–	0	0	0
Other capital items	0	0	–	0	0	0
Maintenance	250 000	0	–	0	0	0
Process efficiency	0	0	–	0	0	0
Employment	0	0	–	0	0	0
Training	0	0	–	0	0	0
Capacity	0	0	–	0	0	0
Output	0	0	–	0	0	0
Profitability	0	0	–	0	0	0
Total number of plants in each group	4	4	1	4	2	2

*Two of the total of the five medium sized German plants ranked to have medium emissions had remarkably lower emissions than their counterparts within the same emission range. These two plants were analysed separately within the German group 2 and called group 2a. The other three medium sized plants in group 2 were called group 2b. For illustration purposes group 2b also includes one large plant with average emissions.

Table 9.4-5 Energy saving measures and their impacts on business

Country	**Germany***		**Spain**		**UK**	
Env. Category	Group 1 and 2a	Group 2b	Group 2	Group 3	Group 2	Group 3
Total number of plants using the initiatives	1	1	1	1	1	1
Driver	Cost	Cost	Cost	Cost	Cost	Cost
Year	Always	Always	1995	n.a.	1996	n.a.
Capital cost in M EUR	n.a.	4	0.04	0.07	3	3.75
Operating cost per t of clinker in EUR	Small	Small	Small	Small	Small	Small
Reduction of energy	25% less energy	Less CO2	50% less energy	n.a.	From 900 to 850 kcal per tonne	Some
Subsidies	0	0.63 M Euros	0	0	0	0
Payback	Positive	Positive	n.a.	n.a.	15%	n.a.
Other capital items	0	0	0	0	0	0
Maintenance	0	0	0	0	0	0
Process efficiency	0	0	0	0	10%	0
Employment	0	0	0	0	0	0
Training	0	0	0	0	0	0
Capacity	0	0	0	0	0	0
Output	0	0	0	0	11%	0
Profitability	0	0	0	0	0	0
Total number of plants in each group	4	4	1	4	2	2

*Two of the total of the five medium sized German plants ranked to have medium emissions had remarkably lower emissions than their counterparts within the same emission range. These two plants were analysed separately within the German group 2 and called group 2a. The other three medium sized plants in group 2 were called group 2b. For illustration purposes group 2b also includes one large plant with average emissions.

Table 9.4-6 Process optimisation and their impacts on business

Country	Germany*		Spain		UK	
Env. Category	Group 1 and 2a	Group 2b	Group 2	Group 3	Group 2	Group 3
Total number of plants using the initiatives	1	2	0	2	1	1
Driver	Process efficiency	Process efficiency	–	Cost/ quality	Cost	Process
Year	n.a.	1993	–	1995-98, 1999	1998	
Capital cost in M EUR	Small	0.15	–	0.9, 0.12	0.01	0.09
Operating cost per t of clinker in EUR	Small	0.015	–	0.015	0	n.a.
Reduction of NO_x in mg	Some, not measured	From 1000 to 800	–	Some, not measured	0	Not meas-ured
Subsidies	0	0	–	0	0	0
Payback	0	0	–	0	0	0
Other capital items	0	0	–	0	0	0
Maintenance	0	0	–	0	0	0
Process efficiency	0	Improved	–	0	0	Improved
Employment	0		–	0	0	1 person less
Training	0	0	–	0	0	Yes
Capacity	0	0	–	0	0	0
Output	0	0	–	0	0	0
Profitability	0	0	–	0	0	0
Total number of plants in each group	5	3	1	4	2	2

*Two of the total of the five medium sized German plants ranked to have medium emissions had remarkably lower emissions than their counterparts within the same emission range. These two plants were analysed separately within the German group 2 and called group 2a. The other three medium sized plants in group 2 were called group 2b. For illustration purposes group 2b also includes one large plant with average emissions.

Table 9.4-7 Expert systems and their impacts on business

Country	Germany*		Spain		UK	
Env. Category	Group 1 and 2a	Group 2b	Group 2	Group 3	Group 2	Group 3
Total number of plants using the initiatives	3	2	1	3	1	1
Driver	Process efficiency (2 times) precondi-tion for alternative fuels	Process eff.	Process	Process	Cost	Process
Year	2000	2000	1995	90-96, 94	1998	1998
Capital cost in M EUR	3, 15	17.5	n.a.	1.2, 1.2	48500	n.a.
Operating cost per t of clinker in EUR	Reduced	n.a.	n.a.	Increased	n.a.	n.a.
Reduction of NO_x in mg	20% less No_x	Yes, not quantified	Small	Small	n.a.	Expected
Subsidies			0	0	0	0
Payback	Positive		0	0	0	0
Other capital items	0		0	0	0	0
Maintenance	0		0	0	0	0
Process efficiency	Improved	Improved	0	0	0	0
Employment	3 less, 4 less		0	0	0	0
Training	Yes	Yes	Some	0	0	Some
Capacity	0	0	0	0	0	0
Output	0	0	0	0	0	0
Profitability	2x in-creased		0	0	0	0
Total number of plants in each group	4	4	1	4	2	2

*Two of the total of the five medium sized German plants ranked to have medium emissions had remarkably lower emissions than their counterparts within the same emission range. These two plants were analysed separately within the German group 2 and called group 2a. The other three medium sized plants in group 2 were called group 2b. For illustration purposes group 2b also includes one large plant with average emissions.

Table 9.4-8 Alternative fuels and their impacts on business

Country	Germany		Spain		UK	
Env. Category	Group 1 and 2a	Group 2b	Group 2	Group 3	Group 2	Group 3
Total number of plants using the initiatives	4	3	0	0 (only trials)	1	1
Driver	Cost	Cost	–	–	Cost	Cost/env.
Year	1980, 1995, 1994/99	1985, 1995	–	–	1994	1995
Capital cost in M EUR	1.5, 3.4	0.1	–	–	0.8	1.5
Operating cost per t of clinker in EUR	Less	12,500	–	–	n.a.	5
Reduction of emissions in mg		1 case: less NO_x	–	–	0	0
Subsidies		10–15%	–	–	0	3 years
Payback	Pos.	Pos.	–	–	Pos.	Pos.
Other capital items	0	Scales	–	–	Small	Feed system
Maintenance	0	0	–	–	0	Some
Process efficiency	0	0	–	–	0	2% increase
Employment	0	0	–	–	0	3
Training	0	0	–	–	0	0
Capacity	0	0	–	–	0	0
Output	0	0	–	–	0	0
Profitability	Pos.	Pos.	–	–	Pos.	3% increase
Total number of plants in each group	4	4	1	4	2	2

*Two of the total of the five medium sized German plants ranked to have medium emissions had remarkably lower emissions than their counterparts within the same emission range. These two plants were analysed separately within the German group 2 and called group 2a. The other three medium sized plants in group 2 were called group 2b. For illustration purposes group 2b also includes one large plant with average emissions.

Table 9.4-9 Noise abatement and their impacts on business

Country	Germany*		Spain		UK	
Env. Category	Group 1 and 2a	Group 2b	Group 2	Group 3	Group 2	Group 3
Total number of plants using the initiatives	2	3	0	2	0	1
Driver	Regulation	Regulation	–	Cost	–	Regulation
Year	1998-2000	1994, over 8 years	–	Recently	–	90ies
Capital cost in M EUR	10	0.2, 0.4, 4	–	0.15	–	0.3
Operating cost per t of clinker in EUR	Negligible	0.01	–	Less energy	–	0
Reduction of noise in dB	8-10	Ca. 5	–	Ca. 5	–	n.a.
Subsidies	0	0	–	0	–	0
Payback	0	0	–	0	–	0
Other capital items	0	0	–	0	–	0
Maintenance	0	0	–	0	–	0
Process efficiency	0	0	–	0	–	0
Employment	0	0	–	0	–	0
Training	0	0	–	0	–	0
Capacity	0	0	–	0	–	0
Output	0	0	–	0	–	0
Profitability	0	0	–	0	–	0
Total number of plants in each group	4	4	1	4	2	2

*Two of the total of the five medium sized German plants ranked to have medium emissions had remarkably lower emissions than their counterparts within the same emission range. These two plants were analysed separately within the German group 2 and called group 2a. The other three medium sized plants in group 2 were called group 2b. For illustration purposes group 2b also includes one large plant with average emissions.

Table 9.4-10 Examples of future environmental investment

Plant type	Type of measure	Effects on business
Germany Group 2 medium	Possibly SNCR, when NO_x limit < 500	Expensive
Germany Group 2 medium	Possibly SNCR, when NO_x limit < 500	Expensive
Spain Group 3 large	filters clinker storage raw material storage improved process transport	Partly detrimental. Partly beneficial; but overall upgrading is necessary
Spain Group 3 medium	precalciner	Only if it can be done with a capacity enlargement
Spain Group 3 large	For 800 mg NO_x: low NO_x burner; For 500 mg NO_x: SNCR	No problem Problematic
UK Group 3 medium	Bag filters; Computer system	Impact on profitability depends on electricity price Improves performance
UK Group 3 Large	Bag filter	Neutral

9.5 Competitive advantages and disadvantages

Plants were asked for their competitive advantages and disadvantages concerning the production and sale of cement. This was done in order to place environmental issues relative to the general economic conditions in which the plants operate. Again plants were matched by size and plants with a favourable environmental performance were matched with those with a less favourable one.

Competitive advantages

The small German plant in group 1 stated that its main competitive advantage was low energy costs. The Spanish counterpart said that the quality of production, short delivery times and the highly skilled labour force was advantageous for the plant (see Table 9.5-1 for an overview of all facets of competitive advantages and their frequencies).

In the medium sized group the German plants mainly stressed variety and quality of products, being part of a multinational group and the ability to reach strict environmental standards. In Spain and the UK location and quality of raw materials and products was seen as the main advantage.

The large German plants stressed their modern process technology, the flexibility of workers and the variety of products as their main competitive advantages. The Spanish and British plants stressed the quality of products and their location including low transport costs. In Britain the large plant size was considered to be a competitive advantage. One Spanish plant thought of its planned use of alternative fuels that it would reduce energy costs in the future, a British plant mentioned its low borrowings and their increasing exports as competitive advantages.

Overall, plants of each environmental category were preoccupied with similar competitive advantages. However, German plants most frequently referred to elements which were also connected to favourable environmental performance like ability to achieve low energy costs, modernity of plant and up to date equipment which enabled them to fulfil strict environmental standards.

Competitive disadvantages

Concerning competitive disadvantages the small German plant in group 1 stressed the higher costs for spare parts because of being an independent plant and not getting any discounts like members of a group would get. The Spanish counterpart complained about missing economies of scale because of its small size, high labour and environmental costs (see Table 9.5-2 for an overview).

In the medium sized category German plants complained most heavily about strict environmental standards. Spanish plants complained about location, small size, variable quality of raw material and in one plant classified as group 3 about more stringent environmental legislation than the average plants in Spain. In a British plant it was thought that its independence and lack of experience with chemistry and also its limited market were the most important competitive disadvantages.

One of the large plants in Germany classified to have a very favourable environmental performance said that its vicinity to Poland forced it to keep prices low. Also the fact that it was not a global player was seen as a disadvantage. Another large German plant grouped as having a less favourable performance complained about the gaps in the local infrastructure, having three production lines and thirdly about high environmental costs. In Spain the plants complained about competition at the coast from Turkey and low prices for export. Also, one plant thought that its location close to a touristic zone had led to more stringent environmental standards. In the UK a plant complained about the tight market, another one about lack of flexibility and rail transport which made the shipments very slow.

To summarise, none of the German group 1 performers complained about high environmental costs, although environmental requirements were said to be high. Only one large German plant classified as group 2 performer felt that it was suffering from a competitive disadvantage through environmental costs, but had stressed problems related to infrastructure and plant design as more important. Several other plants, mainly medium sized ones in group 2 (and from those only plants with higher emissions) thought that the environmental requirements were very high. In Spain and the UK current environmental requirements and costs were hardly mentioned as a competitive disadvantage.

Table 9.5-1 Competitive advantages: main factors and number of responses

Country	Germany	Germany	Spain	Spain	UK	UK
	Group 1 and 2a	Group 2	Group 2	Group 3	Group 2	Group 3
Quality of raw materials				1		1
Quality of production			1	1		1
Nearness to raw materials				2		1
Location				3		1
Age of Plant	2					
Size of Plant						1
Cost to manufacture	1 (low energy costs)			2 (low transport costs)		2 (low transport costs)
Substitute (waste) fuels				1		1
Consistency of Product Quality	2					3
Variety of products	3					1
Flexibility of production		1				
Keeping delivery times			1			1
Env. regulation		1				
Env. costs						
Labour costs			1	1		
Labour quality	1					
Service		1	1			
Part of int. group		1				
Growing market						1
Low borrowings						1
No. of plants	4	4	1	4	0	4

Table 9.5-2 Competitive disadvantages: main factors and number of responses

Country	Germany	Germany	Spain	Spain	UK	UK
	Group 1 and 2a	Group 2b	Group 2	Group 3	Group 2	Group 3
Price	1 (vicinity to Poland)			1		
Quality of raw materials				1		
Quality of production						
Nearness to raw materials		1				
Location		1		2		
Age of Plant		3				
Size of Plant			1	1		1
Cost to manufacture	1 (spare parts)					1 (high transport costs)
Substitute (waste) fuels						
Consistency of Product Quality						
Variety of products						
Flexibility of production						1
Keeping delivery times						
Environmental regulation	2	1		1		
Environmental costs		1 (due to cont. measuring)	1			
Labour costs			1			
Labour quality					1	
Service						
Not part of int. group	1				1	
Long time to get permit		1				
Limited market					2	
No. of plants	4	4	1	4	0	4

9.6 Conclusions on the impact of stringent regulation on the German cement industry

The analysis of the interview data collected in German cement factories shows that the relatively stringent legislation implies relatively high compliance costs, but hardly an impact on the competitiveness of German plants.[35] There are several explanations to this. First of all, German cement plants can offset the higher capital costs by the use of alternative fuels and through general reductions in energy consumption. Although German plants were not entirely high productivity plants, they have implemented the highest number of costly abatement measures in the sample. In implementing these measures, German plants could in several cases achieve economically and environmentally more beneficial effects than elsewhere. Even concerning the purely legislation driven measures which are undertaken only in the German sample and nowhere else, no plant reported a directly negative impact on competitiveness. The ability to achieve low energy costs and the installation of modern equipment were stated as competitive advantages which also enabled the plants to fulfil strict environmental standards. Although environmental requirements were said to be high, the top environmental performers did not complain about environmental costs. The detailed findings will be presented as answers to the hypotheses presented at the outset of this study.

Productivity and cost of compliance: The proportional cost of environmental compliance relative to turnover incurred by the firms was hypothesised to be a negative function of the productivity level because firms with the capabilities to achieve high productivity will also find it easiest to respond to environmental challenges without the penalty of reduced output and employment.

Findings: There is an overall trend that German cement plants have to some extent the highest productivity levels, but – in contrast to the expectation related to the above hypothesis - also incur the highest compliance costs caused by tight environmental standards. Moreover, Spanish plants also achieve high productivity (mainly due to very low employment figures and high capacity utilisation), which is in several plants even higher than in comparable German plants, and have the lowest compliance costs in the sample due to the currently soft regulation.

---

[35] This finding is also confirmed for plants in Sweden and Austria visited by the author in 1999 and 2000 where environmental regulation is at least partially even more stringent than in Germany.

Number and type of initiatives, ability to absorb compliance costs, skills and R&D, environmental awareness: A favourable environmental performance is hypothesised to be linked to characteristics of high productivity (best practise) plants. Still, these environmentally friendly plants are hypothesised to be able to keep compliance costs low because they are assumed to show a high degree of environmental awareness and would be in a good position to use clean technologies and not end-of-pipe technologies through their skill and R&D inputs.

Findings: Indeed, on a matched basis German plants were often high productivity plants and very clearly showed the most favourable environmental performance in the sample implying both the achievement of lowest emission levels and the installation of the highest number of pollution reduction techniques. Contrary to the assumption, their compliance costs were highest in the entire sample and this has to do with the type of abatement technologies used in the cement industry. The majority of effective pollution abatement techniques is of an end-of-pipe nature and the potential for the use of clean technology with associated cost savings is limited. German plants have exploited this potential to the largest extent possible, partially the strong skill base was helpful for this. However, in order to decrease emissions down to the German standards the use of secondary measures with additional compliance costs is inevitable.

Simultaneously, German plants undertook the largest efforts to compensate high compliance costs through the use of alternative fuels and general reductions in energy consumption. Also the UK plants regularly used alternative fuels. In the German and British plants where compliance costs were highest, also the use of alternative fuels measured as a percentage of total energy consumption tended to be largest. A comparison of the time sequence of important environmental investments showed that permits for the use of alternative fuels were always obtained in times with increasing environmental compliance costs. Moreover, German plants have reduced their energy consumption drastically relative to other countries and need on average about 12 per cent less energy per tonne of clinker than for instance their British counterparts.

A direct relationship between environmental awareness (measured through the availability of an environmental management system) and environmental performance was not found in any sample country. Neither were compliance costs related to environmental awareness.

Plant age: The age of a plant and its machinery is likely to have an impact on its environmental performance and on compliance costs. Modern plants can reach better environmental standards because the newest technology also embodies best environmental technology and allows to achieve environmental goals at a lower cost than with old equipment which would need expensive retrofitting.

Findings: New kilns imply a high environmental performance and produce high productivity levels. Within the sample Germany had the highest number of new kilns in place. But in Germany also relatively old kilns can reach a favourable environmental performance and high productivity levels. This is related to high general investment levels both in the past and in the future which implies that the German plants are maintained to a very high standard. Also the system of national emission limit values plays a role. Compliance costs are higher in the new German plants than in the old ones, mainly because the new plants use more secondary abatement techniques.

Plant size: Large plants are hypothesised to undertake more environmental initiatives because they often have more financial and human resources for the solution of environmental problems than small and medium sized plants and have also better opportunities to research cost efficient environmental measures. Thus the proportional cost of compliance relative to turnover is likely to be a negative function of the size of plants/firms.

Findings: Both in Germany and the UK compliance costs rise with plant size. In the German sample there was no single case where a small plant had invested in many (expensive) secondary environmental measures. In Germany and the UK compliance costs do not seem to depend on environmental performance (measured through level of emissions and number of environmental initiatives), whereas in Spain the pattern of compliance costs is such that they clearly fall with plant size and rise with environmental performance reflecting the poor environmental performance of large Spanish sample plants. This result can be expected because especially the large plants in Germany and the UK were recently modernised and upgraded. Consequently, differences in emissions levels within the German and British sample were much smaller than in the Spanish sample.

History of regulation: Plants in countries with a history of stringent environmental legislation will be in a better position to develop or gain access to environmentally and economically beneficial solutions than their counterparts in other countries with a history of less stringent regulation. This is because their supplier network is more experienced with solutions to environmental problems and can offer a more differentiated supply of environmental goods including clean technology.

Findings: The long tradition of environmental regulation has certainly stimulated the supply of environmental goods in Germany and has also made the

German cement industry more experienced in the use of pollution abatement measures. German firms have on average implemented more abatement measures and have done this earlier than their counterparts in Spain and the UK. In several cases (electro filters, energy initiatives, expert systems) the German plants have achieved economically and environmentally more beneficial effects than elsewhere. Concerning the purely legislation driven measures SNCR, MSC and absorbent addition which are undertaken only in the German sample and nowhere else, it does not seem that the German plants suffer from a loss in competitiveness. With regard to SNCR and absorbent addition those plants with the highest investment costs achieved the largest reduction in emissions. All plants using these secondary abatement techniques, also used alternative fuels and could compensate any increase in production costs due to secondary measures with a decrease in energy costs. In none of the plants a directly negative impact on competitiveness was reported.

Moreover, the impact of the history of regulation was also captured when the plants were asked to estimate the importance of environmental factors influencing their competitive performance relative to other factors. Concerning competitive advantages German plants most frequently referred to aspects connected to favourable environmental performance like ability to achieve low energy costs, modernity of plant and up to date equipment which enabled them to fulfil strict environmental standards. With respect to competitive disadvantages German plants stressed that environmental requirements were said to be high. But the top environmental performers did not complain about environmental costs, only one large German plant classified as group 2 performer felt that it was suffering from a competitive disadvantage through environmental costs. Simultaneously this plant stated problems related to infrastructure and plant design as more important. In Spain and the UK current environmental requirements and costs were hardly mentioned at all.

Green market pressure in the supply chain relations: Not only regulatory pressure forces firms to react, but also environmental pressure coming from the market place. Relations within the supply/production chain are likely to be both an influence upon, as well as being influenced by, the level of environmental standards, e.g. a manufacturer may find it easier to increase the environmental standards of its products if it has a reliable and competent base of suppliers to draw on. A manufacturer may be forced to upgrade product and process environmental standards by pressure coming from the customers of plants in the three sectors under study.

Findings: This did not seem to play a role at all in the cement industry. In Germany it was only mentioned that customers wished a large variety of products, larger than in Spain and the UK. All countries stressed that quality issues were very important.

Green consumer pressure: Plants which operate in a market where final demand for clean products and processes is large and consumers are prepared to pay a premium, will be more involved in environmental measures than other firms which do not have similar customer groups and are not rewarded for their environmental activities.

Findings: Since sales to outlets are very small, direct consumer contact has so far not been a subject for the cement industry, also not in Germany. E.g. no German plant used an eco-label for its product.

Market structure: In an oligopoly like the cement industry individual incentives for renewal and updating always depend on what other members of the oligopoly will undertake. Adjustment to new environmental requirements may therefore be procrastinated in an oligopolistic market. On the other hand, if an industry achieves super-normal profits, it is to a certain protected against additional costs arising from environmental compliance and the pursuit of wider management goals which includes environmental protection is possible.

Findings: The analysis has indeed shown that costly secondary measures which really bring emissions down, have been introduced only very recently in the German industry, i.e. the industry has succeeded in avoiding these measures for a long time. On the other hand the measures do not show a significant impact on businesses. Any super-normal profits would play a role in this. Data on profit margins on a plant level were hard to obtain during the in-depth interviews. But during later interviews with headquarter chief executives of multinational cement companies it was said that German plants have always been profitable despite their environmental investments.

Ownership and location: Multinational companies are likely to show better environmental performance than other enterprises because they export both the environmental and the efficiency standard of their home country. Concerning location, plants in peripheral sites are likely to be less advanced on environmental matters, to undertake less environmental initiatives and to be potentially subject to higher environmental costs because of lack of access to suppliers, language barriers, higher transport costs etc. than plants in centrally located regions.

Findings: In Germany it is also possible for independent plants to reach a high level of environmental performance. It of course also reflects the German legislation which is uniform across different types of ownerships. Plants abroad owned by German factories have a somewhat lower environmental standard. Although multinationals perform better than their national counterparts abroad, they do not apply exactly the same standards as at home and can thus lower their compliance costs. Location only mattered in terms of differences in environmental regulation, but not in terms of access to suppliers or anything similar.

Employment impact: The direction of the employment implication of adjustment to stringent environmental regulation is not clear a priori. The available macro-econometric evidence has shown that the overall employment effects of environmental policies are likely to be small.

Findings: There was hardly any employment impact through environmental measures in this industry. The little impacts were dominated by small negative impacts caused through automatisation.

High standards and strict enforcement: Given the above hypotheses, although high environmental standards and strict enforcement may represent a short term cost and burden to the firm, they could in the medium and longer term push firms on to a higher growth path by forcing them to make product and process changes which yield higher competitiveness. If this happened it would represent part of the so-called "double dividend", i.e. gains in environmental performance would also be accompanied by increased economic performance.

Findings: To summarise all answers above, high standards and strict enforcement in the cement industry have caused a cost to the individual plants, which can be partially compensated by process changes and a reduction of energy costs yielding higher competitiveness. Overall, gains in environmental performance have not damaged economic performance, but have also not led to an increase in economic performance.

10. Summary and conclusions

10.1 Introduction

Environmental regulation in Germany has been considered to be more demanding in comparison to many other industrialised countries. In the European Union only countries like Sweden, Austria and the Netherlands have reached such environmental performance. The central aim of this research has been to determine the effects on competitiveness of the adjustment which German firms make in response to this stringent environmental legislation. This adjustment was measured in comparison to other Member States of the European Union where environmental regulation is on average still less demanding than in Germany. To this end three international case studies were carried out: German environmental legislation was compared with Irish, UK and Spanish regulation in various industries which at the time of the study were less stringently regulated than the corresponding German industry branches. The basic research question was whether the more stringent regulation puts Germany at an economic disadvantage relative to the other case study countries. Methodologically, three different ways of measuring a possible disadvantage were applied. In the following a summary of the main findings, the general conclusions as well as policy implications are presented.

10.2 Summary

In three case studies it was examined whether German plants suffer from a negative impact on competitiveness caused by stringent environmental legislation. For the food (dairy and meat) industry the impact of stringent waste water legislation was examined; in the supply chain of the packaging, food and retail industry the influence of packaging regulation was researched and finally the impact of clean air regulation was examined in the cement sector.

In order to answer the question whether Germany is put at an economic disadvantage, three different test methods were applied in each case study. First of all a set of hypotheses concerning the possible relationship between regulation and competitiveness was developed and statistically tested. Thereby the focus was on a detailed measurement of compliance costs and the factors which influence compliance costs and thereby competitiveness (see chapters two and five, and also section 10.2.1 below). Second, a measurement of the detailed economic effects of each environmental initiative was carried out in order to check the net effect of each initiative. Lastly, managers in each country were directly asked to rank environmental aspects in relation to other competitive advantages and disadvantages.

In the following sections the findings to the three test methods are presented in a comparative way for all case study countries and industries. First of all, short summaries of the main findings are given, then in the sections 10.2-1 to 10.2-4 detailed results are presented. Results are as follows:

Summary on test of hypotheses

As a central hypothesis, the economic performance of plants (mainly productivity, but also the skill and R&D level of a plant, the age of machinery, size of plant and similar factors) was assumed to have an impact on the ability of a firm to adjust efficiently to stringent regulation and hence influence the level of compliance costs and competitiveness. It was found in the test of the hypotheses for all case studies that although Germany is a high productivity country, compliance costs are high as well. Thus, German firms can hold a competitive position without being favoured by nationally or internationally low compliance costs. It was also found that Germany takes a leading position in terms of the number of environmental initiatives used. There was some support for a positive relationship between productivity and the number of environmental initiatives undertaken with the evidence being strongest in the food case. In the case studies no consistent relationship between number of environmental initiatives and compliance costs was found. However, German sample plants had undertaken the larg est efforts to implement clean technologies, benefiting from their cost reducing and productivity enhancing effect.

The hypothesis that small plant size is linked with higher compliance costs could not be confirmed for any case study. But there was a tendency for small plants to undertake fewer environmental initiatives than large plants. No generally valid relationship between skills, R&D and compliance costs could be established. Also, in general, lower plant age did not influence compliance costs positively in Germany.

Concerning location, East German plants tended to be faced with higher compliance costs, particularly concerning waste water treatment. Moreover, for East German packaging plants it was found that the number of environmental initiatives was lower than in West Germany. East German cement plants were the most modern plants in the sample. Ownership mattered insofar as multinationals had company-wide environmental policies in place and as large firms they undertook a high number of environmental initiatives. However, German cement multinationals operating abroad did not apply as stringent environmental standards as at home and could therefore lower compliance costs abroad. German cement plants have also benefited from their oligopolistic industry structure insofar as they have only recently implemented tight environmental standards.

Summary on the effects of the implementation of environmental initiatives

The second test method on the implementation of environmental initiatives showed that more economic benefits were achieved both in East and West Germany than in the other sample countries. In the food case small, positive productivity impacts were caused by environmental initiatives. In the packaging study, the majority of initiatives was cost driven and presented the largest source for general financial savings; solid waste initiatives caused by external pressure in Germany produced more unit cost savings and positive competitiveness effects than elsewhere. In the cement study additional capital requirements for legislation driven measures were offset through the use of alternative fuels. Thereby German cement plants could compensate any increase in production costs due to end-of-pipe measures with a decrease in energy costs. These financial savings prevented a loss in competitiveness. German cement plants have also voluntarily invested in environmental measures and could achieve more profitable solutions for their business than the plants in Spain and the UK which undertook the investment because of legislative pressure.

Summary on environmental pressure relative to other competitive pressures

In the third test, German managers were asked to rank environmental aspects next to other competitive factors relevant for their business. In all case studies, environmental factors were neither seen as a competitive advantage nor as a competitive disadvantage.

The findings to the three tests explain why hardly any negative impact of environmental regulation on competitiveness was discovered, although compliance costs in Germany are higher than in the other sample countries.

10.2.1 Test method (1): Hypotheses on possible relationship between environmental regulation and competitiveness: compliance costs and influencing factors

As a principal task the study sought to measure the cost of environmental compliance and find a relationship with the economic performance characteristics of the sample plants. In the following, first of all the results on the detailed measurement of compliance costs are presented. Then the factors which influence compliance costs and thereby competitiveness are discussed.

Results on level of compliance costs

It was attempted to measure compliance costs as the sum of capital and running costs of all environmental initiatives. This was not always fully possible; within each case study, however, an identical measure of compliance costs for each country was used in order to make comparisons meaningful. In all three case studies compliance costs tended to be higher for German sample plants. Compliance costs were found to be more significant in the cement case than in the food and packaging study. Sometimes the cost differentials between Germany and the other sample countries were only small. The largest cost differentials between countries were discovered for sewer charges in the food industry, DSD costs in the packaging case study and for capital costs in the cement sector (see Table 10.2-1 for an overview). The detailed results are as follows:

Table 10.2-1 Overview of compliance costs in each case study

Case study/sub sector/type of compliance cost		Type of compliance cost	Germany	Ireland/UK	Spain
Food: capital and running costs		Range of capital cost of effluent plant in % of plant turnover	2.7-5.5	2.4-2.7	–
		Range of running cost of effluent plant in % of plant turnover	0.14-0.39	0.1-0.5 (but average around 0.2)	–
		Range of absolute sewer charges in Euro	0.78-3.42	0.17	–
Packaging: running costs measured as sum of recycling and landfill costs	Dairy/drinks	Mean as % of plant turnover	0.050 (0.61 incl. DSD)	0.040	–
		Matched Mean as % of plant turnover	0.08 incl. DSD	0.020	–
	Retailing	Mean as % of plant turnover	0.11	0.10	–
		Matched Mean as % of plant turnover	0.10	0.07	–
	Packaging	Mean as % of plant turnover	0.36	0.12	–
		Matched Mean as % of plant turnover	0.28	0.12	–
Cement: capital costs		Average compliance costs as % of turnover	8.4	6	1.6

Source: Own data collection.

In the food case individual firms undertaking the same activity in each country face a range of waste water treatment costs with German firms incurring the highest costs per m^3. Concerning direct dischargers using an effluent plant capital costs of these effluent plants were recorded where available as percentage of turnover, they varied for Germany between 2.7 and 5.5 per cent of current turnover, in Ireland the variation was with 2.4 to 2.7 per cent much smaller. Estimates concerning running costs were made of labour costs, chemical and detergent costs, electricity costs for pumping, maintenance and repairs; also depreciation costs were included where it was still necessary. Average effluent plant running costs were at 0.39 per cent of turnover highest in the German meat industry and at 0.14 per cent lowest in the German dairy industry. In the Irish industries costs varied around 0.20 per cent of turnover in both industries, although the range in costs varied from 0.1 to 0.5 per cent of turnover. Where possible costs per m^3 to firms treating their own effluent were compared with sewer charges. Costs for discharging to municipal sewers were variable and were not necessarily greater than the costs of self treatment, although the usual reason given for installing an effluent plant instead of using the municipal sewer was to reduce costs. In Germany sewer charges amounted to a 49 per cent higher cost although there was a wide variation of between a 5 and a 128 per cent difference between sewer charges and average effluent plant costs within Germany. Absolute sewage fees per m^3 reached the highest values in East Germany. Also in Ireland sewage fees were variable. E.g. in the Republic of Ireland costs per m^3 for two dairies discharging to sewers averaged Euro 0.17 compared with Euro 0.54 for seven firms sampled discharging directly to rivers and so on. The sewer price differential arises from differing charging systems and differing pricing structures between countries. In Germany direct dischargers have to pay a waste water charge depending on their pollution load, indirect dischargers are subject to sewage fees and occasionally also to a strong polluter surcharge. Also, in Northern Ireland and the Republic of Ireland a mixture of licences and charges is used. At the time of the study, however, in contrast to Germany direct discharg ers are not being charged by a waste water charge and sewer fees for indirect dischargers are much less developed than in Germany. E.g. in the Republic of Ireland in the mid nineties, fees for the use of the public sewers are only in a few cases paid by industry. It is reckoned that these capital cost contributions vary widely and globally cover only about one-third of total capital costs. There are also differences between firms in the extent to which they pre-treat or clean their waste water and which influences the cost of water purification.

Also in the packaging supply chain case study, average solid waste costs measured as a percentage of turnover and consisting of net recycling costs and landfill costs were found to be higher in Germany than in Ireland or the UK. According to the German Packaging Ordinance of 1991, 80 per cent of all package waste had to be collected separately and 64 to 72 per cent (depending on the materials used) had to be recycled. Manufacturers, distributors and retailers were obliged early onwards to take back their package and founded the DSD ("Duales System Deutschland GmbH") in 1990 - a private disposal company which is responsible for the collection, sorting and reprocessing of used sales packaging materials of its participating companies. The DSD makes the participants pay a licence fee for every product disposed of the licence symbol is the so-called green dot. In Germany waste costs inclusive of green dot fees raise total solid waste costs, particularly in the dairy/drinks sector.[36] This is the case because DSD costs fall on the packer/filler – in our case this is the dairy/drinks industry – and to some small extent on the retailer in the form of secondary packaging passed on to the customer. Dairies use a high percentage of throwaway plastic packaging for their products for which a high DSD fee must be paid. Excluding DSD costs the mean cost differential between Germany and the Republic of Ireland/UK is small for the dairy/drinks sector and retailing, but greater in the packaging sector. Results based on matched pairs of firms also support the findings. In Germany a shift away from landfill to recycling has taken place in the new waste policy. Since landfill costs in absolute terms have risen greatly in Germany, recycling activities are still cheaper for an individual firm than sending the waste to landfill. In the Republic of Ireland/UK samples net recycling costs were negligible or yielded revenues.

German cement sample plants incurred with an a average of 8.4 per cent of turnover the highest compliance costs (measured as investment costs during the last 5 to 10 years converted into current prices) in comparison with plants in the UK and Spain where average compliance costs were 6 and 1.6 per cent respectively. German dust emission limits are among the most stringent emission limit values in the EU, and the relative stringency is particularly conspicuous in comparison to Spain where dust legislation is in many provinces still quite soft. German NO_x limits are the second most stringent limit values, only in Denmark and Sweden is regulation tighter. SO_2 regulation in Germany is also tighter than the average in the European Union. However, SO_2 is on average not as much a widespread problem in Germany as e.g. in the UK where different raw materials (chalk) produce much higher SO_2 levels. As a consequence of the stricter German limits pollution abatement investment in the German sample was higher than in the other case study countries:

Which factors influence compliance costs and competitiveness? Answers to hypotheses

Compliance costs are influenced by various factors (see hypotheses developed in chapter 2). As a central hypothesis, the economic performance of plants (mainly productivity) was assumed to have an impact on the ability of a firm to adjust

[36] Similar costs were not yet paid in the other sample countries.

efficiently to stringent regulation and hence influence the level of compliance costs and competitiveness. The productivity level was also assumed to be related to the number and type of environmental initiatives in response to environmental regulation. The environmental measures themselves have an impact on compliance costs and competitiveness. Furthermore, other productivity related factors like the skill and R&D level of a plant, the age of equipment, size of plant as well as location/ownership can have an influence on compliance costs. Finally, the non-regulatory pressures a plant is exposed to in the market place as well as market structure itself are not negligible.[37] In the following all these factors are examined and it is worked out whether any of these factors puts Germany in a particular negative or positive situation compared with the other case study countries. Concerning presentational matters in this paragraph each factor of influence is first illustrated as hypothesis. Then a general answer from all three case studies is given followed by the detailed findings of the individual case studies.

Influence of productivity on compliance costs

It was hypothesised that the proportional cost of environmental compliance relative to turnover incurred by the firms is likely to be a negative function of the productivity level. This would be because firms with the capabilities to achieve high productivity will also find it easiest to react to environmental pressures without the penalty of reduced output and employment.

The findings in the case studies indicate no clear evidence in support of this hypothesis. It can be concluded for Germany that a firm can hold a competitive position without being favoured by nationally or internationally low compliance costs. The detailed findings are as follows:

As stated above waste water costs in the food study were found to be higher in Germany. Based on a statistical analysis of the relation between the cost of waste water and measures of firm competitive performance the hypothesis that competitive plants incur low environmental costs is not generally confirmed. Whereas competitive Irish plants faced lower waste water costs than their less well performing national counterparts, competitive German plants did not incur lower waste water costs.

---

[37] Of course, also the type of regulatory system and related policy instruments plays a role for the level of compliance costs. This was already discussed in the paragraph on the level of compliance costs above.

In the packaging case, again no statistically significant relationships have been found. While in many cases there is support for the hypothesis in terms of a relationship between low average waste costs and top performing firms, there are also many cases where low waste costs are associated with relatively poor economic performance. In Germany, only top performers in the retail sector have lower waste costs than the average, but this is statistically not significant. By and large, in dairy/drinks and also packaging the German top performers have even higher costs than the average.

There is a trend in German cement plants to reach high productivity levels, but – in contrast to the hypothesis above - they incur the highest compliance costs caused by tight environmental standards. Therefore our hypothesis is not confirmed.

Influence of productivity on number of environmental initiatives; relationship between number of environmental initiatives and compliance costs

A favourable environmental performance is hypothesised to be linked to characteristics of high productivity (best practise) plants. Just as these high productivity plants would be more aware of any gains in efficiency than the average, their management and workforce can be expected to undertake a higher number of environmental initiatives because they show a higher environmental awareness. The consequence on compliance costs can a priori be twofold: On the one hand compliance costs could rise with the number of initiatives because of their capital and running costs, on the other hand a higher number of environmental initiatives can also reduce compliance costs via financial savings generated through the minimisation of the use of certain resources.

In all three samples German plants had implemented the highest average number of environmental initiatives. There was some support for a positive relationship between productivity and the number of environmental initiatives undertaken by the firm with the evidence being strongest in the food case. No consistent relationship between number of environmental initiatives and compliance costs was found. Evidence of a positive relationship was strongest in the cement sector. In all industries but cement a higher score of environmental awareness was correlated with a higher number of environmental initiatives. The detail is as follows:

In the food industry sample, German plants undertook the highest number of initiatives with West German plants having a slightly higher number in place than their East German counterparts. There was some support for a positive association between achieving average or above average competitiveness and average or above average use of waste water initiatives both in the dairy and meat industry and both in Germany and the Republic of Ireland/Northern Ireland. Linking waste water costs to waste water initiatives in the food industry case study, a positive relationship emerged across all sample countries and a correlation of 0.3 was calculated. Similar small and positive correlations were found for the relationship between average water prices and average number of waste water initiatives. However, the statistically significant correlation disappeared, when the data were disaggregated to within country comparisons for any of these measures. Consequently, there was no evidence that a higher number of waste water initiatives leads to either higher or lower waste water costs.

In the packaging case the larger average number of initiatives undertaken by German firms arises from a greater percentage of firms engaged in each individual waste activity. It is noteworthy that in four areas, especially, there is more activity undertaken by German firms. This is in packaging design (particularly German retail and packaging companies are more active than their Irish/UK counterparts), material switching, packaging recovery and waste segregation which both are particularly important across all sectors compared with the two Irish economies. Overall, the support for a relationship between firm performance (productivity) and the number of solid waste initiatives undertaken by the firm, is generally inconsistent and not statistically significant. The strongest linkage was found in the German dairy/drinks industry where top performers were able to implement a higher number of solid waste initiatives than the overall German sample. Moreover, also Germany's top retailers used a much larger number of initiatives than the average German retail firm. The relationship between numbers of solid waste initiatives undertaken and solid waste costs showed no statistically significant results and in Germany some positive linkage between high waste costs and high number of waste initiatives in dairy/drinks and retailing was observed.

Also German cement plants were on average the environmentally most active plants in the sample, using with on average eleven initiatives per plant the highest number of environmental measures. Spanish sample plants had on average 5.4 measures per plant installed and in the UK it was 5.8 initiatives per plant. From the list of 18 relevant environmental measures German plants used 12 measures more frequently than the plants in the other sample countries. Differences were most marked for the use of expert systems and automatic quality control, in the prevention of fugitive dust, the use of continuous measurement and updating of electrostatic precipitators. In addition, from the list of these 12 initiatives, three measures were used in Germany only and nowhere else (multi stage combustion, SNCR and absorbent addition). Concerning the relationship between the number of environmental initiatives and economic performance the German plants with low emissions and many pollution abatement measures show in some respects a better company performance than their counterparts with less favourable environmental performance, but not in all aspects. Especially, the German small and large plants with high environmental performance reach a higher productivity level. However, production costs, price per tonne, sales per head and capacity utilisation do not give them an economic advantage over the plants in other countries. There was a positive relationship between the level of compliance costs and the number of environmental initiatives. This has to do with the type of abatement technologies used in the cement industry. The majority of effective pollution abatement techniques is of an end-of-pipe nature and the potential for the use of clean technology with associated cost savings is limited (see next paragraph below).

Type of environmental initiative and compliance costs

In general, compliance costs could be assumed to increase with the number of environmental initiatives. High productivity plants, however, will not only react more actively to environmental pressure with a higher number of environmental initiatives than other plants with lower productivity, but also prefer clean technologies to end-of-pipe technologies. This is because the latter is normally associated with additional costs and losses in productivity, whereas clean technology can improve productivity, reduce compliance costs and simultaneously bring emissions down.
In all three case studies it was found that Germany operates more clean technology solutions than the other sample countries. The detail is as follows:

In response to the relatively strict regulation German food plants undertook more clean technology initiatives than their Irish counterparts. Especially, recycling/reuse[38] of water incl. water monitoring and recovery of valuable resources from waste water was more widespread in Germany. As a consequence of these measures German plants used a lot less water per litre of milk or kg of meat produced than their Irish counterparts in the sample. But since waste water fees and charges per m^3 are relatively high in Germany, waste water costs were still highest in the German sample.

From the ten packaging initiatives in the packaging study seven can be classified as clean technologies and only three as end-of-pipe technologies. The percentage of German plants involved in six of the seven clean technology initiatives

[38] Recycling cannot easily be subsumed under the categories cleaner or end-of-pipe technology. Process-internal recycling can be understood as cleaner technology while process-external recycling is an end-of-pipe technology (see Rennings/Zwick 2001)). In the food case study water recycling is clearly undertaken in a process-internal way.

was much larger than in the Republic of Ireland or the UK[39]. This indicates that there is more pressure in Germany to avoid and reduce packaging waste in a cost efficient way. Also in waste segregation which is an end-of-pipe technology and a precondition for recycling, 80 per cent of German firms were engaged, whereas only 20 per cent of the Irish/UK plants undertook this initiative. The reason is that recycling activities are cheaper for an individual German firm than sending the waste to landfill. Still, the total of recycling and DSD costs makes solid waste costs in the German sample higher than elsewhere.

As in the other two case studies, German cement plants have exploited the potential for the use of clean technology with associated cost savings to the largest extent possible: 13 of the total number of 18 environmental measures are reckoned to be primary measures with some potential for cost savings. These measures include general reductions in energy consumption: For instance, German plants have reduced their energy consumption drastically relative to other countries and need on average about 12 per cent less energy per tonne of clinker than for instance their British counterparts.

However, in order to decrease emissions down to the German standards the use of secondary measures with additional compliance costs is inevitable. Simultaneously, German cement sample plants have the longest tradition to compensate high compliance costs through the use of alternative fuels. Thereby large savings in energy costs could be achieved which came close to the level of compliance costs. A comparison of the time sequence of important environmental investments showed that permits for the use of alternative fuels were always obtained in times with increasing environmental compliance costs.

Influence of skill and R&D level on compliance costs

The type of environmental activity is also anticipated to be connected with the skill and R&D endowment of plants. Plants with a well trained labour force and a strong R&D basis will also integrate environmental activities in their work routines and come up with a higher number of individual, clean technology solutions and are likely to be more successful in reducing environmental costs than their counterparts lacking these skills. Regulation could lead firms to increase

---

[39] These six initiatives were reuse, special design, lightweighting, material switching and packaging recovery. In the packaging industry recycling was also counted as clean technology initiative because it is a process-internal measure in this industry. In the food and retail industry recycling efforts cannot be integrated in the production or service process and are thus process-external.

their R&D efforts, thus leading to a decrease in production costs and stimulating productivity growth.

No generally valid relationship between R&D and compliance costs could be found.

In the food case study no clear relationship between the skill and R&D level and environmental costs was found. Thus it can be argued that low compliance costs do not depend on high skill and R&D levels.

In the packaging study some instances of a cost reducing influence of skills and R&D on waste costs were found in each country, but results were stronger in Germany. Especially, packaging plants with a strong skill and R&D base had lower than average waste costs. However, none of these relationships is supported by statistically significant correlations.

In the cement study no direct influence of skills on compliance costs could be measured for German sample plants. However, the German plants reported that higher skill levels and training were necessary for the introduction of expert systems, low NO_x burners, multi stage combustion and SNCR. For expert systems a positive influence of skills on environmental performance could be proven for Germany and this initiative was more frequently used than in the other sample countries.

Influence of plant age on compliance costs

The age of a plant and its machinery is likely to have an impact on its environmental performance and on compliance costs. Modern plants with younger machines are more productive and less polluting than plants using older machines. They can reach better environmental standards because the newest technology also embodies best environmental technology and are allowed to achieve environmental goals at a lower cost than with old equipment which would need expensive retrofitting.

In general, lower plant age did not influence compliance costs positively in Germany.

Concerning the hypothesis that low plant age would be connected to lower waste water costs, in the food study the opposite was found for Germany and results were mixed for the Republic of Ireland and Northern Ireland.

While the hypothesised direction of the relationship between younger plant age and lower waste costs is confirmed for the German dairy/drinks and packaging industry, the relationship is not found to be statistically significant.

Within the cement sample Germany had the highest number of new kilns in place. Compliance costs are higher in the new German plants than in the old ones, mainly because the new plants use more secondary abatement techniques. But in Germany also relatively old kilns can reach a favourable environmental performance and high productivity levels. This is related to high general investment levels both in the past and in the future which implies that the German plants are maintained to a very high standard.

Influence of plant size on compliance costs and linkage with number of environmental initiatives

Plant size is assumed to have an influence on the adjustment to regulation and related environmental pressure. Large plants are hypothesised to undertake more environmental initiatives because they often have more financial and human resources for the solution of environmental problems than small and medium sized plants and have also better opportunities to research cost efficient environmental measures. Large multiplant firms could possess certain economies of scope across several markets in complying with a regulation; that is large diversified companies may be able to spread compliance costs across several markets. Thus the proportional cost of compliance relative to turnover is likely to be a negative function of the size of plants/firms.

The hypothesis that small plant size is linked with higher compliance costs could not be confirmed for any country and case study. But there was a tendency that small plants undertake fewer initiatives than large plants.

No clear influence of plant size on waste water costs was found in the food case study. In dairying there was a clearly positive relationship between size and number of initiatives in all sample countries, but less so in meat processing.

In the packaging study there was no general evidence of a relationship between waste costs and plant size, although there was some evidence of higher costs associated with smaller size for firms in the German dairy/drinks sector. In retailing some positive relationships between waste costs and shop size were found. In this case study it became particularly clear that small firms undertake fewer initiatives than their larger national counterparts, especially in the German retailing sector.

In the cement study both in the German and the UK sample compliance costs rise with plant size. In the German sample there was no single case where a small plant had invested in many (expensive) secondary abatement measures. In Spain the pattern of compliance costs is such that they clearly fall with plant size reflecting the poor environmental performance of large Spanish plants.

Influence of non-regulatory pressure on compliance costs

Not only regulatory pressure forces firms to react, but also environmental pressure coming from the market place. Relations within the supply/production chain are likely to be both an influence upon, as well as being influenced by, the level of environmental standards, e.g. a manufacturer may find it easier to increase the environmental standards of its products if it has a reliable and competent base of suppliers to draw on. Also a manufacturer may be forced to upgrade product and process environmental standards by pressure coming from the customers of plants in the three sectors under study.

In the three case studies non-regulatory pressure was least important for Germany. Legislatively driven initiatives were always more frequent in Germany than in the other countries and the compliance costs for these measures were higher than for initiatives motivated by other drivers.

The motivation for waste water investments was principally reported to be found in legislative requirements, with German sample plants expressing more of this type of pressure than Irish plants. The second source of driver was found in cost reduction aspirations with an almost even distribution between industries and sample countries. However, often the cost considerations followed legislative requirements and consequent costs and charges. Customer pressure and public image were far less frequently quoted as reasons for undertaking environmental initiatives.

In the packaging study the most important driver in each of the countries was cost, followed by legislation and customer pressure. In Germany legislation was responsible for significantly more initiatives than in either of the other countries. Health and safety, technology and other drivers were relatively unimportant. In Germany recycling, special design, material switching, packaging recovery, and packaging reduction (especially removal and reduction of labels and removal of outer packaging, use of larger packs and the elimination of excess packaging) were important initiatives driven by legislation. External pressures gave rise to other initiatives, particularly increased material reuse, but also special design, material switching, light weighting and the removal of outer packaging. Firms were also asked about the importance of direct environmental pressures from customers or suppliers in comparison to other pressures. While customer/supplier environmental pressures on packaging were present they were only one of many pressures. The main pressures firms faced were with respect to the more usual competitive aspects of price, quality and service.

In the cement study, there are some purely legislation driven measures (SNCR, MSC and absorbent addition) which are undertaken in the German sample only and nowhere else. All plants using these costly abatement techniques, also used alternative fuels and could compensate any increase in production costs due to secondary measures with a decrease in energy costs. In none of the plants a directly negative impact on competitiveness was explained. In several cases where initiatives were in all countries undertaken voluntarily, motivated by process efficiency and cost reductions (electro filters, energy initiatives, expert systems), the German plants have achieved economically and environmentally more beneficial effects than elsewhere. There was no major evidence of customer/supplier environmental pressure in the cement industry.

Influence of market structure (packaging and cement study only)

Market structure can influence the level of compliance costs and/or the number of environmental initiatives. In a relatively competitive market like the food industry, constant strive for new cost efficient environmental solutions is required because chances to pass on costs to customers are small. In an oligopoly like the cement industry individual incentives for renewal and updating always depend on what other members of the oligopoly will undertake. On the one hand, adjustment to new environmental requirements may therefore be delayed in an oligopolistic market. On the other hand, if an industry achieves super-normal profits, it is to a certain extent protected against additional costs arising from environmental compliance and the pursuit of wider management goals which includes environmental protection is possible. Thus, market power will allow to pass on environmental cost increases. Also, where a firm is acting in a protected local market, it can pass on any higher cost arising from environmental regulation than if it was active in a global market with competition from countries with laxer regulation. Monopsony power in the retail sector (packaging supply chain study) could be used to exert environmental pressure and force suppliers to adopt environmental initiatives.

No evidence of environmental pressure due to monopsony power in the retail sector was found. Concerning the oligopolistic situation in the cement industry there was some evidence for the hypothesis.

In the packaging case the hypothesis that those firms supplying major retailers would face a greater environmental pressure and would therefore carry out more environmental initiatives was tested. For the firms supplying major retailers there was no evidence of a greater number of environmental initiatives.

The analysis for cement has indeed shown that costly secondary measures which really bring emissions down, have been introduced only very recently in the German industry, i.e. the industry has succeeded in avoiding these measures for a long time. On the other hand the measures do not show a significant impact on businesses. Any super-normal profits would play a role in this. Data on profit margins on a plant level were hard to obtain during the in-depth interviews. But during later interviews with headquarter chief executives of multinational cement companies it was said that German plants have always been profitable despite their environmental investments.

Influence of location and/or ownership on compliance costs and number of environmental initiatives

Location: Location can affect the cost of compliance. The assumption that environmental regulation has differential impacts between regions and countries would be related to differences in competitiveness between Member States of the EU which are based on the underlying differences in productivity and skills/R&D. These factors are related to the efficient adjustment to regulation. Therefore plants in peripheral locations are likely to be less advanced on environmental matters, to undertake less environmental initiatives and to be subject to higher environmental costs because of lack of access to suppliers, language barriers, higher transport costs etc. than plants in centrally located regions.

Ownership: Ownership is likely to be of influence on the environmental response a plant or a firm gives to environmental regulation. The industries under concern in this study are to varying degrees composed of multinational enterprises. Multinational companies are likely to show better environmental performance than other enterprises because they export both the environmental and the efficiency standard of their home country (at least in their new production facilities, see Epstein and Roy, 1997 and 1998). Therefore, where multinational branch plants are sampled in those parts of the EU with lower environmental standards/enforcement, they are expected to undertake a higher number of environmental initiatives and are in a better position to reduce environmental costs than indigenous companies producing the same products.

There was some evidence on the above hypothesis concerning location. East German plants were faced with higher compliance costs both in the food and the packaging case. It was also found that the number of environmental initiatives was lower in East German packaging plants than in West Germany. Ownership mattered insofar as multinationals had company-wide environmental policies in place. However, German multinationals operating abroad did not apply as stringent environmental standards as at home and could therefore lower compliance costs.

In the food case location mattered insofar as Germany with the most stringent legislation in place had higher waste water costs than either the Republic of Ireland or Northern Ireland. Moreover, sewer charges were highest in East Germany. This had to do with the newly built municipal treatment plants after reunification. No influence of ownership on compliance costs was found.

The packaging study showed that in East German dairies DSD costs tended to be higher because more non-reusable packaging was used. A clear lag was also found concerning the number of solid waste initiatives, for example, between indigenous East German producers in the paper industry compared with their western counterparts. On average German PLCs undertook more initiatives than limited companies. Since PLCs were larger firms than their limited liability counterparts (though not necessarily larger plants) the data lend support to a positive relationship between firm size and total numbers of environmental initiatives adopted. In addition PLCs tended to have environmental policies. About half those plants sampled within the PLC group were branch plants of multinational companies. Many of these plants reported that their environmental policy was based on strong company-wide policies to implement good environmental practices.

In Germany it is also possible for independent cement plants to reach a high level of environmental performance. It of course also reflects the German legislation which is uniform across different types of ownerships. Plants abroad owned by German factories have a somewhat lower environmental standard. Although German multinationals perform better than their national counterparts abroad, they do not apply exactly the same standards as at home and can thus lower their compliance costs. Within the sample location only mattered in terms of differences in environmental regulation, but not in terms of access to suppliers or anything similar. In East Germany the most modern cement plants with the best environmental technology were in place. Permits for a the use of high percentage of alternative fuels reduced their net compliance costs.

10.2.2 Test method (2): Detailed economic effects of each environmental initiative

In each case study the detailed economic effects of the relevant environmental initiatives were recorded in order to explore the net effects of each measure. The detailed economic effects included any productivity impacts, financial savings and competitiveness effects. Overall, although German sample plants tended to incur higher capital costs as a consequence of regulation, they were more often able to generate positive (but often only marginal) productivity effects, financial savings and beneficial competitiveness effects from undertaking environmental initiatives than their sample counterparts in less stringently regulated countries. The detailed findings are as follows:

In the food case study more German than Irish firms were able to record positive productivity effects triggered by waste water activities, but these impacts were only marginal. These productivity effects were based on the recognition of plants that more work was carried out by existing staff in order to carry out an environmental initiative with the extra work being absorbed within the company. Also, environmental initiatives gave rise to better understanding of how the company worked and productivity enhancing spin-offs were a by-product of these initiatives. However, there were little implications by way of productivity for the core business.

In the packaging case study comparisons of the effects of solid waste initiatives between countries showed that initiatives undertaken by German firms were in total more likely to require additional capital costs, were less likely to bring about unit cost savings, but were more likely to trigger general financial savings and more likely to give rise to positive competitiveness effects compared with those undertaken in the other sample countries. A differentiation according to regulatory and non-regulatory pressures showed that German initiatives driven by regulation were more likely to require additional resources (except labour), were less likely to create financial savings and were less likely to improve the competitiveness of the firm compared with non-regulatory drivers. The additional costs for German plants mainly consisted of recycling costs. Initiatives driven by external pressure (e.g. consumer pressure) were also more likely to require additional resources but they were also identified as having a higher probability of generating unit cost savings or positive competitiveness effects compared with other non-regulatory drivers. Most expenditures were cost neutral and few required capital expenditure. The majority of solid waste initiatives, however, was cost driven in every country and these initiatives triggered the largest number of general savings effects for the plants under consideration.

Comparisons of the effects of clean air initiatives in the cement study showed that the additional capital requirements for legislation driven and uniquely in Germany undertaken measures were offset through the use of alternative fuels. Thereby German plants could compensate any increase in production costs due to end-of-pipe measures with a decrease in energy costs. These financial savings prevented a loss in competitiveness. Furthermore, German plants have also voluntarily invested in dust reducing measures (electrostatic precipitators) and could achieve more profitable solutions for their business than the plants in Spain and the UK which undertook the investment because of legislative pressure. Also, German plants with particular energy initiatives have achieved a greater reduction of their energy consumption than plants in the other countries. Also their share of energy costs in total production costs is lowest in the sample. No difference between sample plants concerning the implementation of process optimisation measures could be detected. Only with regard to expert systems German plants spent more and achieved more beneficial results both with respect to economic and environmental consequences of this measure.

10.2.3 Test method (3): Managers' opinion about competitive advantages and disadvantages

Managers were asked to estimate the importance of environmental factors influencing the firm's competitive performance relative to other factors. Overall, environmental factors were neither seen as a competitive disadvantage in Germany nor as a competitive advantage. Only in the packaging case study several German packaging companies reported that their packaging initiatives have generated competitive advantages. One German cement plant declared its ability to reach strict environmental standards to be of positive influence. Several East German food plants saw high waste water treatment costs as a competitive disadvantage, but the necessary investment for effluent plants was not regarded as a constraint on growth. In the packaging case factors like overcapacity, fierce competition or price were considered to be stronger and more important pressures than environmental factors. In cement manufacturing several German plants complained about high environmental requirements, but at the same time it was said that pressures like the need for better plant design and infrastructure were more pressing. The detailed findings are as follows:

Competitive advantages

The main competitive advantages in the food sector were related to product quality, plant location and labour related factors, in particular skills and costs. A variety of other advantages were associated with prices, costs, fiscal and credit advantages. Several firms also mentioned the technical aspects of their production processes to be favourable aspects for competing on the national and international markets.

In the packaging case study most advantages were also related to product quality and variety, but emphasis was also put on design and presence in niche markets as competitive advantage. Several German packaging companies stated that their environmental initiatives triggered important competitive advantages. No Irish or UK firm quoted environment as a competitive advantage.

Concerning competitive advantages only one German cement plant clearly declared the ability to fulfil strict environmental standards as a competitive advantage. Several other German plants stated low energy costs and modernity of plant as most important competitive features. Indirectly, however, these aspects are connected to favourable environmental performance. In Spain and the UK nearness to raw material, low transport costs and consistency of product quality were the most frequently stated competitive advantages.

Competitive disadvantages

Concerning competitive disadvantages in food companies pressures on price, location (for peripheral firms), labour availability, quality and labour relations were amongst the most important. Eight plants reported environmental costs to constitute a competitive disadvantage for them, seven of these plants were located in East Germany, the other in West Germany. The concern arose in new plants in East Germany where managers had made considerable capital investments to meet the new environmental regulation and the complaints arose about these investments and high treatment costs of waste water. However, in relation to the overall investment requirements at new green field sites in East Germany, the environmental investments were not large. Managers did not state that the environmental investment represented a constraint on future growth.

In German packaging firms environmental factors were mentioned on very few occasions as a competitive disadvantage. Again no Republic of Ireland/UK firm reported that environmental matters constituted a competitive disadvantage. No packaging firm cited environmental regulation or costs as a constraint on growth. In Germany overcapacity, raw material costs, fierce competition, low margins and transport costs were more frequently cited. In addition, German sample firms were shown to export more in both the retailing and packaging sectors, despite having the highest compliance costs (but less in the dairy/drinks industries than their Irish counterparts). With respect to external pressures, firms were asked to list those direct pressures they faced from customers and suppliers and those they demanded of suppliers. Price, quality and service were not only more frequently cited, they were also considered to be stronger and more important pressures than environmental factors.

With respect to competitive disadvantages German cement plants stressed that environmental requirements were said to be high. But the top environmental performers did not complain about environmental costs, only one large German plant felt that it was suffering from a competitive disadvantage through environmental costs. Simultaneously this plant stated problems related to infrastructure and plant design as more important. In Spain and the UK current environmental requirements and costs were hardly mentioned at all.

10.2.4 Overview of all results

In summary, the main explaining factors for the result that German plants do not suffer from a negative impact on competitiveness caused by stringent environmental legislation varied from case study to case study. There were factors with more and less explanatory value for the ability to adjust efficiently to stringent environmental legislation (see Table 10.2-2 below for a differentiated overview).

A number of factors which seemed to be essential shall be mentioned first. The frequent use of clean technology solutions which is less costly both on the resource and financial side than the use of end-of pipe technology was a conspicuous characteristic of the German sample. Also, the micro level analysis of the implementation of environmental initiatives confirmed that there were both more environmental and economic benefits in the German sample than elsewhere. High environmental awareness went hand in hand with high number of environmental initiatives in two of the three case studies (food and packaging). Furthermore, the relative unimportance of environmental pressure in comparison with other pressures on competitiveness presents a further explanation why competitiveness is not endangered by environmental legislation. Finally, the low absolute levels of compliance costs measured as a percentage of turnover, at least in the food and packaging case study, do explain why environmental regulation does not have an influence on competitiveness in these sectors. The cement case has shown that compliance costs can be high in Germany. The industry, however, is in a position to compensate these costs through the use of alternative fuels. This finding does not imply that other pollution intensive industries without a similar scope for compensation of compliance costs, are also protected against a potential competitive threat from regulation in Germany.

There were also factors which seemed to be less important for the relationship between environmental regulation and competitiveness. Thus, high German productivity levels were not among the essential factors explaining our findings. Only in the retail branch top performers had lower waste costs than the sample average. It implies for our case studies that also plants with lower productivity can withstand high compliance costs. There was also hardly any evidence found for a relationship between low age of plant and low compliance costs. Neither did high skill and R&D levels help to produce a high number of individual or cost saving environmental solutions. Only in branches where design changes were required (e.g. in the packaging industry), a direct influence of skills was observed. Small plant size was not found to be a disadvantage when complying with environmental regulation. This is especially important for Germany where industry consists to a vast extent of small and medium sized enterprises.

Table 10.2-2 Overview of results on the impact of stringent environmental regulation in the three case studies of the German manufacturing industry

Test method	Result in food case study: low absolute compliance costs	Result in packaging case study: low absolute compliance costs	Result in cement case study: high absolute compliance costs
Test method (1): Hypotheses			
High productivity and low compliance costs	Not confirmed	Confirmed only for retailing	Not confirmed
High productivity and high no. of environmental initiatives	Some support	Confirmed only for dairy/drinks and retailing, but not statistically significant	Inconsistent findings, support only for small and large plants
High environmental awareness and high no. of env. initiatives	Confirmed	Confirmed	Not confirmed
No. of environmental initiatives and compliance costs (relationship could a priori be both positive and negative)	High no. of initiatives and high level of compliance costs, but not statistically significant	High no. of initiatives and high level of compliance costs (only for dairy/drinks and retail), but not statistically significant	High no. of initiatives and high level of compliance costs, but not statistically significant
Type of environmental initiative and compliance costs	3 of 7 initiatives are clean technologies, still: high waste water fees and charges	7 of 10 initiatives are clean technologies, still: DSD fees	13 of 18 initiatives are clean technologies, still: expensive end-of-pipe tech.
High skills/R&D and low compliance costs	Not confirmed, inconsistent	Confirmed, but only for packaging plants	Not confirmed
Young plant age and low compliance cost	Opposite was found	Confirmed for dairy/drinks and packaging, but not statistically significant	Not confirmed
Large plant size and low compliance costs	Unclear relationship	Confirmed: Higher costs for small firms (but dairy/drinks only)	Not confirmed, waste costs rise with plant size
Large plant size and high no. of environmental initiatives	Confirmed for dairy industry, less so in meat sector	Confirmed for retail sector: Small firms undertake less init.	Confirmed
Non-regulatory pressure and compliance costs	Legislation is main driver for initiatives, followed by cost; higher costs for regulatory driven measures	Legislation is main driver for initiatives, followed by cost and customer pressure; higher costs for regulatory driven measures	Legislation is main driver for costly initiatives; compensation through use of alternative fuels

Test method	Result in food case study: low absolute compliance costs	Result in packaging case study: low absolute compliance costs	Result in cement case study: high absolute compliance costs
Market structure and no. of environmental initiatives and level of compliance costs	Not examined	No evidence of monopsony power in the retail sector	Evidence of late tightening of regulation in the industry; probably supernormal profits
Location and no. of environmental initiatives and level of compliance costs	Higher compliance costs in East Germany; but: no constraint to growth	Lower no. of environmental initiatives in East Germany; higher compliance costs in East Germany; but: no constraint to growth	Location did not matter for no. of environmental initiatives
Ownership and no. of environmental initiatives and level of compliance costs	No influence	PLCs undertake more initiatives	Also independent plants reach a high no. of environmental initiatives; multinationals have lower compliance costs abroad than at home;
Test method (2): Effects of environmental initiatives on a micro level			
Implementation of environmental initiatives	More productivity effects than elsewhere	More financial savings and more positive competitiveness effects; a lot of measures were cost neutral	Legislation driven measures: compensation through alternative fuels and energy saving activities; Voluntary actions: Sometimes more profitable solutions than elsewhere
Test method (3): Environmental pressure relative to other competitive pressures			
Competitive advantages	Quality, location, skills etc.	Several plants in the packaging industry report environmental aspects as source of competitive advantage; otherwise: quality, variety	One time environmental stringency was cited as competitive advantage; otherwise: Low energy costs, modern plants

Test method	Result in food case study: low absolute compliance costs	Result in packaging case study: low absolute compliance costs	Result in cement case study: high absolute compliance costs
Competitive disadvantages	In East Germany several plants mentioned high environmental costs; but environmental investment was small in comparison of overall investment in Greenfield sites. Otherwise: price, location, labour availability	Overcapacity; fierce competition on price	Despite high environmental requirements, top environmental performers did not complain about compliance costs. Need for better plant design and infrastructure was reckoned to be more important.

10.3 General conclusions

This study employs a methodology that measures how environmental regulation influences plant competitiveness. Viewing the method as a test of the Porter hypothesis, it has shown little evidence to fully support neither the "win-win" side of the debate nor the traditional school. Our results show that environmental policy with respect to the regulation of waste water, packaging waste and clean air has led neither to an improvement nor to a loss in overall competitiveness in the German manufacturing industry.

Specifically, in the food case study environmental costs facing a firm varied according to the local authority pricing structure, use and cost of water, existence and age of effluent plant. Plant location was strongly determined by cost considerations concerning raw material supply, the level of environmental cost arising from plant location was not critical to the firm's success. It was found that productivity varied widely and also fresh and waste water costs depending on whether a firm owned its own water resources and the way of effluent treatment. Environmental cost were not found to be an important factor influencing the survival or growth of sample firms. Running costs measured as percentage of turnover were very small and insignificant.

In the packaging case study at the micro level there was some evidence of competitiveness benefits arising from regulation and external pressures in Germany, but these advantages were not evident at the firm level anymore. Probably the most important reason is that the impact of packaging regulation and the cost of complying with this regulation is relatively insignificant in relation to other factors which influence the competitive performance of firms.

High clean air standards in the cement case study have caused a cost to the individual plants, which distract resources for other investment projects, but which can be partially compensated by process changes and a reduction of energy costs. Overall, gains in environmental performance have not damaged economic performance, but have certainly not led to an increase in economic performance. This is because compliance costs are relatively high for the sector due to the type and technical scale of abatement techniques available.

How do these results compare with other studies? This question is not easy to answer since we do not know of other studies treating the same sectors or using the identical methodology. However, there are some studies examining similar research questions for Germany. In an international comparative study on the relationship between environmental and economic performance in the European paper industry in Germany, Italy, the Netherlands and the UK Wagner/Ngyen/Azomahou (2001) find a better environmental performance of firms in Germany, but a worse economic performance relative to other countries. The poorer economic performance is linked to the relatively rigid German type of regulatory system. This latter finding was in our study only found to some degree in the cement industry. In the food and packaging sectors German plants were performing well economically. Wagner/Ngyen/Azomahou (2001) also generally confirm that firm size is not found to have any significant effect on either environmental or economic performance. Consequently, according to their findings the influence of environmental regulation on economic performance is not as negative for small firms as often argued in the literature. In our case studies there was some evidence that small plants undertake fewer environmental initiatives than large firms, but only one instance that small firms incur higher compliance costs than large firms. Therefore our findings follow Wagner/Ngyen/Azomahou (2001) partially, but not fully because the non-confirmation of higher compliance costs for small firms could be due to the lower number of environmental initiatives used by small firms. Of course, the eventual cost impact of these initiatives depends on their nature as clean or end-of-pipe technology. Therefore variability of findings according to different sectors can be expected.

In another recent international study on the impact of best available techniques on the European paper industry which includes Germany in the white line chipboard segment it is found that top environmental performers also have strong economic performance (Hitchens/Farrel/Lindblom/Triebswetter (2001)). The study used just like this one the matched plant technique. Therefore results are methodologically very well comparable, although a different sector was researched.

What can this study add to the literature? This study adds to the existing literature on the relationship between environmental regulation and competitiveness insofar it analyses three factors which are hardly addressed in other studies irrespective of the type of method used: the first is the role of the age of a plant, the second is the type of environmental technology used and the third is market pressure as additional factor of influence important for a firm's adjustment next to regulation.

Concerning the age of a plant it was found that low plant age does not generally help to bring compliance costs down. This is probably due to the fact that modern plants used a high number of environmental initiatives which can be cost increasing. In relation to this it has to be considered that the type of abatement technology used is decisive for the actual extent of compliance costs. Although each case study sector provided the potential for clean technology solutions, it was found that often the costs of end-of-pipe measures forced by stringent regulation overcompensate possible cost reductions of clean technologies used at the same time. The costs of end-of pipe technologies also seem to dominate the low age effect from which it was hypothesised at first hand that modern machinery would embody the most up-to-date and cost minimising technology. Finally, the study showed that non-regulatory environmental pressure stemming from the market place gives an incentive to plants to search for cost minimising and sometimes cost reducing environmental measures early onwards.

We close this section with a word of caution about the weaknesses of this study. First of all, environmental and economic performance could be influenced by many more factors than included in the theoretical framework. Furthermore, the mutual relationships between the different variables used, could be researched in greater depth. To this end, the research method could be extended by a more mathematical approach using simultaneous equation systems. However, it could also be imagined to expand the current approach with more qualitative research methods like policy evaluation. This would imply more emphasis on the political process concerning the negotiation of environmental policy and also on the impact of different policy instruments. On both these levels important implications for competitiveness are anchored during early stages of negotiation of environmental regulation, long before firms are affected by environmental regulations.

10.4 Implications for policy

As was found in this study, the relationship between environmental regulation and competitiveness is a story with many facets and as such it should be treated by policymakers. Although environmental costs varied widely both at the national and international level and from sector to sector, there was no evidence that existing levels of regulation prevented any firm from achieving internationally competitive performance. Irrespective of country there was a positive correlation between the number of environmental initiatives and high productivity levels. That implies that best practice firms will try to be environmental forerunners as well. But there was no evidence that above average productivity firms achieved the lowest compliance costs, although good management may well seek to minimise these costs. All in all, it was found that the existing levels of environmental regulation in Germany do not harm competitiveness. The implication for other countries besides Germany is that the current softer national regulation is not harmful either. More stringent EU policy is likely to require a transition period (as granted in the EC Directives under concern for this thesis).

In this study it was found that environmental policy generated only very small traces of a double dividend: some positive employment and productivity effects were reported together with environmental benefits. However, the economic effects never touched on the core business. In this context it could be asked whether new, less rigid and more flexible policy instruments could be used for better results. Also, both national and EU environmental policy should promote environmental awareness and the use of cleaner technology. Lastly, small firms should be treated with care by policy. Although small firms clearly contribute to pollution, it was found in our study that they undertake substantially fewer environmental initiatives than large firms. Policymakers should therefore try to mobilise their potential for pollution reduction without burdening their competitiveness.

Appendix A.3: The issue of alternative fuels in the cement industry

Alternative fuels that are fed through the main burner, will be decomposed in the primary burning zone, at temperatures up to 2000 °C. But waste fed to a secondary burner, preheater or precalciner will be burnt at lower temperatures, which not always are enough to decompose halogenated organic substances and bind volatile components in the cement clinker. When improperly used the use of waste containing volatile metals (mercury, thallium) or volatile organic compounds can result in an increase of the emissions of mercury, thallium or VOCs.

Therefore the burning of alternative fuels leads in most countries to additional maximum emission limits for heavy metals. These differ widely between countries and regions (see Table A 4.3-1 for individual examples). Many authorities have also restricted the contents of heavy metals in alternative fuels. This affects plants in the choice of alternative fuels which they can utilise (Kehl, without year).

Table A.3-1: Examples of maximum limits imposed by the authorities for the levels of heavy metals in secondary fuels, in mg/kg (individual plants in each country)

	Germany, example 1	Germany example 2	UK	Belgium	Sweden
Cd	5	10	20	100	No regulation
TI	2	5	20	100	
Hg	1	1.5	10	10	
As	20	5	50	200	

Source: information obtained during interviews

Until recently the (old) EU Waste Incineration Directive did not cover NO_x emissions from the so-called co-incineration (i.e. burning of alternative fuels) in cement plants. But with the revised Directive of November 2000[40] cement plants burning waste (hazardous and/or non-hazardous waste) are subject to NO_x regulation as well. Two types of emission limit values are foreseen in the new legislation: One type applies to cement plants, and other co-incinerators, where the heat input comes from non-hazardous waste or where less than 40 per cent of the heat input comes from hazardous waste. Another type of emission limit values will apply to dedicated incinerators, or co-incinerators, burning untreated

[40] This legislation must be transposed by the Member States into national law by 1 January 2003. Existing cement plants with a permit to burn waste by 2003, will have until 2006 to comply with the legislation. Other plants which start burning waste before 2005 will have until 2006 to comply with the legislation.

mixed municipal waste or where more than 40 per cent of the heat input comes from hazardous waste.

In general the new emission limit values are the same for cement plants and for incinerators – except for dust and NO_x. Higher values for dust from cement plants are accepted as the cement dust originates primarily from the raw materials whereas incinerator dust originates entirely from waste. The higher values for NO_x are accepted as NO_x emissions from cement plants do not increase when waste is used as alternative fuel.

Furthermore, the higher gas temperatures in cement kilns (2000° C) result inevitably in higher NO_x levels than in incinerators where temperatures between 900 – 1100° C are used.

The following emission limit values are provided for cement plants burning non-hazardous waste or less than 40 per cent hazardous waste:

Table A.3-2 Emission limit values for cement plants burning non-hazardous waste or less than 40 per cent hazardous waste as to European Waste Incineration Directive of November 2000

Pollutant	**Limit value measured as daily average, 10% O_2, dry, mg/m³ (dioxins ng/m³)**
Total dust	30
Hydrogen chloride (HCl)	10
Hydrogen fluoride (HF)	1
NO_x for existing plants NO_x for new plants	800 500
Cadmium (Cd) Thallium (TI)	0.05
Mercury (Hg)	0.05
Antimony (Sb), arsenic (As), lead (Pb), chromium (Cr), cobalt (Co), copper (Cu), manganese (Mn), nickel (Ni), vanadium (V)	0.5
Dioxins and furans	0.1
Sulphur dioxide (SO_2)	50
Total Organic Carbon (TOC)	10

Concerning monitoring operators will have to ensure continuous monitoring of NO_x, dust, TOC, HCI, HF and SO_2. Continuous monitoring will also be required for combustion and exhaust gas temperature, O_2 and H_2O levels, and pressure. Heavy metals and dioxins and furans would be measured twice a year and all measurement results presented in a standardised format.

Appendix A.6 Solid waste and environmental awareness initiatives in Germany

Table A.6-1 Description of most important solid waste initiatives for Germany

Reuse materials	**Most striking details**
Pallets	About four fifths of all companies participate in an exchange system for reusable pallets. Costs are very low and usually consist of replacement costs for broken palettes.
Bottles	Reuse of bottles takes mainly place in the dairy and drinks industry. Both glass and PET bottles are reused. The standardised pool system facilitates take back of empty bottles. All of the retailers offer soft drinks and milk in reusable bottles.
Bins	Plastic packaging producers reuse the cardboard octabins in which the granule is supplied.
Steel containers	Drinks companies reuse steel containers in which fruit ingredients are delivered. Metal manufacturers reuse the containers in which ink for spraying cans is delivered.
Plastic containers/big bags	Dairies reuse plastic containers for detergents and big bags (made from strong plastic fibres) in which dry fruit ingredients for milk products are delivered.
Recycle materials	
Foil	Foils, in which inputs are wrapped, are collected in all industries.
Plastics	Plastic waste like lids, ribbons etc. is collected across industries.
Process scrap	Recycling of process scrap is undertaken in the packaging industry. Manufacturers collect process scrap and send it to recycling companies. However, it is not allowed to use recycled material for the production of food containers.
Glass	Glass producers frequently have their own recycling plants.
Paper/board	Paper and board recycling is undertaken in nearly 70 per cent of all companies.
Wood	About a third of retailers recover and recycle wood, to a certain extent even in their own recycling plants.
Cork	Some of the retailers provide containers where consumers can bring cork.
All materials	About 9 per cent of companies recycle all materials.
Recycling plant	Four retailers and 3 packaging companies (1 paper and 2 glass) have their own recycling plants.. One retailer for instance had just contracted for a period of 60 months with a recycling company which provided a 24-hour service on the premises of a retail depot. The papermill and the glass producers use a very high percentage of secondary material in their production process. Own recycling plants are quite profitable for them.

Reuse materials	**Most striking details**
Special design	20 per cent of all companies developed special designs in order to reduce packaging. For instance one carton producer developed a new folding technique which decreased the amount of paper necessary for the individual carton.
Material switching	About 30 per cent of all firms switched the type of material and the majority tried to use more environmentally friendly materials. A drastic example was the production stop of PVC food containers in three companies due to health reasons prior to the Packaging Ordinance; PVC was replaced by PP. One dairy introduced UHT milk in reusable glass bottles instead of gabletop packaging; but it was too expensive and was listed out by retailers. One drinks company switched from glass to PET for economic reasons.
Packaging recovery	
General recovery	98 per cent of all companies recovered packaging waste.
Bottle banks	All retailers provide bottle banks on their premises.
Reducing labels/volumes	Many of the retailers urged their suppliers to reduce volumes of packaging.
Eliminate excess packaging	Due to the Packaging Ordinance retailers reduced as much transit packaging as possible
Waste reduction: Compactor	46 per cent of all companies compact their waste in order to save transport costs.
Waste segregation	About four fifth of all companies segregated waste in different containers before it was taken away by recycling companies.
Waste targeting: Food waste	Some retailers introduced special programmes to reduce food waste.

Table A.6-2 Description of environmental awareness initiatives for Germany

Environmental audits	Companies with an environmental management scheme – either ISO 14000 or EMAS – regularly carry out environmental audits.
Environmental manager	Companies with an environmental management scheme usually employ an engineer who had the explicit position of an environmental manager. In the other companies, frequently production managers also cover the environmental area. In small companies the general manager takes care of the environment.
Environmental team	Companies with a strong environmental profile had installed an environmental team or department which could include up to 10 employees.
Environmental policy	As part of the environmental management scheme companies had developed an environmental policy. It was generally a broad statement and part of the environmental report of companies.
Environmental information service	Some companies provide information on environmental variables for their employees; for example a dairy shows its effluent data for BOD and COD on a daily basis on a monitor. The measure was thought to have an awareness rising potential for employees.
Staff training	All companies which had implemented environmental management systems had also provided environmental training for their staff. In some cases, business managers found staff training important anyway, i.e. without implementing an environmental management scheme and regularly sent all employees to environmental courses. However, in uncertified companies frequently only the environmental managers get environmental training.
Sponsorship	One company reported that it sponsored the World Wildlife Fund.
Lifecycle analysis	Life cycle analysis was carried out in order to research the environmental impacts of certain products. A drinks company looked at the environmental effects of reusable PET bottles vs. reusable glass bottle.
Packaging Guidelines	All big retailers developed packaging guidelines which specify characteristics of the packaging which retailers wanted from their suppliers after the German Packaging Ordinance had been implemented.
Improved working conditions	One company stated in its environmental report that environmental initiatives in the area of noise protection had also improved working conditions.

Appendix A.8 Solid waste projects and their effects in German sample plants

A.8-1 Solid waste projects with additional capital and running costs in German plants

Large-scale projects of significantly more than DM 500,000 were undertaken in just a few cases. One dairy invested DM 30 million in a new bottling line for UHT milk; two drinks companies spent about DM 40 million each for new equipment to fill reusable PET bottles. A few other new machines were bought to make material switch, reuse or light weighting possible. The related investment costs were between DM 70,000 and DM 500,000.

In the packaging industry the two largest investments took place in the glass industry, where two companies set up their own recycling plants and invested about DM 2 million each. One East German paper processing plant invested DM 6 million for a complete equipment exchange in order to facilitate material switch from grey board to cardboard. However, this had more to do with product quality and ability to compete in the market than as a direct consequence of the Packaging Ordinance. Several machine adjustments in the plastics industry as a consequence of light weighting also triggered investment effects, more or less around DM 100,000.

The largest investment which one retailer undertook was DM 33 million for a central disposal hub; however, most retailers have outsourced these recycling centres and just pay running costs to its operator. Other investments were, for instance, for reusable fruit containers and trays (DM 200,000 and DM 500,000 respectively). Investment for a compactor was DM 90,000 in one case.

However, there were are also increased running costs associated with solid waste initiatives and they accounted with a share of almost 78 per cent for the vast majority of all cost effects in this area. Usually the running costs stayed below DM 50,000 and were related to recycling activities, repair of palettes and so on.

A.8-2 Costs arising from legislation in Germany

About half of the legislation driven initiatives in the area of solid waste in Germany did not trigger any cost effect at all. The majority of the costs incurred through legislation could be classified as running costs, i.e. recycling costs for various materials or costs for segregation of waste (container rental and so on). Only in eight cases did the new legislation require additional investment and in five of these the initiatives were also undertaken to reduce costs and in response to external pressures.

A few details will be described. A glass factory invested DM 2.1 million in an on-site recycling plant; this activity was undertaken due to legislation, but it was also cost driven because the costs of used glass had increased on the market and the company decided it was more profitable to recycle glass by itself. Two plastics companies had to move away from the use of PVC for food containers and switched to PP/PE. This was motivated by legislation which was in place prior to the Packaging Ordinance and had to do with health and safety. Respective investments were DM 0.5 and 1.0 million. Two manufacturers of packaging machinery developed new machinery as a direct consequence of the packaging legislation. Investment was DM 30,000 in one case for a new machine to fill UHT milk in reusable bottles (which failed on the market in the end) and DM 50,000 in the other case for a machine adjustment in order to light weight reusable PET bottles. This latter example was not only motivated by legislation, but also by cost savings for both the machine manufacturer and its client, the drinks industry, which can save DSD fees if bottles are lighter. A dairy invested in new equipment in order to facilitate recycling. Finally two retailers reported investment in compactors and a recycling plant. In one of these cases it was specified that the initiative was also influenced by the increase in costs (both for landfill and recycling) which resulted from the new legislation. In the other case it was said that the initiative was also related to external market pressure. One retailer also said that recycling was seen as a business opportunity.

A.8-3 Externally driven solid waste initiatives and costs in Germany

Solid waste initiatives driven by customer/supplier pressure or public image had in more than 67 per cent no cost effect across all industries. The dairy/drinks industry, however, seems to be an exception in this respect. There were six examples where external pressure required the industry to invest in new bottling lines. In two of these cases in the drinks industry the investments went hand in hand with material switch away from glass to PET and also with larger packs (1-litre instead of the traditional 0.7-litre bottles).

Running costs occurred only in the packaging industry. Two glass factories reported labour costs of DM 50,000 for adjustment of machinery for the production of lighter bottles due to customer pressure. A plastics factory spent an additional annual DM 15,000 for the wage costs of a person who was in charge of washing reusable containers. This was undertaken because the supplier wished it. Another plastics factory also reported wage costs of DM 100,000 for recycling used plastic containers produced by the firm.

All externally driven solid waste initiatives in the retail sector were classified as reuse of bottles or trays and were in most cases reported as consumer driven; in most cases they were reported as cost neutral. One retailer said that the initiative was in parallel driven by the quota for reusable bottles in city supermarkets, whereas in Greenfield sites (where consumers come by car) it was a general rule.

A few solid waste initiatives in the packaging industry were also without any costs because the supplier offered new systems for the delivery of inputs, e.g. the use of steel containers for inks in the metal can industry.

A.8-4 Sources of employment effects of German firms created through solid waste initiatives

In the following the sources of employment effects of German firms created through solid waste initiatives are analysed sector by sector.

In the *dairy/drinks industry* a total of 10 employment effects were counted in 15 firms. Two of them were negative and occurred in the dairy industry, one was a consequence of reuse of big bags, in another case light weighting of cardboard boxes for cheese caused a machine change and went hand in hand with rationalisation. As a consequence two employees less were needed to operate the line. Five of the remaining eight positive employment effects could be precisely quantified. In one case segregation of waste in a dairy required an additional one and half to two persons. A large drinks company employed three people for recycling all materials; in a small drinks company one person was in charge of recycling glass. Another large drinks company needed 30 and 15 people, respectively, for washing reusable steel and plastic containers. In two cases the reuse of pallets and recycling efforts were said to produce a positive employment effect.

Changes in employment in the *packaging industry* due to solid waste initiatives gave rise to 14 effects found in 29 firms: the lowest per-firm change in the overall sample. Seven effects were negative; three of these took place in the plastic industry where two different light weighting procedures for yoghurt pots led to small decreases in employment. In another case the banning of PVC for food containers cost 600 jobs (that was already prior to the Packaging Ordinance). Another two cases with negative employment effects were reported in the packaging machinery industry. A new coffee filling machine was developed where less vacuum is needed and also less packing material. The machine is 'simpler' than the previous model. It is cheaper in production for the machinery producer because fewer people are involved and it is also cheaper for the user. It was not a brand new development, but it was speeded up by the Packaging Ordinance. In general, all machinery is light weighted by using lighter construction elements. This is a time advantage during construction. Depending on the size of the machine, sometimes fewer people are needed for construction. Material switching in one East German paper processing plant caused the redundancy of 34 people. Downspeccing board in a West German plant caused a decrease of the respective shop floor employment of 10 per cent. Positive employment effects of one additional job were found three times (operation of compactor, washing of reusable plastic boxes, preparation of paper pieces as transport protection). Two glass factories, each employing 30 people, had their own recycling plants. One producer of reusable transport packaging estimated that the total increase of 800 employees was due to the fact that the company could produce lighter crates and thus gained market share. One East German paper plant reported a positive employment effect arising from recycling of paper waste.

A total of 23 employment effects in the *retail sector* were discovered in the area of solid waste. Two of them were negative; the extent could not be exactly specified. Both of the negative employment effects were associated with reuse initiatives: the introduction of reusable fish trays in the stores of a large retailer made handling more efficient, thus fewer people were required; in the same company the reuse of plastic containers for fruit and vegetables instead of throwaway wooden crates facilitated handling and triggered a negative employment effect. Positive employment effects were connected in 50 per cent of the cases to the reuse of bottles, the sorting of which and dealing with deposit refunds is labour intensive. The other half of the effects were a consequence of recycling efforts in the shops. Quantifiable employment effects were rarely reported. In one case two people were responsible for operating a waste compactor. There were three cases in which about 30 people fulfilled waste segregation tasks. One of these employment effects took place in a company to which a retailer had centrally outsourced waste disposal for all shops in a larger region. Another two retailers had central disposal plants, but they were outsourced as well. One retailer had set up its own recycling plant where 1 000 persons are employed; 80 per cent of all company waste is sorted (employees integrate it in their work in the shops) and recycled in the recycling plant, e.g. 1 000 t of different plastic materials.

A.8-5 Competitiveness effects arising from solid waste initiatives in the German sample

In *dairy/drinks* reuse of materials, mainly the reuse of palettes, bottles and crates, produced employment and productivity effects as well as costs and savings. For instance, in one large drinks company there was a DM 40 million investment for a new bottling line for 1-litre, reusable PET bottles which led to an extension of market share, higher productivity and savings of energy and transports costs. This company and also a smaller one that had invested in the same new bottling technique experienced positive competitiveness effects. According to the companies these competitiveness effects could be measured in terms of increased market share in comparison to the traditional glass bottling method. One medium-sized dairy company stressed that its glass bottling technique led to a slight decrease in market share. Despite its proximity to customers, which had been specified as an important competitive advantage for a glass bottler, it was also prepared to install new bottling lines for plastic packaging. Moreover, about half of the recycling activities triggered costs and savings. In one company where all materials were recycled it was stressed that the recycling centre was organised as a profit centre and operated in a cost covering manner. This was seen as a positive competitiveness effect. In five cases a material switch was undertaken. One case induced a positive productivity effect and an increase of market share (the filling of parmesan cheese in glasses instead of using plastic pouches) which was seen as positive for competitiveness; in the other case (at least physical) productivity was negatively affected (switch from aluminium butter foil to paper wrapping). However, in the end costs equalled savings of this activity. Light weighting efforts (through pressures on packaging suppliers) led to savings in material costs and positive impacts on competitiveness. One dairy had saved almost DM 3 million of both material costs and DSD fees through down gauging the thickness of plastic containers for yoghurt (the factory used massive amounts of this kind of packaging because it was the only type it used). This resulted in an improvement of cost competitiveness; another firm that had light weighted glass packaging had the same experience. One East German dairy had light weighted the cardboard boxes for cheese; this caused a machine change (investment of DM 100,000). However, the new machine had a higher output and this was seen as beneficial for competitiveness. The driver was the cost of packaging and investment in better production technology. The company hoped that 80 per cent of the cost of the new machine could be paid back quickly. In only one case had a solid waste initiative triggered a negative competitiveness effect. A dairy did not achieve sufficient demand for its UHT milk sold in reusable glass bottles. As a consequence a bottling line bought for DM 30 million is unused at the moment. The remaining solid waste efforts in the dairy/drinks industry had not very large effects and in particular no effects on competitiveness.

In the *packaging industry* most of the productivity effects were centred in the cluster of solid waste initiatives of the plastics industry. About one-half of these effects were positive and frequently had to do with light weighting. However, many companies stressed that no further light weighting of plastics could be done with existing technology or material. Some of the producers faced quality problems because the lighter plastics pots were harder to unpack and then customers would complain about machine stops; this was estimated to be a competitive disadvantage for the plastics industry. Negative productivity effects mainly had to do with material switch (no more PVC). In one packaging machinery company two negative productivity (and competitiveness) examples were given as a consequence of packaging legislation. About 30.000 hours of R&D (i.e. 20 full-time jobs within one year) were devoted to the development of a machine which fills UHT milk in reusable bottles. But consumers did not buy the milk because it is much more expensive than in throwaway containers. Also the search for monomaterials has cost a lot of R&D time in order to invent, for example, a new sealing technique for plastic pots. Plastic pots for UHT coffee milk are made of three different plastic layers: PE, a protection layer and a stability layer. For recycling reasons the search for monomaterials started around the time when the packaging ordinance was written down. In the meantime it has been stopped and composite plastic foils are still in use. They are more economic because they need less packing material. But usually there was no productivity effect from producing equipment for processing thinner material. Cost reductions for the filler vary between 5 and 30 per cent of the packing material; the employment effect for the machine producer depends on the size of the machine. The major environmental efforts are directed towards reduction of material.

Packaging companies could gain market share in only a few cases; for instance in the area of solid waste through the introduction of reusable paper boxes, the installation of a recycling plant for paper or the production of clean packaging technology. Price and unit costs decreased only when major savings occurred. This was primarily the case for light weighting efforts which also required machine adjustments and some investments. Competitiveness effects were in a few instances negative, such as when PVC was abolished; or some plastics firms also complained about frequent machine stops due to light weighting.

But there were also a few examples of positive competitiveness effects. In the glass industry own recycling plants were seen as a positive competitive effect because they give a price advantage. Light weighting metal cans and lids had a positive effect for the industry because material savings can be achieved without a decrease in quality. Downspeccing board was regarded as a cost saving activity, but not all of those companies undertaking it, would label it as a competitive advantage. Several companies argued that everybody is doing it and it does not give a competitive lead in the industry.

Both legislation and cost-driven solid waste initiatives triggered positive competitiveness effects and cost savings in the *retail sector*. Necessary machinery usually means containers, compactors and waste interceptors. About half of the legislation-driven initiatives triggered a cost effect. The cost driven initiatives required, in over 60 per cent of the cases, an investment or at least an initial cost. Positive competitiveness effects were gained through reuse efforts and also through recycling. The savings (of not shipping the garbage from throwaway packaging to the landfill, but recycling it) gives the retailers a competitive advantage which, however, was said to be small and non quantifiable. Reuse of bottles supports an image gain.

11. References

Abwasserabgabengesetz, 1976, letzte Änderung 1998 Bundesgesetzblatt I. Bonn.

Aden, J., Kyu-Hong, A., 1999. What is Driving the Pollution Abatement Expenditure Behaviour of Manufacturing Plants in Korea? World Development, 27(7): 1203–1214.

Adler, U., 1997. Integrierter Umweltschutz als Beispiel zukunftsfähiger Innovation. ifo Schnelldienst, 17/18: 44–52.

Alanen, L., 1996. The Impact of Environmental Cost Internalisation on Sectoral Competitiveness: A New Conceptual Framework. UNCTAD DISCUSSION PA PER, UNCTAD/OSG/DP/119.

Albrecht, J., 1998. Environmental Regulation, Comparative Advantage and the Porter Hypothesis. Working Paper 98/50, Gent University, Faculty of Economics and Applied Economics.

Alpay, S., 2001. Environmental Regulations, Innovation and International Competitiveness: Some New Insights. Department of Economics, Bilkent University, mimeo.

ARGUS in association with ACR and Carl Bro a/s, 2001. European Packaging Waste Management Systems. Main Report, Final Report for the European Commission, Brussels.

Ark, B. van, 1996. Productivity and Competitiveness in Manufacturing: A Comparison of Europe, Japan and the US. In: K. Wagner and B. van Ark (Editors), International Productivity Differences: Measurement and Explanations. Elsevier Science, Amsterdam, pp. 23–52.

Arnold, Frank S., 1995. Economic Analysis of Environmental Policy and Regulation. John Whiley & Sons, Inc. New York.

Barbera, A.J., McConnell, V., 1990. The Impact of Environmental Regulations on Industry Productivity: Direct and Indirect Effects. Journal of Environmental Economics and Management, 18: 50–65.

Barrass, R., Sprenger, R.-U., Triebswetter, U., 1997. Cohesion and the Environment: Policy Integration at European Level. European Environment, 7(4): 101–109.

Barret, B.F.D., Therivel, R., 1991. Environmental Policy and Impact Assessment in Japan. Routledge, London and New York.

Barrett, A., Lawlor, J., Scott, S., 1997. The Fiscal System and the Polluter Pays Principle: A Case Study of Ireland. Ashgate Publishing Ltd., England.

Bauer, J., 1992. Integrierte Umwelttechnik - Abwasser. Landsberg/Lech.

Baumol, W.J., Blackman, S.A.B., Wolff, E.H., 1989. Productivity and American Leadership. MIT Press, Cambridge, MA.

Baumol, W.J., Panzar, J. C., Willig, R.D., 1982. Contestable Markets and the Theory of Industry Structure. San Diego.

Berg, J.P., 1999. The Cohesion Fund & the Environment: Ireland. European Commission, Directorate-General for Regional Policy and Cohesion, Brussels.

Biglaiser, G., Horrowitz, J.K., 1995. Pollution Regulation and Incentives for Pollution Control Research. Journal of Economics and Management Strategy, 3: 663–684.

Birdsall, N., Wheeler, D., 1992. Trade Policy and Industrial Pollution in Latin America: Where Are the Pollution Havens? In: P. Low (Editor), International Trade and the Environment. World Bank Discussion Paper 159, World Bank, Washington D.C.

Bojkow, E., 1992. Getränkeverpackung und Umwelt, Daten und Fakten zur Verpackungsdiskussion. Springer Verlag, Wien/New York.

Boyd, D.A., McClelland, J.D., 1999. The Impact of Environmental Constraints on Productivity Improvement in Integrated Paper Plants. Journal of Environmental Economics and Management, 38: 121–142.

Bradford, D., 2000. Motivating SMEs Towards Improved Environmental Performance. The IPTS Report, 41(2): 25–29.

Breitenacher, M., 1997: Mit Innovationen gegen Verbrauchsrückgang und Überkapazitäten, Getränkekonsum 1996 abermals rückläufig. ifo Schnelldienst, 5–6: 39–41.

Breitenacher, M., Franke, A., Grefermann, K., Sprenger, R.-U., Triebswetter, U., Wackerbauer, J. with the help of Dressel, K., 1997. Förderung von ökologisch sinnvollen Getränkeverpackungen. Forschungsbericht 103 10 904 UBA- FB 97–038, Texte 17/97, ifo Institut, München.

British Cement Association (BCA), without year. UK Cement Manufacture and the Environment. Crowthorne, UK.

British Cement Association, 1998. Annual Report. Crowthorne, UK.

Brüggemeier, F.-J., Rommelspacher, T., 1992. Blauer Himmel über der Ruhr, Geschichte der Umwelt im Ruhrgebiet 1840–1990. Klartext-Verlag, Essen.

Bundesgesetzblatt (BGBl.), 1990. Bundesimmissionsschutzgesetz, 14/5/90, BGBl. Teil I: 880.

Bundesgesetzblatt (BGBl.), 1991. Verordnung über die Vermeidung von Verpackungsabfällen, 20/6/91, BGBl. Teil I.

Bundesgesetzblatt (BGBl.), 1998. Verordnung über die Vermeidung und Verwertung von Verpackungsabfällen (Verpackungsverordnung – VerpackV), BGBl. Teil I: 2379, letzte Änderung per Verordnung vom 28.8.2000, BGBl. I: 1344.

Bundeskartellamt, 1993. Wettbewerbspolitische und kartellrechtliche Probleme der deutschen Entsorgungswirtschaft. Berlin.

Bundesministerium für Umwelt, Naturschutz und Reaktorsicherheit (BMU) und Bundesministerium für Wirtschaft (BMWi), 2001. Bericht an das Bundeskanzleramt, Schätzung der Kosten bei Einführung eines Pflichtpfandes auf Einweg-Getränkeverpackungen, Berlin.

Bundesministerium für Umwelt, Naturschutz und Reaktorsicherheit (BMU), 2001. Entwurf Zweite Verordnung zur Änderung der Verpackungsverordnung, Bonn.

Bundesministerium für Umwelt, Naturschutz und Reaktorsicherheit und der Länderarbeitsgemeinschaft Wasser (BMU) (Editors.), 1994. Mindestanforderungen an das Einleiten von Abwasser in Gewässer § 7a Wasserhaushaltsgesetz; Milchverarbeitung. Bonn.

Bundesverband der Deutschen Zementindustrie e.V., 1998. Zahlen und Daten 97/98. Köln.

Bundesverband der Deutschen Zementindustrie e.V., 2000. Zahlen und Daten 99/00. Köln.

Bundesverband der Gas- und Wasserwirtschaft e.V. (Editors), 1998. Abwassergebühren 1997. Bonn.

Buttermann, H.G., 1997. Ein Modell zur Erklärung des Faktoreinsatzes in der deutschen Zementindustrie, Rheinisch-Westfälisches Institut für Wirtschaftsforschung. RWI-Papiere, 48, Essen.

Carlton, D.W., Perloff, J.M., 1990. Modern Industrial Organisation. Little Brown, Glenview, IL.

Carson, R., 1962. Silent Spring. Houghton Mifflin, Boston.

Cembureau - The European Cement Association, 1996. World Cement Directory. Brussels.

Cembureau - The European Cement Association, 1997. "Best Available Techniques" for the Cement Industry. A Contribution from the European Cement Industry to the Exchange of Information and Preparation of the IPPC BAT Reference Document for the Cement Industry on BAT. Brussels.

Cembureau - The European Cement Association, 2000. www.cembureau.be.

Cemento-Hormigon, 1997. www.cemento-hormigon.com

Central Statistics Office (Republic of Ireland), 1998. Census of Industrial Production, Dublin.

Christie, I., Rolfe, H., Legard, R., 1995. Cleaner Production in Industry, Integrating Business Goals and Environmental Management. Policy Studies Institute, London.

Coenen, R., Klein-Vielhauer, S., Meyer, R., 1995. TA-Projekt Umwelttechnik und wirtschaftliche Entwicklung: Integrierte Umwelttechnik - Chancen erkennen und nutzen. Büro für Technikfolgenabschätzung beim Deutschen Bundestag. TAB-Arbeitsbericht 35, Bonn.

Cohen, W.M., Levinthal, D.A., 1989. Innovation and Learning: The Two Faces of R&D. The Economic Journal, 99: 569–596.

Commission of the European Communities, 1992. Green Paper "Industrial Competitiveness and the Protection of the Environment", Brussels.

Commission of the European Communities, 1996. Benchmarking the Competitiveness of European Industry, Internet Version, 9 October, Brussels.

Commission of the European Communities, 1999. Interim Report According to Article 6.3 (a) of Directive 94/62/EC on Packaging and Packaging Waste. COM, 1999, 596 final, Report from the Commission to the Council and the European Parliament, Brussels.

Conrad, K., Morrisson, C., 1989. The Impact of Pollution Abatement Investment on Productivity Change: An Empirical Comparison of the U.S., Germany and Canada. Southern Economic Journal 55: 684–698.

Costa, Ch.; Franke, A., 1995. Handelsunternehmen im Spannungsfeld umweltpolitischer Anforderungen – Der Weg von der Abfall- zur Kreislaufwirtschaft in der Distribution. ifo Studien zu Handels- und Dienstleistungsfragen 48, ifo Institut, München.

Cramer, J., Schot, J., 1989. Problemen rond innovatie en diffusie van milieutechnologie: Een onderzoeks programmeringsstudie verricht vanuit een technologiedynamica perspectief. RMNO publication series 44. Rijswijk, Netherlaands: RMNO.

Daly, A., Hitchens, D.M.W.N., Wagner, K., 1985. Productivity, Machinery and Skills in a Sample of British and German Manufacturing Plants: Results of a Pilot Enquiry. National Institute Economic Review, 111: 48-61.

Dean, J., 1992. Trade and the Environment: A Survey of the Literature. In: P. Low (Editor), International Trade and the Environment. Washington, World Bank Discussion Paper 159.

Dean, T.J., Brown, R.L., 1994. Pollution Regulation as a Barrier to New Firm Entry: Initial Evidence and Implications for Future Research. Academic Management Journal, 38: 288–303.

Department for Environment, Food and Rural Affairs, 2001. Special Study of Meat and Meat Products, www.maff.gov.uk.

Department of the Environment, 1997. Producer Responsibility Obligations (Packaging Waste) Regulation Compliance Cost Assessment. London.

Department of the Environment, Transport and the Regions, 2000. News Release 712.

Dertouzous, M.L., Lester, R.K., Solow, R.M., 1989. Made in America: Regaining the Productive Edge. MIT Press, Cambridge, MA.

Dielemann, H., de Hoo, S., 1993. Toward a Tailor-made Process of Pollution Prevention and Cleaner Production: Results and Implications of the PRISMA Project. In: K. Fischer and J. Schot (Editors), Environmental Strategies For Industry, International Perspectives on Research Needs and Policy Implications. Island Press, Washington D.C., pp. 245–275.

Dobilas, G., Macpherson, A., 1997. Environmental Regulation and International Sourcing Policies of Multinational Firms. Growth and Change, 28: 7–23.

Dosi, G., 1988. Sources, Procedures and Microeconomic Effects of Innovation. Journal of Economic Literature, 26(3): 1120–1171.

Dosi, G., Freeman, C., Nelson, R., Silverberg, G., Soete, L. (Editors.), 1988. Technical Change and Economic Theory. London, Pinter Publishers.

Dreißigacker, H.-L., Surendorf, F., Weber, E. (Editors.), 1974. Technische Anleitung zur Reinhaltung der Luft (TA-Luft). Köln.

Dresdner-Kleinwort-Benson Research, 1998. Pan-European Building Book.

Dr. Holländer, Fraunhofer Institut für Lebensmitteltechnologie und Verpackung, personal communication, 1997.

Duales System Deutschland GmbH (DSD), 1995a. Lizenzentgeltliste. Köln.

Duales System Deutschland GmbH (DSD), 1995b. Wertstoffrecycling in Zahlen – Techniken und Trends. Köln.

Duales System Deutschland GmbH (DSD), 1996. Annual Report. Köln.

Duales System Deutschland GmbH (DSD), 1999. Annual Report. Köln.

Duales System Deutschland GmbH (DSD), 2001. www.gruener-punkt.de

Dufour, C., Lanoie, P., Patry, M., 1998. Regulation and Productivity. Journal of Productivity Analysis, 9: 233–247.

Dumez, H., Jeunemaître, A., 2000. Understanding and Regulating the Market at a Time of Globalisation: the Case of the Cement Industry. Macmillan Press Ltd, Houndmills, Basingstoke, Hampshire RG21 6XS and London.

Eastwood, D., Eaton, M., Monds, F., Stark, T., 1995. Northern Ireland's Green Economy: An Examination of Environmentally Based Employment Opportunities. European Environment, 5: 134–144.

Ehrlich, P., Ehrlich, A., 1972. Bevölkerungswachstum und Umweltkrise. Die Ökologie des Menschen. S. Fischer, Frankfurt am Main.

EIPPC (European Integrated Prevention and Pollution Control) Reference Document on Best Available Techniques in the Cement and Lime Manufacturing Industries, 2000. Institute for Prospective Technological Studies, Seville.

Environment Agency in Reading, personal communication, 1999.

Environmental Protection Agency, 1996. BATNEEC Guidance Note for the Slaughter of Animals, Integrated Pollution Control Licensing. Dublin.

Environmental Protection Agency, 2001. Ireland's Environment – A Millennium Report. Dublin.

Epstein, M.J., Roy, M.J., 1997. Strategic Learning Through Corporate Environmental Management: Implementing the ISO 14001 Standard. INSEAD Working Papers 97/61/AC, Fontainebleau.

Epstein, M.J., Roy, M.J., 1998. Managing Corporate Environmental Performance: A Multinational Perspective. INSEAD Working Papers 98/12/AC, Fontainebleau.

Erdmann, G., 1993. Elemente einer evolutorischen Innovationstheorie. Tübingen.

Eskeland, G., Harrison, A., 1997. Moving to Greener Pastures? Multinationals and the Pollution-Haven Hypothesis. Washington, World Bank Policy Research Working Paper 1744.

Ettlinger, N., 1997. An Assessment of the Small Firm Debate in the United States. Environment and Planning A, 29: 419–442.

Europäisches Handelsinstitut (EHI), 1996. Handel aktuell '96. Köln.

Ewringmann, D., Scholl, R., 1996. Zur fünften Novellierung der Abwasserabgabe: Meßlösung und sonst nichts? Finanzwissenschaftliche Diskussionsbeiträge, Finanzwissenschaftliches Forschungsinstitut an der Universität zu Köln, Köln.

Feldhaus, G., Vallender, W. with collaboration of Hansel, H.D., 1984. Bundesimmissionsschutzrecht. Band 1 A: Bundesimmissionsschutzgesetz: Text und Kommentar, Band 1 B: Ausführungsvorschriften des Bundes zum Bundesimmissionschutzgesetz. Zweite, völlig neu überarbeitete Auflage, Wiesbaden.

Ferrantino, M., 1995. International Trade, Environmental Quality and Public Policy. Washington, US International Trade Commission, Office of Economics Working Paper.

Ferraro Garcia, F.J., 1982. La industria del cemento en España. Anales de la Universidad Hispalense, Publicaciones de la Universidad de Sevilla, Serie: Ciencias Económicas y Empresariales 13, Sevilla.

Fiederer, H.-J. et al., 1994. Freihandel trotz Wettbewerbsverzerrungen? Zu den Auswirkungen der Öffnung der Ostmärkte. Institut für Angewandte Wirtschaftsforschung, Tübingen.

Financial Times, 23 October 1996. 'Aluminium: a survey'.

Fischer, P., 1989. Umweltschutz durch technische Regelungen – Zur Bedeutung der Grenzwertfestsetzungen und Verfahrensbeschreibungen des Immissions- und Atomrechtes. Berlin, Duncker & Humblot.

Fisher, A.C.; Peterson, F.M., 1976. The Environment in Economics: A Survey, in: Journal of Economic Literature, 14(1): 1–33.

Forbairt (Editor), 1996. Environmental Licensing and Control in Ireland, A Guide for Industrialists. Dublin.

Frank, A., Swarte, H., 1986. Milieutechnologie, toepassing in kleine en middelgrote ondernemingen. Erasmus Center for Environmental Studies, publication series 1, Rotterdam.

Gabel, H.L., Sinclair-Desgagne, B., 1997. The Firm, Its Routines and the Environment. INSEAD Working Papers 97/05/EPS, Fontainebleau.

Geisendorf, S., Gronemann, S., Hampicke, U., Immler, H., 1996. Die Bedeutung des Naturvermögens und der Biodiversität für eine nachhaltige Wirtschaftsweise. Möglichkeiten und Grenzen ihrer Erfaßbarkeit und Wertmessung. Kurzfassung, F&E Vorhaben Nr. 101 03 165/02 im Auftrag des BMU/UBA, Universität Gesamthochschule Kassel.

Gollop, F.M., Roberts, J.M., 1983. Environmental Regulations and Productivity Growth: The Case of Fossil-Fuelled Electric Power Generation. Journal of Political Economy, 91: 654–674.

Gray, W.B., 1987. The Cost of Regulation: OSHA, EPA and the Productivity Slowdown. American Economic Review, 12: 998–1006.

Gray, W.B., Shadbegian, R.J., 1995. Pollution Abatement Costs, Regulation and Plant Level Productivity. NBER Working Paper 4994, National Bureau of Economic Research, Washington D.C.

Griliches, Z., 1986. Productivity, R&D and Basic Research at the Firm Level in the 1970s. American Economic Review, 76(1): 141–154.

Griliches, Z., 1995. R&D and Productivity: Econometric Results and Measurement Problems. In: P. Stoneman (Editor), 1995, Handbook of the Economics of Innovation and Technological Change. Chapter 3: pp. 52–89, Blackwell, Cambridge, Mass.

Halstrick-Schwenk, M., Horbach, J., Löbbe, K., Walter, J., 1994. Die umwelttechnische Industrie in der Bundesrepublik Deutschland. Untersuchungen des Rheinisch-Westfälischen Instituts für Wirtschaftsforschung (RWI) 12, Essen.

Hampicke, U., 1996. Perspektiven umweltökonomischer Instrumente in der Forstwirtschaft insbesondere zur Honorierung ökologischer Leistungen. Stuttgart, Materialien zur Umweltforschung 27, hrsg. vom Rat von Sachverständigen für Umweltfragen.

Hartje, V.J., 1990. Zur Struktur des ökologisierten Kapitalstocks: Variablen und Determinanten umweltsparender technologischer Anpassungen in Unternehmen. In: K. Zimmermann., V. J. Hartje and A. Ryll (Editors), Ökologische Modernisierung der Produktion, Berlin, pp. 135–198.

Hartje, V.J., Lurie, R.L., 1984. Adopting Rules for Pollution Control Innovations: End-of-pipe versus Integrated Process Technology. International Institute for Environment and Society (IIUG), Berlin.

Haskoning Consulting Engineers and Architects, 1996. Study on the Implementation of Article 13 of Directive 91/271/EEC on Urban Waste Water Treatment – Comparison of the National Legislation of the EC Member States. Nijmegen, The Netherlands.

Hegenbart, B.; Jasch, C., 1992. Ökologie und Ökonomie in der österreichischen Milchwirtschaft. Schriftenreihe 7/1992 des Instituts für ökologische Wirtschaftsforschung Wien.

Hemmelskamp, J., 1999. Umweltpolitik und technischer Fortschritt, Eine theoretische und empirische Untersuchung der Determinanten von Umweltinnovationen. Physica Verlag, Heidelberg.

Henriques, I., Sadorsky, P., 1996. The Determinants of an Environmentally Responsive Firm: An Empirical Approach. Journal of Environmental Economics and Management, 30: 381–395.

Hettige, H. et al., 1996. Determinants of Pollution Abatement in Developing Countries: Evidence from South and Southeast Asia. World Development, 24(12): 1891–1904.

Hicks, J., 1932 and 1962. The Theory of Wages. Macmillan, London.

Hitchens, D.M.W.N., Wagner, K., Birnie, E., 1990. Closing the Productivity Gap: A Comparison of Northern Ireland, Britain and Germany. Avebury, Aldershot.

Hitchens, D.M.W.N., Wagner, K., Birnie, E., 1993. East German Productivity and the Transition to the Market Economy. Avebury, Aldershot.

Hitchens, D.M.W.N., Birnie, J.E., Hamar, J., Wagner, K., Zemplinerova, A., 1995. Competitiveness of Industry in the Czech Republic and Hungary. Avebury, Aldershot.

Hitchens, D.M.W.N., Birnie, E., McGowan, A., Triebswetter, U., Cottica. A., 1998. The Firm, Competitiveness and Environmental Regulations, A Study of the European Food Processing Industries. Edward Elgar, UK.

Hitchens, D.M.W.N., 1999. The Implications for Competitiveness of Environmental Regulations for Peripheral Regions in the EU. Omega International Journal of Management Science, 27: 101–114.

Hitchens, D.M.W.N., Birnie, E., Thompson, B., Triebswetter, U., Bertossi, P., 2000. Implementation of the EC Directive on Packaging and Packaging Waste and Implications for the Environmental Performance and Competitiveness of Firms along the Food Industry Supply Chain. Edward Elgar, Cheltenham.

Hitchens, D.M.W.N., Farrel, F., Lindblom, J., Triebswetter, U., 2001. The Impact of Best Available Techniques (BAT) on the Competitiveness of the European Pulp and Paper Industry. IPTS Seville.

Hohmeier, O., Schäfer, W., Sieber, A., 1988. Systematische Darstellung der Luftreinhaltepolitik einzelner OECD-Mitgliedsländer. Fraunhofer-Institut für Systemtechnik und Innovationsforschung (ISI), Karlsruhe.

Hülsemeyer, F., 1997. Aspekte der Milchvermarktung in Deutschland. Kieler Milchwirtschaftliche Forschungsberichte, 49: 35–45.

Jacobson, D., Andréosso-O'Callaghan, B., 1996. Industrial Economics and Organisation. McGrawHill, Maidenhead.

Jaeckel, U., 1996. Novellierung der Verpackungsverordnung und Behandlung von Getränkeverpackungen. Unterlagen zum Seminar "Mehrwegschutz und Binnenmarkt" am 12.November 1996 im Hotel Maritim, Nürnberg.

Jaffe, A.B., Peterson, S.R., Portney, P.R., 1995. Environmental Regulation and the Competitiveness of U.S. Manufacturing: What does the Evidence Tell Us? Journal of Economic Literature, 33: 132–163.

Jaffe, A.B., Stavins, R.N., 1995. Dynamic Incentives of Environmental Regulation: The Effects of Alternative Policy Instruments on Technology Diffusion. Journal of Environmental Economics and Management, 30: 95–111.

Jaffe, A.B.; Palmer, K., 1996. Environmental Regulation and Innovation: A Panel Data Study. National Bureau of Economic Research. Working Paper 5545, Cambridge (MA).

Jänicke, M., Weidner, H. (Editors), 1995. Successful Environmental Policy. A Critical Evaluation of 24 Cases. Edition sigma, Berlin.

Jenkins, R., 1999. Environmental Regulation and International Competitiveness – A Framework for Analysis. Research Project "Environmental Regulation, Globalisation of Production and Technological Change", Background Report 20.

Jones, H.R., 1974. Pollution Control in Meat, Poultry and Seafood Processing. Noyes Data Corporation. Park Ridge, New Jersey.

Jorgensen, D.W., Wilcoxen, P.J., 1990. Environmental Regulation and U.S. Economic Growth. Rand Journal of Economics, 21: 314–340.

Kalt, J.P., 1988. The Impact of Domestic Environmental Regulatory Policies on US International Competitiveness. In: M. Spence and H.M. Hazard (Editors), International Competitiveness. HarperRow, Cambridge, MA, pp. 221–226.

Kemp, R., 1993. An Economic Analysis of Cleaner Technology: Theory and Evidence. In: K. Fischer and J. Schot (Editors), 1993. Environmental Strategies For Industry, International Perspectives on Research Needs and Policy Implications. Island Press, Washington D.C., pp. 79–113.

Kemp, R., 2002. Synthesis report of 1st BLUEPRINT Workshop, "Environmental Innovation Systems", Brussels, Jan 23 to 24, 2002.

Kemp, R., Soete, L. 1990. Inside the "Green Box": On the Economics of Technological Changes and the Environment. In: C. Freeman and L. Soete (Editors), New Explorations in the Economics of Technical Change. London.

Kennedy, P., 1994. Innovation stochastique et coût de la réglementation environnementale. Actualité économique, 70: 199–210.

Klöckner, H., 1992: Verpackung – Fluch oder Segen? Verpackungswirtschaft 1950–2000, Gladenbach.

Kosz, M., 1997. Integrierter Umweltschutz und Arbeit, Erste Erfahrungen und langfristige Perspektiven. Umwelt und Arbeit III, Informationen zur Umweltpolitik III, Studie im Auftrag der Kammer für Arbeiter und Angestellte für Wien, Wien.

Krugman, P., 1991. The Age of Diminished Expectations, MIT Press, Cambridge, MA.

Kuckartz, U., 2000. Umweltbewußtsein in Deutschland 2000, Ergebnisse einer repräsentativen Bevölkerungsumfrage. Studie im Auftrag des Umweltbundesamtes, Berlin.

Kuhn, M., 1996. Produktdifferenzierung und Sustainable Development – Was kann die Industrieökonomik zu den neueren Ansätzen der Umweltökonomie beitragen? In: J. Köhn and M.J. Welfens (Editors), Neue Ansätze in der Umweltökonomie. Metropolis-Verlag, Marburg, pp. 479–505.

Lange, B., 1999. National Environmental Regulation? A Case-Study of Waste Management in England and Germany. Journal of Environmental Law, 11(1): 59–86.

Langowski, H.-Chr., 1997. Kunststoffverpackungen für die Molkereiindustrie. Lebensmittelindustrie und Milchwirtschaft, 18: 756–766.

Lanjouw, J.O., Mody, A., 1993. Stimulating Innovation and the International Diffusion of Environmentally Responsive Technology: The Role of Expenditures and Institutions. Mimeo.

Lawlor, J., 1996. The Use of Economic Instruments for Environmental Services in Irish Local Authorities. Economic and Social Research Institute, Dublin.

Leonard, J.H., 1984. Are Environmental Regulations Driving US Industry Overseas? The Conservation Foundation, Washington, D.C.

Leonard, J.H., 1988. Pollution and the Struggle for the World Product: Multinational Corporations, Environment and International Comparative Advantage. Cambridge University Press, Cambridge.

Levêque, F., 1993. "How Can Environmental Policymakers Tackle Industrial Diversity?" In: OECD, Environmental Policy and Industrial Competitiveness, Paris.

Levinson, A., 1996. Environmental Regulations and Manufacturers' Location Choices: Evidence from the Census of Manufacturers. Journal of Public Economics, 62: 5–29.

Low, P., Yeats, A., 1992. Do "Dirty" Industries Migrate? In: P. Low (Editor), International Trade and the Environment. World Bank Discussion Paper 159: 89–103, World Bank, Washington D.C.

Lucas, R.E., 1988. On the Mechanics of Economic Development. Journal of Monetary Economics, 22: 3–42.

Lucas, R.E.B., Wheeler, D., Hettige, H., 1992. Economic Development, Environmental Regulation and the International Migration of Toxic Industrial Pollution: 1960–1988. In P. Low (Editor), International Trade and the Environment. World Bank Discussion Paper 159, World Bank, Washington D.C.

Maas, C., 1990. Determinanten betrieblichen Innovationsverhaltens. Theorie und Empirie, Berlin.

Mason, G., van Ark, B., Wagner, K., 1994. Productivity, Product Quality and Workforce Skills: Food Processing in Four European Countries. National Institute Economic Review, 147: 62–83.

Maurer, O., Drescher, K., 1995. Evolution of the Dairy Market in Germany. In: D. Rama, and R. Pieri (Editors), The European Dairy Industry: Consumption Changes, Vertical Relations and Firm Strategies. Franca Angeli Publishers, Milano, pp. 37–73.

Meadows, D.H., Meadows, D.L., Randers, J. Behrens W.W., 1972. The Limits to Growth: A Report for the Club of Rome's Project on the Predicament of Mankind Universe, New York.

Meffert, H., Kirchgeorg, M., 1992. Marktorientiertes Umweltmanagement, Grundlagen und Fallstudien. Stuttgart.

Meier, W., 1997. Wie stark belasten die Grüne-Punkt-Gebühren den Milchpreis? Milch-Fettwaren-Eier-Handel, 64: 497-498.

Michaelis, P., 1993. Ökonomische Aspekte der Abfallgesetzgebung. Kieler Studien 254, Tübingen.

Ministry of the Environment in Catalunia, personal communication, 2000.

Minsch, J., 1991. Kausalität und externe Effekte. Ein Beitrag zu einer problemorientierten Weiterentwicklung der Externalitätenkonzeption. In: F. Beckenbach (Editor), Die ökologische Herausforderung für die ökonomische Theorie (Ökologie und Wirtschaftsforschung), Marburg, 2: 121–125.

Müller, E., 1986. Innenwelt der Umweltpolitik. Sozial-liberale Umweltpolitik - (Ohn)Macht durch Organisation? Opladen, Westdeutscher Verlag.

Müller, E., 1989. Sozial-liberale Umweltpolitik. Von der Karriere eines neuen Politikbereiches. Aus Politik und Zeitgeschichte, 47–48: 3–15.

Murty, M.N., Kumar, S., 2001. Win-win Opportunities and Environmental Regulation: Testing of Porter Hypothesis for Indian Manufacturing Industries. Delhi University Enclave, Institute of Economic Growth, India, mimeo.

Nelson, R.R., Winter, S.G., 1977. In Search of Useful Theory of Innovation. Research Policy, 6: 36–76.

Nelson, R.R., Winter, S.G., 1982. An Evolutionary Theory of Economic Change. Harvard University Press, Belknap Press, Cambridge, MA.

Neumann, L.F.; Pastowski, A., 1994. Umweltpolitik und internationale Wettbewerbsfähigkeit, Theoretische Überlegungen, empirischer Befund und qualitative Abschätzung möglicher Wirkungen der EG-Umweltpolitik. Leistungen und Grenzen des EG-Umweltschutzes, Bonn, 1–43.

Newell, R.G., Jaffe, A. B., Stavins, R.N., 1996. Energy-Saving Technology Innovation: The Effects of Economic Incentives and Direct Regulation. Discussion paper, Harvard University, Cambridge, MA.

Oficemen, personal communication, 1999.

O'Leary, G., Carty, G., 1998. Urban Waste Water Discharges in Ireland. A Report for the Years 1996 and 1997. Environmental Protection Agency, Wexford.

O'Mahony, M., 1993. Capital Stocks and Productivity in Industrial Nations, National Institute Economic Review, 145: 108-117.

Oates, W.E., Palmer, K., Portney, P.R., 1993. Environmental Regulation and International Competitiveness: Thinking about the Porter Hypothesis. Discussion Paper 94–02, Resources for the Future.

OECD, 1985a. The State of the Environment. Organization for Economic Co-operation and Development, Paris.

OECD, 1985b. Environment Policy and Technological Change. Organization for Economic Co-operation and Development, Paris.

OECD, 1987. The Promotion and Diffusion of Clean Technologies. Organization for Economic Co-operation and Development, Paris.

OECD, 1992. Technology and the Economy: The Key Relationships. Organization for Economic Co-operation and Development, Paris.

OECD, 1993. Environmental Policies and Industrial Competitiveness. Organization for Economic Co-operation and Development, Paris.

OECD, 1994a. OECD Environmental Performance Reviews. Organization for Economic Co-operation and Development, Paris.

OECD, 1994b. Recovery and Recycling of Packaging Material, Trends and Industry Impact. OECD Working Papers 30, Organization for Economic Co-operation and Development, Paris.

OECD, 1995. Technologies for Cleaner Production and Products. Organization for Economic Co-operation and Development, Paris.

OECD, 1996. The Global Environmental Goods and Services Industry. Organization for Economic Co-operation and Development, Paris.

OECD, 1998. The Effects of Government Environmental Policy On Costs and Competitiveness – Iron and Steel Sector. OECD Working Paper, Organization for Economic Co-operation and Development, Paris.

OECD, 2000. Environmental Performance Reviews – Ireland. Organization for Economic Co-operation and Development, Paris.

OECD, 2001. Environmental Performance Reviews – Germany (2nd cycle), Conclusions and Recommendations. Organization for Economic Co-operation and Development, Paris.

Office for National Statistics, 1997. Regional Trends, London.

Office for Official Publications of the European Communities, 1999. EU Focus on Waste Management. Directorate-General Environment, Nuclear Safety and Civil Protection, Luxembourg.

Oficemen (Spanish Cement Association), 1992. Memoria Anual, Madrid.

Oficemen (Spanish Cement Association), 1996. Memoria Anual, Madrid.

Oficemen (Spanish Cement Association), 1997. Memoria Anual, Madrid.

Öko-Institut e.V., 1995. Jenseits vom "Grünen Punkt", Alternativen zur Verpakkungsverordnung und Grünem Punkt. Langfassung, Darmstadt.

Orsenigo, L., 1991. The Dynamics of Competition in a Science Based Technology: the Case of Biotechnology. In: D. Foray and C. Freeman (Editors), Technology and the Wealth of Nations. London, pp. 41–65.

Palmer, K., Oates, W., Portney, P., 1995. Tightening Environmental Standards: the Benefit-cost or the No-cost Paradigm? Journal of Economic Perspectives, 9(4): 119–132.

Pashigian, P., 1984. The Effect of Environmental Regulation on Optimal Plant Size and Factor Shares. Journal of Law and Economics, 27: 1–28.

Pavitt, K., 1984. Sectoral Patterns of Technical Change: Towards a Taxonomy and a Theory. Research Policy, 13: 343–373.

PLM Recklinghausen Dosenwerk GmbH, personal communication, 1998.

Poguntke, T., 1992. Between Ideology and Empirical Research: The Literature on the German Green Party. European Journal of Political Research, 21(4): 337–356.

Porter, M., Esty, D., 1998. Industrial Ecology and Competitiveness. Strategic Implications for the Firm. Journal of Industrial Ecology, 2(1): 35–43.

Porter, M., van der Linde, C., 1995. Toward a New Conception of the Environment-Competitiveness Relationship. Journal of Economic Perspectives, 9: 97–118.

Portney, P.R., 1981. The Macroeconomic Impacts of Federal Environmental Regulations. In: H.M. Peskin, P.R. Portney and A.V. Kneese (Editors), Environmental Regulation and the U.S. Economy. Johns Hopkins University Press, Baltimore.

Pott, G., 1999. Reducing Emissions at Slite. World Cement Review, 8.

Prais, S.J., 1981. Vocational Qualifications of the Labour Force in Britain and Germany. National Institute Economic Review 98: 47-59.

PRVR, 1969. Projektgruppe Regierungs- und Verwaltungsreform beim Bundesminister des Innern. Erster Bericht zur Reform der Struktur von Bundesregierung und Bundesverwaltung, Bonn.

Ratnayake, R., 1996. Do Stringent Environmental Regulations Reduce International Competitiveness? Evidence from an Inter-Industry Analysis. Working Paper Series 160, University of Auckland, Auckland Business School.

Rehbinder, E., Stewart, R., 1985. Environmental Protection Policy. Integration Through Law: Europe and the American Federal Experience. Berlin and New York.

Rennings, K., Zwick, T., 2001. Employment Impact of Cleaner Production on the Firm Level – Empirical Evidence from a Survey in five European Countries. Centre for European Economic Research (ZEW).

Repak, 2001. www.repak.ie

Repetto, R., 1995. Jobs , Competitiveness and Environmental Regulation: What Are the Real Issues? World Resources Institute, Washington, D.C.

Rogers, E.M., 1995. Diffusion of Innovations. 4.Auflage, New York.

Romer, P.M., 1990. Endogenous Technological Change. Journal of Political Economy, 98: 71–102.

Rosenberg, N., 1974. Science, Innovation and Economic Growth. Economic Journal, 84: 90–108.

Roustan, M., Laigo, S., Caille, J., Roulph, Ch., 1996. Basic Concepts of Cleaner Technologies. In: K.B. Misra (Editor), Clean Production, Environmental and Economic Perspectives. Springer, Berlin/Heidelberg, pp. 171–182.

Rügemer, W., 1995. Staatsgeheimnis Abwasser. Zebulon Verlag, Düsseldorf.

Ruttan, V.W., 1997. Induced Innovation, Evolutionary Theory and Path Dependence: Sources of Technical Change. The Economic Journal, 107(444): 1520–1529.

Salzwedel, J. (Editor), 1982. Grundzüge des Umweltrechtes, Berlin.

Schiller, B., Ellerbrock, H.-G., 1992. Mahlung von Zementen mit mehreren Hauptbestandteilen. Zement-Kalk-Gips, 45: 325.

Schmalensee, R., 1994. The Costs of Environmental Protection. In: M.B. Kotowski (Editor), Balancing Economic Growth and Environmental Goals. American Council for Capital Formation, Center for Policy Research, Washington D.C., pp. 55–75.

Schneider, M., Hoenig, V., Hoppe, H., 2000. Application of "Best Available Techniques" (BAT) in the German Cement Industry. Zement-Kalk-Gips International, 1: 1–11.

Schumpeter, J.A., 1987. Theorie der wirtschaftlichen Entwicklung. Eine Untersuchung über Unternehmergewinn, Kapital, Kredit, Zins und den Konjunkturzyklus. 7. Auflage, Berlin.

Scott, S., Lawlor, J., 1994. Waste Water Services: Charging Industry the Capital Cost. Policy Research Series 22, Economic and Social Research Institute, Dublin.

Shenoy, Chako, McGovern, 1997. International Cement Review.

Smith, J.B., Sims, W. A., 1985. The Impact of Pollution Charges on Productivity Growth in Canadian Brewing. Rand Journal of Economics, 16(3): 411–423.

Solow, R.M., 1957. Technical Change and the Aggregate Production Function. Review of Economics and Statistics, 39: 312–320.

Sorsa, P., 1994. Competitiveness and Environmental Standards: Some Exploratory Results. Washington, World Bank, Policy Research Working Paper 1249.

Spenner, D., 1996. Preiskampf und Wettbewerb in der rheinisch-westfälischen Zementindustrie. Europäischer Verlag der Wissenschaften, Frankfurt.

Sprenger, R.-U., Pupeter, M., 1980. Evaluierung von gesetzlichen Maßnahmen mit Auswirkungen im Unternehmensbereich dargestellt am Beispiel der ökonomischen Auswirkungen des Abwasserabgabengesetzes auf industrielle Direkteinleiter. Gutachten im Auftrag des Bundeskanzleramtes, ifo Institut, München.

Sprenger, R.-U.; Hartmann, M.; Wackerbauer, J.; Adler, U. unter Mitarbeit von Bernd Lemser, 1991. Umweltschutz in den neuen Bundesländern: Anpassungserfordernisse, Investitionsbedarf, Marktchancen für Umweltschutz und Handlungsbedarf für eine ökologische Sanierung und Modernisierung. ifo Studien zur Umweltökonomie 16, ifo Institut, München.

Sprenger, R.U., Körner, J., Paskuy, E., Wackerbauer, J., 1994. Das deutsche Steuer- und Abgabensystem aus umweltpolitischer Sicht – eine Analyse seiner ökologischen Wirkungen sowie der Möglichkeiten und Grenzen seiner stärkeren ökologischen Ausrichtung, ifo Studien zur Umweltökonomie 18, ifo Institut, München.

Sprenger, R.-U., 1996. Environmental Policy and International Competitiveness: the Case of Germany. Paper presented to Workshop on Environmental Policies and International Competitiveness, Cambridge, 28–29 October.

Sprenger, R.-U., Barrass, R., Foeth, M., Triebswetter, U. with contributions from Chiara Aluffi-Pentini and Claudia Wenzl, 1996. Economic and Social Cohesion in the EU: The Environmental Dimension. Project on behalf of the European Commission, DG XVI, Draft Final Report, ifo Institute, Munich.

SRU, 1987. Umweltgutachten 1987. Kohlhammer, Stuttgart und Mainz.

SRU, 1994. Umweltgutachten 1994. Für eine dauerhaft-umweltgerechte Entwicklung. Metzler-Pöschel, Stuttgart.

Statistical Office of the European Communities, 1981. Structure and Activity of Industry 1976. Statistical Office of the European Communities, Luxembourg.

Statistical Office of the European Communities, 1984. Structure and Activity of Industry 1979/80. Statistical Office of the European Communities, Luxembourg.

Statistical Office of the European Communities, 1993. Structure and Activity of Industry 1989/90. Statistical Office of the European Communities, Luxembourg.

Statistical Office of the European Communities, 1994. Comparison in Real Terms of the Aggregates of the ESA. Office for Official Publications of the European Communities, Luxembourg.

Statistical Office of the European Communities, 1994a. Panorama of European Industry. Office for Official Publications of the European Communities, Luxembourg.

Statistical Office of the European Communities, 1995. Panorama of European Industry. Office for Official Publications of the European Communities, Luxembourg.

Statistical Office of the European Communities, 1995a. Panorama of European Enterprises. Office for Official Publications of the European Communities, Luxembourg.

Statistical Office of the European Communities, 1995b. Structure and Activity of Industry 1989–1992. Office for Official Publications of the European Communities, Luxembourg.

Statistical Office of the European Communities, 1995c. Comparison in Real Terms of the Aggregate of ESA Results for 1992 and 1993. Office for Official Publications of the European Communities, Luxembourg.

Statistical Office of the European Communities, 1997. Panorama of European Industry. Office for Official Publications of the European Communities, Luxembourg.

Statistical Office of the European Communities, 1999. Eurostat Yearbook, Data from 1987 – 1997. Office for Official Publications of the European Communities, Luxembourg.

Statistical Office of the European Communities, 2000. Panorama of European Enterprises 1999, Data 1988-1998. Office for Official Publications of the European Communities, Luxembourg.

Statistical Office of the European Communities, 2000a. Eurostat Yearbook. Office for Official Publications of the European Communities, Luxembourg.

Statistical Office of the European Communities, 2002. Eurostat Yearbook 2000. Office for Official Publications of the European Communities, Luxembourg.

Statistics Netherlands, Statistical Office of the Federal Republic of Germany, Centre d'Economie et d'Ethique pour l'Environnement et le Développement, Wuppertal Institute for Climate, Environment and Energy, Institut für ökologische Wirtschaftsforschung, 1996. Methodological Problems in the Calculation of Environmentally Adjusted National Income Figures. Study for the European Commission Directorate General XII, Final Report, Introduction and Part One: Theoretical Perspectives.

Statistisches Bundesamt, 1994. Statistisches Jahrbuch für die Bundesrepublik Deutschland 1994. Wiesbaden.

Statistisches Bundesamt, 2002. Investitionen für den Umweltschutz im Produzierenden Gewerbe. Sonderauskunft auf Anfrage, Wiesbaden.

Steedman, H., Wagner, K., 1987. A Second Look at Productivity, Machinery and Skills in Britain and Germany. National Institute Economic Review, 122: 84–95.

Steedman, H., Wagner, K., 1989. Productivity, Machinery and Skills: Clothing Manufacturing in Britain and Germany. National Institute Economic Review, 128: 40–87.

Steger, U., 1993. The Greening of the Board Room: How German Companies are Dealing with Environmental Issues. K. Fischer and J. Schot (Editors), Environmental Strategies for Industry. Island Press, Washington D.C.

Stewart, R.B., 1993. Environmental Regulation and International Competitiveness. The Yale Law Journal, 102(8): 2039–2106.

Stoneman, P., 1983. The Economic Analysis of Technological Change. Oxford.

Stoneman, P., Kwon, M.J., 1994. The Diffusion of Multiple Process Technologies. Economic Journal, 104: 420–431.

Strasser, K. A., 1997. Cleaner Technology, Pollution Prevention and Environmental Regulation. Fordham Environmental Law Journal, 9(1): 1 - 106.

Süddeutsche Zeitung, 15 May 1997. "Mehr Abfall verbrennen?"

Süddeutsche Zeitung, 21 May 1997. "Verpackungsverordnung, Müll vermeiden."

Süddeutsche Zeitung, 7 Oct. 1997. "Welsch warnt vor einer "Müll-Anarchie".

Süddeutsche Zeitung, 23 Sept. 1997. "Entsorgungswirtschaft fordert Bau von 30 neuen Müllverbrennungsanlagen".

Süddeutsche Zeitung, 21/22 April 2001. "Duales System verstößt gegen EU-Recht".

Süddeutsche Zeitung, 3 May 2001. "Sammelfleiß der Verbraucher ungebrochen".

The Canmaker Magazine, June 1995.

The Cement Industry, 8 Dec. 1998

The Global Cement Report 1998.

Umweltbundesamt, August 2000. Getränkeverpackungen II, Berlin.

United Nations, 1992. World Investment Report, New York.

United Nations Environment Programme Cleaner Production Working Group for the Food Industry, 1999. Cleaner Production Checklists for the Food Industry, Internal Document.

United Nations Environment Programme, 2000. Cleaner Production Assessment in Meat Processing. Prepared by COWI Consulting Engineers and Planners AS, Denmark.

Valpak, 2001. www.valpak.co.uk

Vassilopoulos, M., 1999. Clean Technology Adoption by Firms – An Investigation of Current Literature, Appendix no. 3 of Appendices Group 1 to: The Impact of Best Available Technologies (BAT) on the Competitiveness of European Industry. Interim Report, IPTS, April 2000, Seville.

Verordnung über die Vermeidung und Verwertung von Verpackungsabfällen, 21.08.1998.

Wackerbauer, J., 1998. Umweltschutzwirtschaft in Deutschland: Wettbewerbsposition auf internationalen Märkten. ifo Schnelldienst, 51(13): 23–27.

Wackerbauer, J., 2002. The German Eco-Industry: Country Summary. ECOTEC, Research and Consulting Ltd.

Wagner, K., 1979. Ein Modell der Preisbildung in der Zementindustrie. Bielefeld.

Wagner, K. and Vassilopoulos, M., 2000. The European Cement Industry, Appendix C1 of Appendices Group 2 of: The Impact of Best Available Techniques (BAT) on the Competitiveness of the European Cement Industry (Interim Report), Report to IPTS, Seville.

Wagner, K., Triebswetter, U., 2001. The Impact of Best Available Techniques (BAT) on the Competitiveness of the European Cement Industry. Draft Final Report to IPTS, Seville.

Wagner, M., Nguyen, V.P., Azomahou, T., 2001. Determinants of the Relationship between Environmental and Economic Performance at the Firm Level: An Empirical Analysis of the European Paper Industry. Mimeo.

Wallace, D., 1995. Environmental Policy and Industrial Innovation, Strategies in Europe, the US and Japan. Earthscan, London.

Walley, N., Whitehead, B., May-June 1994. It's not Easy Being Green. Harvard Business Review, 46–52.

Walley, N., Whitehead, B., 1996. It's Not Easy Being Green, in: R. Welford and R. Starkey (Editors.), The Earthscan Reader in Business and the Environment. London, Earthscan.

Wasserhaushaltsgesetz, 1957. Neugefaßt durch Bekanntmachung vom 12.11.1996 Bundesgesetzblatt I 1695; Änderung durch Art. 3 Abs. 1 G vom 27.12.2000 Bundesgesetzblatt I 2048.

Weidner, H., 1992. Basiselemente einer erfolgreichen Umweltpolitik, Eine Analyse und Evaluation der Instrumente der japanischen Umweltpolitik. Edition sigma, Berlin.

Weidner, H., 1995. 25 Years of Modern Environmental Policy in Germany. Treading a Well-Worn Path to the Top of the International Field. Paper of the Science Centre Berlin, FS II 95–301, Berlin.

Weimann, J., 1991. Umweltökonomik, Eine theorieorientierte Einführung. Berlin u.a.

Welford, R., 1995. Environmental Strategy and Sustainable Development: The Corporate Challenge for the 21st Century. Routledge, London and New York.

Wey, K.-G., 1982. Umweltpolitik in Deutschland. Kurze Geschichte des Umweltschutzes in Deutschland seit 1900. Opladen, Westdeutscher Verlag.

World Bank, 1992. World Development Report 1992. Oxford University Press, New York, NY.

Wulf-Schnabel, J., Lohse, J., 1998. Economic Evaluation of NO_x Abatement Techniques in the European Cement Industry. Report compiled for DG XI of the Commission of the European Communities.

www.bmu.de/fset800.php

www.ehsni.gov.uk, Industrial pollution control in Northern Ireland.

Xepapadeas, A., de Zeeuw, A., 1997. Environmental Policy and Competitiveness: The Porter Hypothesis and the Composition of Capital. Journal of Environmental Economics and Management, 37: 165–182.

Zentrale Markt- und Preisberichtstelle GmbH, 2000. Vieh und Fleisch 2000, Deutschland – Europäische Union – Weltmarkt. Bonn.

Peter Lang · Europäischer Verlag der Wissenschaften

Karin Feiler (ed.)
European Forum on Sustainability of the Club of Rome

Sustainability Creates New Prosperity

Basis for a New World Order, New Economics and Environmental Protection. Review by Members of The Club of Rome and International Experts. Preface by Klaus Töpfer.
With Contributions by Martin Bartenstein, Orio Giarini, Hans Küng, Uwe Möller, Patrick M. Liedtke, Mahendra Shah, Franz Josef Radermacher, Walter R. Stahel, Ernst Ulrich von Weizsäcker

Frankfurt am Main, Berlin, Bern, Bruxelles, New York, Oxford, Wien, 2004. 239 pp.
ISBN 3-631-51973-7 / US-ISBN 0-8204-6549-6 · pb. € 19.80*

Since the report to The Club of Rome *Limits of Growth* was published some 30 years ago, the question is still open which approach we must take with our planet. Three factors will have the most influence on the global development of the environment: demographic and climate development and the potential for innovation. The concept of sustainability requires long-term thinking in order to use natural resources in such a way that future generations have the same opportunities we do. This message is confirmed in this review by the members of the Club of Rome's European Forum on Sustainability, which was founded in autumn 2002 in Vienna by the European Support Centre of the Club of Rome in co-operation with the Austrian Federal Ministry for Economic Affairs and Labour and in partnership with the United Nations Information Service (UNIS) Vienna, together with high-level experts and scientists. An immediate and extensive global course correction across the board would, however, be required. Thus far in fact, the price paid for the success of globalisation has been very high – environmental degradation and a global social split. This review presents as potential solutions a new economic model, an international treaty between the North and the South, a fundamental right to food and the principles of a sustainable retirement reform.

Contents: The Significance of Sustainability – Sustainability of Prosperity · Sustainability: Expectations and Reality · Sustainable Development: A Contradiction in Terms or an Economic Necessity · World History from the Viewpoint of Sustainable Development · The Climate Change: Mitigation and Adoption

FSC
www.fsc.org
MIX
Papier | Fördert gute Waldnutzung
FSC® C083411

Kolibri 360 GmbH
Ferdinand-Jühlke-Straße 7
99095 Erfurt, Deutschland
produktsicherheit@kolibri360.de

Druck:
CPI Druckdienstleistungen GmbH
im Auftrag der
Kolibri 360 GmbH
Ferdinand-Jühlke-Str. 7
99095 Erfurt